Basic Petroleum Geology

Basic

Petroleum

Geology

Peter K. Link

OGCI PUBLICATIONS
Oil & Gas Consultants International, Inc.
TULSA

Library of Congress Catalog Card Number: 81-83909
International Standard Book Number: 0-930972-10-4
Second Printing October 1982
Third Printing November 1983

Second Edition, September 1987
Second Printing, July 1990
Third Printing, April 1992
Fourth Printing, September 1996
Fifth printing, May 1998

This Book Is Dedicated To

Lucille

Preface

Perhaps this book should be titled Basics of the Geology of Petroleum because it deals with fundamental geologic concepts in the context of petroleum. It represents the result of a desire to create an experiment, wherein the fundamentals of physical geology are presented in the context of their application to petroleum-related concepts. Most volumes on petroleum geology are directed toward students who, the authors presume, have geological training. They discuss details of petroleum composition, exploration, and drilling without addressing environments of deposition, geologic time, and principles of deformation.

Since petroleum exploration and production make use of many of the same geological parameters, it is essential that geologists, geophysicists, and engineers share a common ground and knowledge, understanding each other as well as what they see in the geologic record. Written for individuals with limited or non-specific geologic experience, *Basic Petroleum Geology* addresses the fundamental concepts of the earth in terms of petroleum occurrence, exploration, and recovery. As a review and different point of view, this book is a reference for the practising geologist, geophysicist and engineer.

It is certain, however, that after this book is complete and in distribution, geologic concepts will have changed and new data become available. That is the way of all study. Geology waxes and wanes in its various aspects that are ephemerally in vogue. We have travelled the tunnels of turbidites, plate tectonics, source rock analysis, and related esoteric jargon. We have distilled the plausible from the morass and applied it to the geologic continuum. It is often difficult to define the plausible because to do so may require discarding old ways or interesting ideas. A good geologist, indeed, a good researcher, discards reluctantly and is never satisfied. Inasmuch as geology is an art, the focusing of diverse concepts upon an interpretation requires imagination and innovation. If this book stirs the imagination and prompts innovation, its objective will have been achieved.

ACKNOWLEDGEMENTS: I learned a great deal in writing the first edition of this book. This, the second edition, has educated me further, because of new information generated in the interim between the first edition and this revision. None of the original writing or this subsequent edition would have been possible without the contributions of many people.

The second edition is largely the work of Verma Hughes who organized and facilitated the revisions, and Loretta Sharp, the artist. Their work is sincerely appreciated, for without them, the second edition would not have materialized.

July, 1987   *Peter K. Link*
                                               Tulsa, Oklahoma

Contents

Basic Petroleum Geology

1　　　　　　　　　　　Earth Structure

Introduction

Geology is the study of the earth, its internal and external composition, structure, and processes by which it develops and is changed. Inasmuch as the earth is constantly changing, its processes and the history of its processes are important in the formation and preservation of economic mineral and hydrocarbon deposits.

As the earth changes, clues are formed that are important in the exploration for and the recovery of hydrocarbons. The history of the earth, and particularly the sedimentary record, is replete with clues, some obvious, some subtle, that provide important information in the search for natural resources and raw materials. Geologists, geophysicists, and engineers who explore for, find, and produce hydrocarbons must be able to interpret the geologic record and to evaluate physical, chemical, and biological parameters toward the successful conclusion of an exploration program. In this manner, the characteristics of the earth and its history tell the explorer and producer how hydrocarbons can be found profitably and eventually recovered and brought to market.

Geology is the primary subject of this book, which discusses geologic principles, concepts, and parameters and how they apply to the much narrower scope of petroleum. Some geologic environments relate directly to petroleum generation, migration, accumulation, and production. Others are not intrinsically petroleum-related but can become so by geologic accident. Geology is the rationale in this volume. Petroleum geology relates the rationale to the specific, which does not exist without geology as the supporting framework. A student cannot know about petroleum accumulations in river channel or desert deposits unless he knows about how river sands and dunes are formed and how substantially different they are. It is not enough to encounter a sand of unknown origin and call it a reservoir. Nor is it sufficient to drill a geologic structure without understanding its deformational character. The sand and the structure have orderly parameters that require specific production treatment different from that needed to produce other sands and structures. Only then does the reservoir become a beach sand deformed into a block-faulted anticline, for example. The beach designates certain geometric and reservoir characteristics, and the block faulting designates certain fracture and deformational factors. Reservoir and structure are defined and produced according to the best related techniques. Orderly development of the geologic rationale is essential. Without it, we cannot hope to be other than random in our approach.

Structure of the Earth

Earth structure is fundamental to an understanding of the study of earth processes and earth history, which are so important in exploring for and producing hydrocarbons. Observation of the earth as a planet illustrates three primary aspects of the crust: the atmosphere, the hydrosphere, and the lithosphere. These are surface manifestations and exist as air masses, oceans, and continental areas, respectively. Superimposed upon the surface of the earth is the biosphere, comprising the organic realm of the earth's life forms that inhabit the three environments and that are often unrestricted as to which of the three spheres they live in. Many life forms experience various stages of development in several different spheres. In some instances they make the transition easily and regularly.

A definition of the earth, then, includes air masses, oceans, continents, and the life in each of them as manifestations of its external characteristics. However, the earth is more than what can be seen externally from a position on its surface. From space, the earth appears to be spherical, or nearly so, and is somewhat flattened at its poles of rotation. It bulges slightly at the equator due to the body forces within the earth caused by the rotation. The flattened poles and increased equatorial girth create a stress field within the earth and cause movements of the different external layers from the atmosphere down to and including the lithosphere. This stress field is important in the development and distribution of continental and oceanic elements of the crust.

Because the earth's surface, oceans, and atmosphere are readily accessible and visible, they are studied relatively easily. Studies of the interior of the earth are not easily accomplished, however, because there has been no direct access to earth materials and structure below six miles beneath the surface. This means that most of what is known about the deep interior of the earth is circumstantial and based upon indirect information. A cross section (Fig. 1) illustrates the various layers, which with their variable compositions are thought to comprise the earth.

Lithosphere

On the outside of the earth is the lithosphere, which consists of rigid, solid rock from 65 to 100 km thick and includes oceanic crust and continental crust in its upper layers and the uppermost part of the mantle (Fig. 2). Oceanic and continental crust consist of silicon and magnesium, and silicon and aluminum-bearing rocks, which are called SIMA and SIAL respectively. Simatic rocks are heavier than sialic rocks and generally underlie the latter when together though some interpretations suggest a lateral transition between them. SIMA and SIAL combine to form the crust. It has a specific gravity of 2.6–2.65, is less dense than the remainder of the earth, and ranges from 5 to 60 km in thickness.

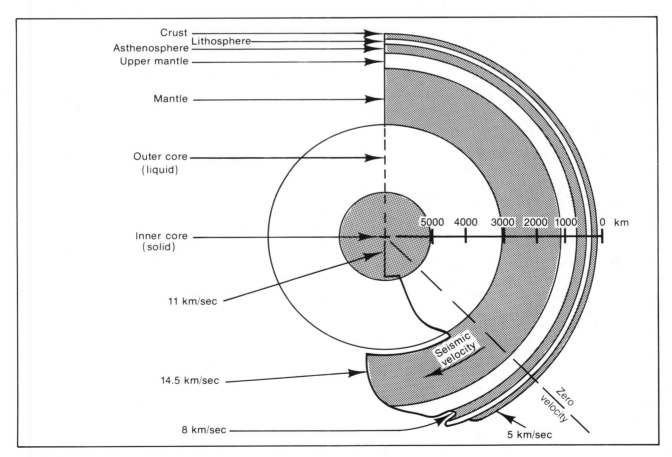

Fig. 1. Internal structure of the earth

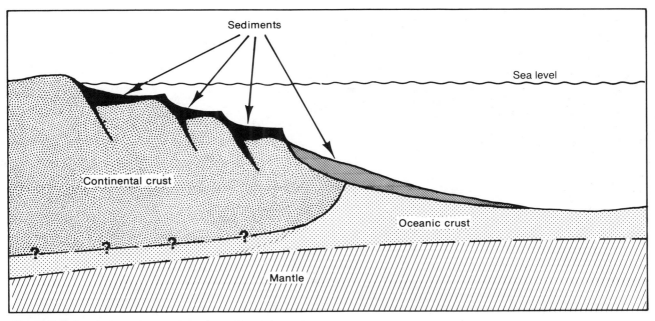

Fig. 2. Relation between oceanic and continental crust

Mantle

Beneath the lithosphere, the mantle is approximately 2,900 km thick and includes in its upper part the asthenosphere, which is about 200 km thick. Temperature and pressure are balanced so that the asthenosphere is very near the melting point and can flow when subjected to stress. Volcanic activity and deformation of the crust result from movement within the asthenosphere.

Iron and magnesium-bearing silicate minerals make up the mantle, which has a higher density than the lighter lithosphere and a specific gravity of 4.5–5.0. Because the mantle is heavier than the lithosphere, seismic waves travel through it at greater velocity. The seismic velocity change occurs between the base of the crust and the upper mantle. Seismic wave velocities increase at this surface, which is called the Mohorovicic Discontinuity after the Yugoslavian seismologist who discovered it.

Core

Iron and nickel are the predominant constituents of the core of the earth. The core is approximately 7,000 km in diameter and consists of an outer liquid portion about 2,200 km thick and a solid inner core with a diameter of 2,600 km. Rotation of the earth is thought to create circulation currents within the liquid core that generate the magnetic field around the earth.

Materials contained in the earth's core are the heaviest found in the earth and have an average specific gravity of 10.5. This is over twice the average specific gravity of 4.5–5.0 of the mantle and almost four times the 2.6–2.65 specific gravity of the lithosphere and crust.

Determination of Earth's Structure and Composition

External characteristics of the earth are easily studied and are important factors in describing the earth's surface structure and composition. Internal characteristics are another matter, since erosion reveals depths up to 25 km below the surface, and drilling has yet to exceed depths much greater than 12 km. Rocks from depths of 200 km come from volcanic activity, and eroded, deep-seated intrusive features may represent original depths of less than 25 km. Beyond these very shallow samples, no other direct information is available to indicate the character of the deeper portions of the earth's interior. Indirect methods are used to determine the character of what we are unable to see.

Earthquake Waves in the Earth

Earthquake (seismic) waves (Fig. 3) consist of primary, secondary, and surface waves that travel on the earth's

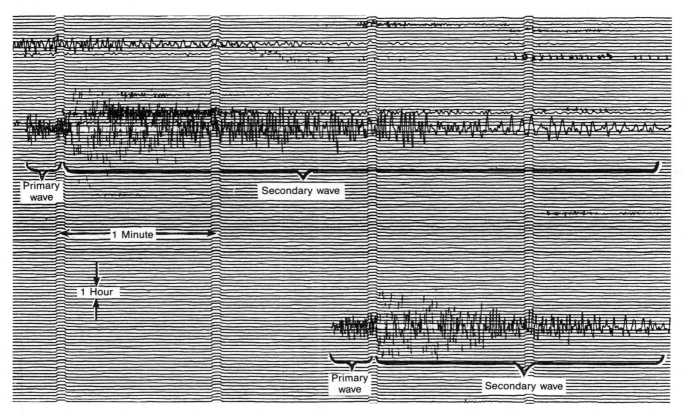

Fig. 3. Seismogram of several of the May 2–3, 1983, earthquakes at Coalinga, California. Record is from the Bakersfield, California seismograph.

surface and through the different densities of the various earth layers at different speeds.

Primary waves vibrate in the vertical plane as compressions and rarefactions and travel the fastest of the wave types. Secondary waves vibrate from side to side in the horizontal plane and arrive after the primary waves. Surface waves move by spherical propagation similar to waves caused by a pebble thrown in a lake and are slowest of the three wave types. They have large amplitudes and cause the most damage to surface structures.

Certain characteristics of earthquake waves are important in discriminating between the various earth layers and establishing what is known about their composition. Therefore, travel times of seismic waves and their ability or lack of ability to be transmitted through various substances are important (Fig. 4).

The Mohorovicic Discontinuity (Moho) at the base of the crust represents a substantial increase in seismic wave velocity in the remaining portion of the underlying lithosphere. A low velocity zone occurs in the asthenosphere below the lithosphere. A steady increase in primary and

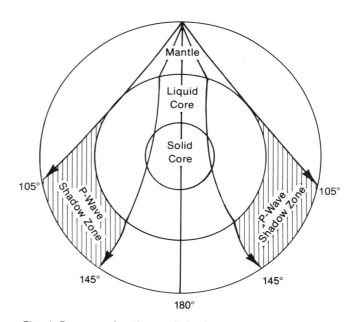

Fig. 4. P-wave refraction and shadow zone

secondary wave velocity persists until the core is reached. Here, the primary waves abruptly slow down and secondary waves stop completely and are not transmitted. Because secondary waves do not pass through liquids, this portion of the core is considered to be liquid.

Beyond this zone, primary wave velocities steadily increase with depth. When the inner core is reached, wave velocities become stabilized below an abrupt velocity increase at the base of the liquid core.

Additional information concerning the earth's core comes from a combination of primary wave refractions and secondary waves not passing through liquids (Figs. 4, 5). Refraction and the slowing of primary waves by the liquid core create a shadow zone where the waves are not received on the side of the earth between approximately 105° and 145° from the earthquake source. Since two areas of refraction are developed in a sphere, two shadow zones result on any given cross-section of the earth.

The slowing of primary waves through the outer core is indicative of its liquid state. An increase in the velocity of waves traveling directly through the central portion of the core indicates that it is solid.

Because seismographs were not developed until the late 1800's, studies of acoustic properties of the earth were not possible until recently. It was not until after the early 1900's that the existence of the core of the earth was confirmed.

Specific Gravity of the Earth

Surface materials of the earth have specific gravities of less than 3.0, which is far too low to account for the 5.5

specific gravity of the entire earth. In order for the average specific gravity of the earth to be higher than that for surface materials, the interior materials must be sufficiently high to create an average of 5.5.

Comparison of Earth Materials with Materials from Space

Indications of what the composition of the interior of the earth might be come from meteorites that fall to the earth's surface. They consist of meteorites composed of nickel and iron, stony meteorites of periodotite (pyroxene and olivine), and stony-iron meteorites that combine peridotite, iron, and nickel. The specific gravities and other characteristics of these materials are compatible with those suggested for various layers within the earth. So the upper and lower mantles and the core are thought to contain much of the same materials found in meteorites from space.

Earth's Magnetic Field

Rotation of the earth is thought to create a slow rotation of the liquid iron core relative to the mantle exterior to it (Fig. 6). A magnetic field may result from the electrical

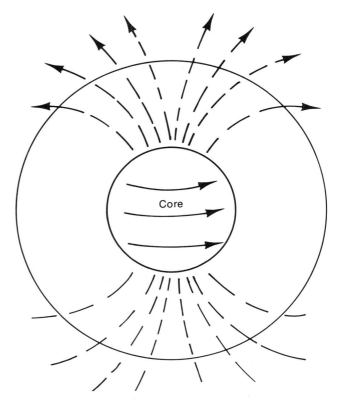

Fig. 6. Earth's magnetic field and rotation within the core

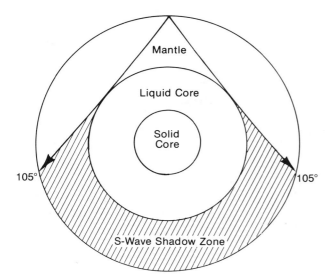

Fig. 5. S-wave shadow zone

currents generated by the relative motion between the liquid core and the mantle. Inasmuch as there is a magnetic field and the earth does rotate, the conclusion that there is a liquid portion of the core is compatible with available data.

Earth's Temperature Gradient

Temperature data from deep wells and mines indicate that a rise of 1° C per 27 meters or about 3° C per 100 meters is typical with increased depth. If the rise continued to the center of the earth, however, an unrealistically high temperature would be achieved. Interior earth temperatures (Fig. 7) are estimated from melting points of rock materials thought to constitute each layer.

Convection is thought to occur in the mantle which, though not liquid, is nearly so, and allows heat transfer by very slow flow. However, the mantle remains solid enough to allow the transmission of secondary earthquake waves. Convection is the probable mechanism by which the temperature of the center of the earth is maintained well below theoretically possible high levels. Melting points are not reached, and various earth zones remain nearly solid because of the convective heat exchange.

Plate Tectonics

Early geological workers noticed that certain portions of the continents would fit together if moved from their present positions. Other workers considered that the continents had probably been together as one mass at one time and had subsequently broken apart.

Alfred Wegener used continental shapes and geological information (Fig. 8) in the formulation of his theory of continental drift. Wegener decided that the continents moved away from a large central continent in response to the earth's rotation (Figs. 9 and 10). Wegener and other early proponents of continental drift conducted their work and proposed their theories in the early 1900's. Their ideas were not widely accepted at the time and eventually languished with little popular support until new data resulted in new ideas in the 1950's and 1960's. Since then, studies of the continents and the ocean floor have proliferated and become oriented toward the concepts of continental drift and plate tectonics. The continental drift theory became respectable, and in the 1970's it has explained much of what is known about tectonics of the earth's surface.

Evidence for Continental Drift

Factors pointing to continental drift are largely circumstantial and therefore interpretive. Relocation of the continents on the surface of the earth provides a very positive

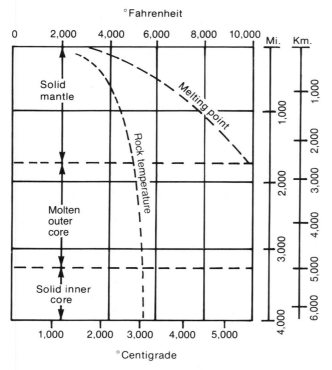

Fig. 7. Internal earth temperature

Fig. 8. Pre-drift continental fit

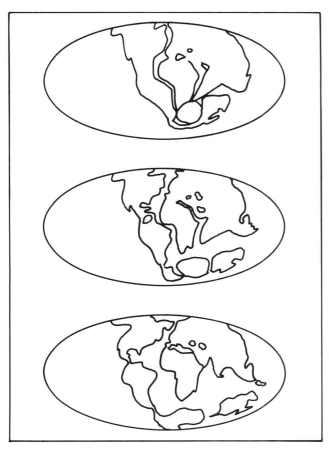

Fig. 9. Continental drift according to Wegener (1915)

fit when the process integrates all of the masses into the nuclear continent or Pangaea of Wegener. The Atlantic margins of Europe, North and South America, and Africa offer remarkable coincidence. The relations between Africa, Antarctica, and India appear to be very close as well.

Climate reconstructions for continents in the past indicate patterns compatible with a central grouping and an associated pattern of climatic zones. Separation of the continents disrupted the climatic continuity and placed climate zones into locations incompatible with present positions. Coal deposits in Antarctica, old glacial features of southern hemisphere continents, lack of old glaciation in the northern hemisphere, and the lack of consistent direction of old glacial grooves strongly suggest continental repositioning.

Fossils of specific land animals and plants occur in same-age rocks in South America, Australia, India, Antarctica, and South Africa (Fig. 11). Distribution of these fossils to each of the continents indicates that their development probably occurred on a common continent that subsequently broke up into segments, which ultimately separated and drifted apart.

Fossils of certain types of ferns and reptiles have been found on the southern continents. Seeds from the ferns are too large to have been wind-carried across the oceans, and the reptiles were exclusively land creatures and could not have swum across large stretches of water. Continental drift conveniently explains the distribution of these fossils.

Flora and fauna developed in Australia, the Seychelles Islands, and on Madagascar, for example, suggest isolation after continental separation. Unique animals and plants in these geographically remote areas indicate lack of faunal and floral communication with other areas during their post-drift evolution.

Several structural trends that extend across portions of some of the continents end abruptly at their coasts and reappear on the continents facing across the ocean. The Appalachian Mountains, for example, terminate at the Newfoundland coast but resume their trends in Ireland and Brittany, indicating that their continuity has been interrupted by continental separation.

Other geophysical and comparative surface data can suggest that continents have moved and continue to move. Direct measurements along major fault zones indicate movements of up to 6 cm per year in some places. Movements of one cm per year or less are considered more common in many localities. Notwithstanding the different types of evidence available in favor of continental drift, there remains evidence against it, and the actuality of the theory remains without conclusive observable proof.

Crust of the Earth

If the earth is visualized as a sphere covered entirely by a dense, tar-like liquid, it is easily imagined that tabular fragments of light-weight wood could be distributed locally about the surface of the sphere. The wood fragments floating on the tar would move about the surface of the sphere in response to forces directed upon its surface. Because of the difference in density between the tar and the wood, the latter floats upon the former as a cork floats on water and moves with forces affecting the water.

The crust of the earth consists of dense oceanic crust, composed of silicon and magnesium, that floats upon the denser lower lithosphere that overlies the mantle. Less dense continental crust consisting of silicon and aluminum floats above and adjacent to the oceanic crust. Forces directed upon the surface cause the continents to move about. They break apart, collide, slide side-by-

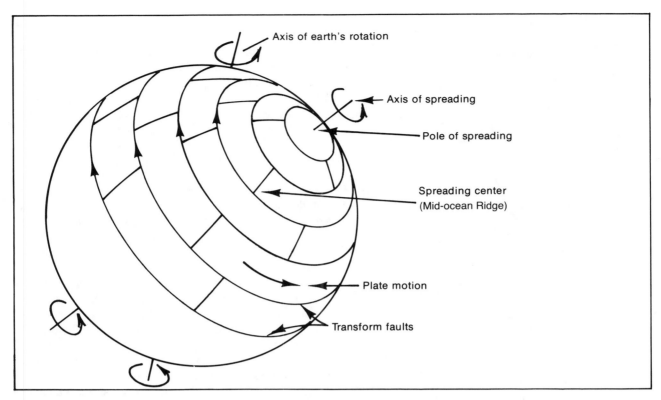

Fig. 10. Plate motion on a sphere

	South Africa	South America	Antarctica	India	Australia
Cretaceous	Marine Sediments	Basalt	None	Volcanics	Continental and Marginal Marine
				Marine Sediments	
Jurassic	Basalt		Basalt	Clastics	Continental and Marginal Marine
				Volcanics	
		Sandstone		Clastics	
Triassic	Clastics	Clastics with reptiles			Continental and Marginal Marine
	Clastics with reptiles		Clastics and Coal with *Glossopteris*	Clastics with reptiles	
Permian	Clastics	Clastics with *Glossopteris*		Clastics with *Glossopteris*	Clastics
	Clastics with *Glossopteris*	Marine Shale		Shale	
		Shale with reptiles	Tillite		Clastics and coal with *Glossopteris*
Carboniferous	Shale with reptiles	Clastics and coal with *Glossopteris*		Tillite	
	Clastics and coal with *Glossopteris*	Tillite	Tillite		Tillite
	Tillite				

Fig. 11. Southern hemisphere stratigraphic columns

side, and apparently have done so several times throughout earth history. It is the movement of the continents that provides the basis of the theory of continental drift and plate tectonics.

Movement of continental masses and tectonic plates occurs in three fundamental ways. They are divergent, convergent, and transform motions and involve continental separation and extension, continental collision and compression, and oblique movement of continents past each other, along a common zone of motion.

Causes of Continental Motion

The fact that continental drift is a theory means that the causes of continental motion must be theoretical as well. One theory for plate motion concerns the development of convection currents within the mantle. Although the mantle is sufficiently viscous to behave as a solid, it may respond to continuous stress applied by heat and develop convection currents. In this manner the mantle would flow at temperatures below that of its melting point. Convection cells (Fig. 12) in the mantle may occur in pairs and provide a mechanism for the separation of continental masses from a common spreading center (Figs. 13 and 14). Movements in the crust, produced by convection, would not exceed a few centimeters per year.

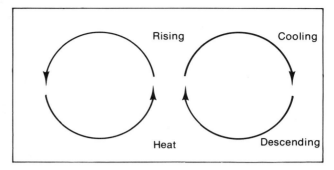

Fig. 13. Convection and heat exchange

Another possible mechanism for continental movements might involve the body forces produced in the earth by rotation on its axis. It is not difficult to imagine a scenario in which the thin, relatively incompetent crustal materials could respond to rotational forces that affect every part of the earth. Rotational forces affect the flow of the hydrosphere and atmosphere and are important in the distribution of ocean currents and air masses, respectively. Light-weight crustal materials floating on denser, highly viscous lithosphere and mantle materials might respond to the same rotational forces and similarly move about on the surface of the earth.

Gravitational effects of the sun, moon, and the planets in the solar system create tides within the earth's crust. These are known as earth tides and have been measured

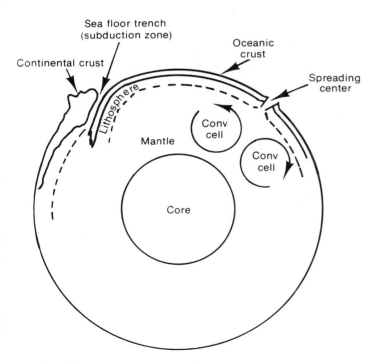

Fig. 12. Convection and crustal motion

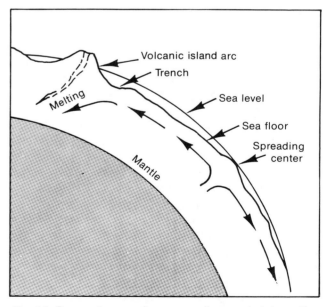

Fig. 14. Convection, seafloor spreading, and plate motion

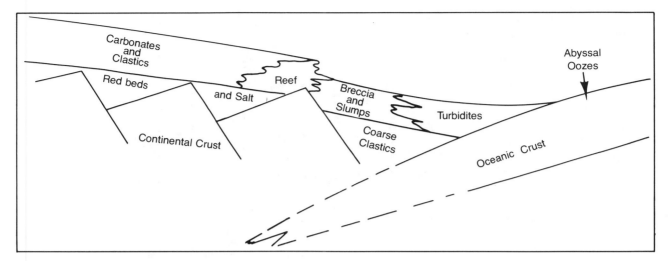

Fig. 15. Divergent continental margin of North Atlantic type

by special instruments for a number of years. It seems possible that earth tides or the forces that cause them might contribute to the motions of crustal plates on the earth.

Continental Margins

As lithospheric plates move about on the earth's surface, certain types of continental margins evolve in response to the nature of the various motions. Each of the different types of margins has particular characteristics that are not only unique but also have bearing on the hydrocarbon potential of the type of margin involved.

Since plate margins involve motion, stresses are produced along them and result in most of the major tectonic activity affecting the crust. Only minor tectonic activity occurs within the central portions of the crustal plates, and it primarily assumes the form of a regional expression of stress bearing upon the margins of the plates.

Continental margins assume three different modes according to the types of motions and related deformation that affect them. *Divergent continental margins* (Figs. 15 and 16) occur where plates have broken apart along a spreading center and moved away from each other. The coasts of North and South America moved away from the western coasts of Africa (Figs. 17 and 18) and western Europe. They originally were joined in the vicinity of the Mid-Atlantic Ridge, which represents the common spreading center for both sides of the north and south Atlantic Oceans. The spreading along the Mid-Atlantic Ridge is apparently continuing since volcanic activity persists along it.

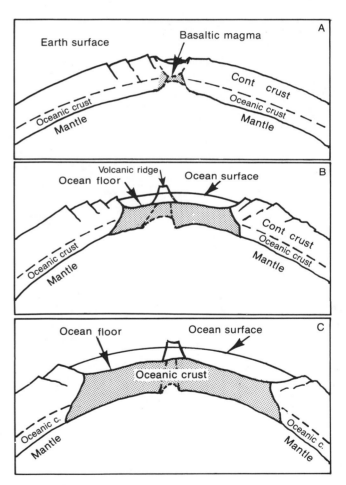

Fig. 16. Continental separation and ocean basin development

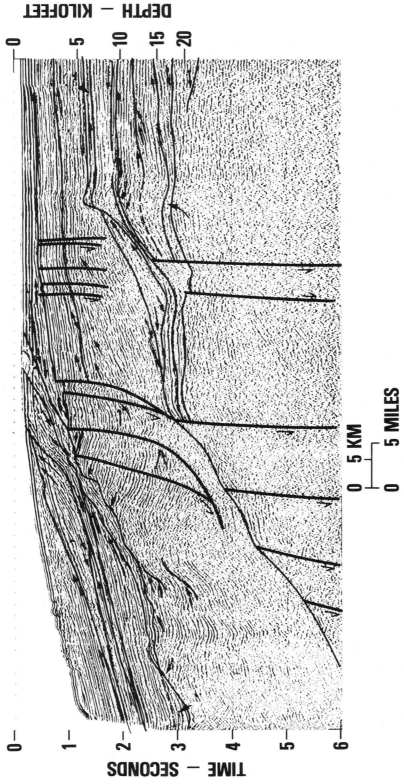

Fig. 17. Seismic line offshore West Africa, showing divergent continental margin. From Todd and Mitchum, 1977. Permission to publish by AAPG

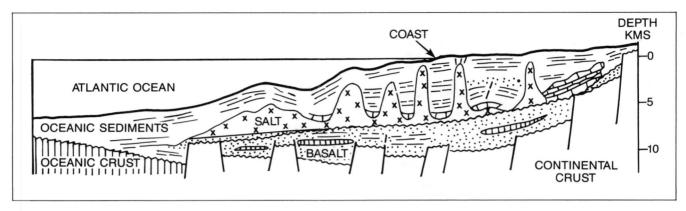

Fig. 18. Offshore Gabon, West Africa showing divergent continental margin, sediments and structure. From Kingston, Dishroon and Williams, 1983. Permission to publish by AAPG.

Divergent continental margins are characterized by extension as the plates move away from each other. Red beds and salt are deposited in the fault-block basins that result. As separation continues, marine sediments are deposited over the salt and red beds. Carbonate banks and reefs form over the margins of the high fault blocks where water depths are suitable. During the Mesozoic break-up of the Gulf of Mexico Basin, red beds, salt, marine sandstone and shale, and Cretaceous reefs were deposited. Petroleum has been found associated with these sediments as well as those in the Tertiary that have been affected by salt dome activity and growth faulting. Divergent continental margin sediments produce petroleum in the North Sea, the Gulf of Mexico, the coasts of west Africa and eastern South America, and northeastern Canada. Characteristics of divergent continental margins include:

1. Block faulting with deformation limited to tilting
2. Red beds deposited as basal sediments in fault-block basins
3. Salt deposited over the red beds
4. Marine sandstone and shale as water depths increase
5. Reef and bedded carbonates
6. Possible petroleum occurrences in or associated with the above sediments and structures

Comparison of areas illustrating progressively greater rifting demonstrates the sedimentary sequences associated with continuing separation (Fig. 19).

Convergent continental margins (Fig. 20) are developed where two continental plates or one continental and one oceanic plate collide. Several types of collisions are possible because several combinations of crustal mate-

rials may be represented in the collisions. In order to accommodate encroachment of one plate by another, a zone of plate consumption is developed whereby one plate overrides the other along a subduction zone. Here the lighter continental material rides over the denser oceanic material, which plunges beneath it, and potential space problems are resolved. Subduction zones only develop where there is a density difference between the crustal materials that are colliding. Where two continental plates come together, subduction continues as long as there is oceanic crust in the space between the continental masses. When the two continental masses meet, the subduction zone ceases to exist because neither mass is capable of being subducted, and a suture or zone of junction between the masses is formed along a resulting mountain range. Collision of two oceanic plates results in the subduction of one of them beneath the other. Where subduction occurs, the oceanic crust being subducted plunges into the asthenosphere where it is melted into the mantle.

A convergent margin consisting of an oceanic plate and a forming continental plate that includes a volcanic island arc will develop a subduction zone along the oceanward side of the arc (Figs. 21 and 22). The subduction zone will dip beneath the island arc and will consume the oceanic crust. Because the two plates are moving together, compression is the regional stress regime. However, because of up-bowing of the island arc due to drag along the subduction zone, extension can occur along the axis of the island arc.

Characteristics of a convergent zone of this type can include the following:

1. Zone of compression within the subduction zone and possible extension along the axis of the island arc.

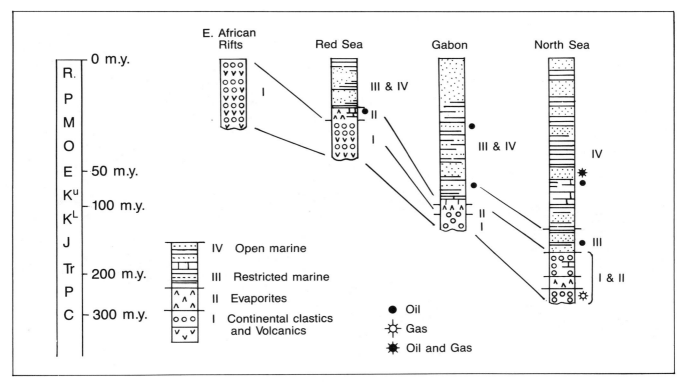

Fig. 19. Sedimentary sequences in rift basins. From Stoneley, 1982.

2. Reverse and normal faulting.
3. Volcanic sediments near the volcanic island arc. Sediments become cleaner and better sorted at distance from volcanic sources.

4. Block faulting in marine basin behind the volcanic island arc.
5. Possible petroleum occurrences in clean marine sediments on high fault blocks in basin behind island arc.

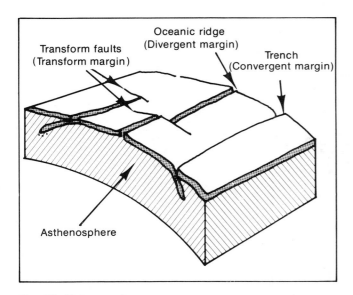

Fig. 20. Plate margins

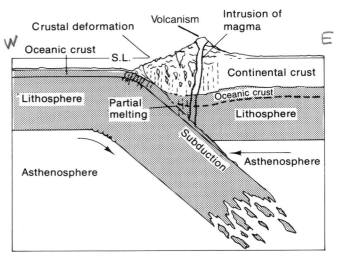

Fig. 21. Convergent trench continental margin

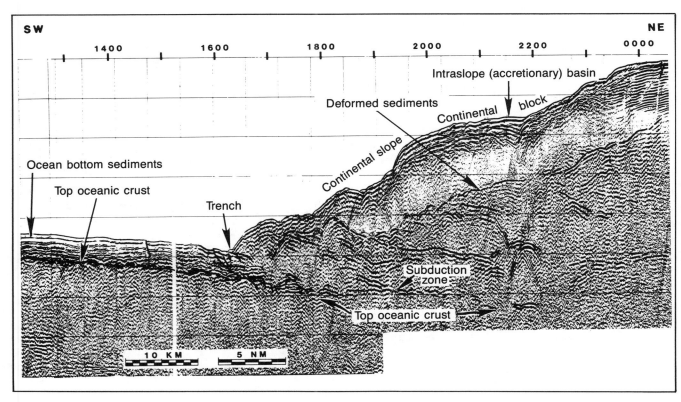

Fig. 22. Seismic section of Middle America trench subduction zone. From Crowe and Buffler, 1983. Permission to publish by AAPG.

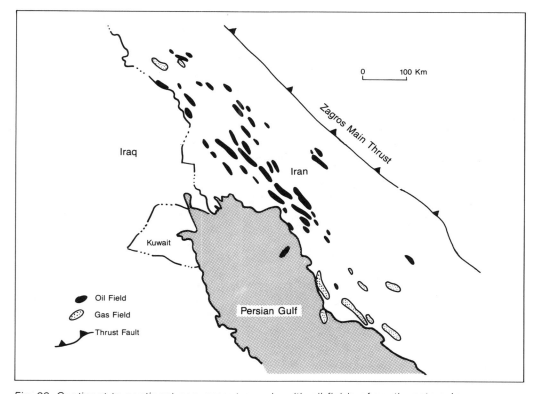

Fig. 23. Continent to continent convergent margin with oil fields of southwestern Iran.

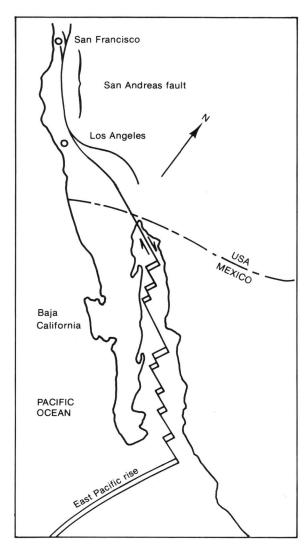

Fig. 24. East Pacific Rise and San Andreas transform margin

Oil and gas fields in Sumatra are of this type. They are developed in marine sediments that are structurally preserved in fault blocks on the Malaysian side of the volcanic mountains of southwest coastal Sumatra. Convergent tectonics are responsible for important production on both sides of the Andes Mountains in Colombia, Equador and Peru. Continent to continent collision formed important petroleum-producing structures along the Zagros zone of southwestern Iran (Fig. 23).

Transform continental margins (Fig. 24) occur where two plates move past each other without converging or diverging. Maximum principal compressive stress is horizontally directed at approximately 30 degrees to the zone of shear along which fault-block structures are formed. Marine sedimentation in the resulting basins occurs simultaneously with deposition and in certain cases traps the petroleum, which is retained in numerous sandstone reservoirs.

Oil fields along the San Andreas fault of southern California and related faults in the system are located on fault blocks containing multiple sandstone reservoirs. These fields are excellent examples of production from structures along a continental transform plate margin. A transform margin between the Caribbean and South American plates produces petroleum from related structures in southern Trinidad and northern Venezuela (Figs. 25 and 26).

A transform continental margin can be characterized by the following:

1. Strike-slip movement along a major shear zone as two continental plates move by each other
2. Fault-block structures in echelon to the major fault zone at approximately 15–45 degrees to it
3. Sediment deposition concurrent with deformation along the transform fault system

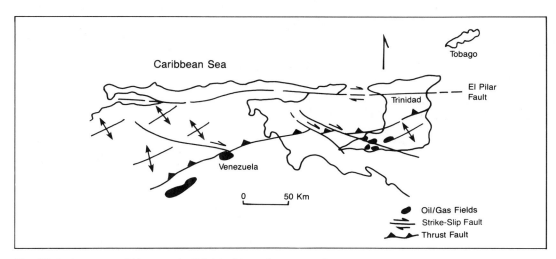

Fig. 25. Index map of Venezuela-Trinidad transform margin.

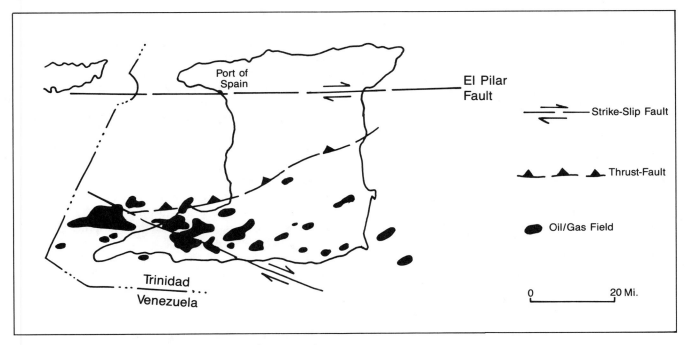

Fig. 26. Index map of southern Trinidad transform margin.

4. Development of multiple reservoir and source beds
5. No volcanic activity

Continental Margins and Petroleum

Oil and gas can occur along virtually any type of continental margin, provided favorable structural and stratigraphic conditions are present. Marine deposition and proper trapping conditions are necessary for petroleum generation and accumulation. Divergent, convergent, and transform continental margins are all capable of providing the necessary depositional and structural conditions described in the following.

1. Oil and gas fields along the Atlantic Coasts of Africa and South America and along the Gulf Coasts of Texas and Louisiana involve fault blocks, salt domes, carbonate banks, and reefs. They are fields typical of divergent continental margins. The Hewett gas field in the British North Sea (Figs. 27 and 28) experienced similar development.

2. Since several types of convergent continental margins are possible, a variety of petroleum fields is possible. Fields in Sumatra on the shelf side of the volcanic Barisan Mountains are excellent examples of production from the convergent continental margin. Figures 29, 30, 31, and 32 illustrate the Minas Field located in east-central Sumatra.

3. Several types of structures can be formed where two continental blocks move past each other along a transform shear zone. The most important of these relative to petroleum occurrences are the faulted folds that are distributed in echelon along the primary transform shear zone. Figure 33 illustrates the Dominguez Field, which is typical of the fields of southern California along the Newport-Inglewood trend.

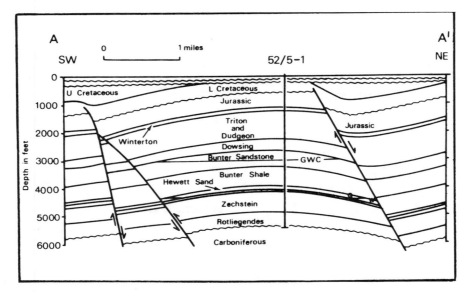

Fig. 27. Hewett Field, geologic section. From McQuillen, Bacon, and Barclay, (1979). Permission to publish by Graham & Trotman Limited.

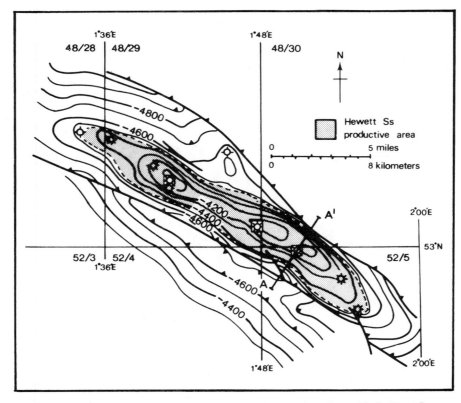

Fig. 28. Hewett Field, structure map on Hewett Sandstone. From McQuillen, Bacon, and Barclay, (1979). Permission to publish by Graham & Trotman Limited.

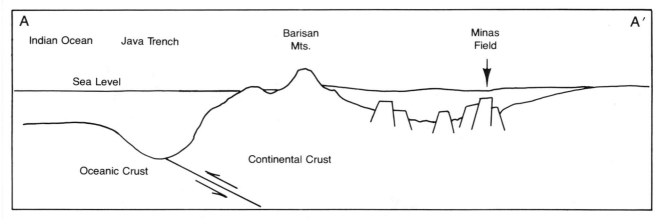

Fig. 29. Indonesian convergent margin and the Minas Oil Field. See Figure 30 for line of section.

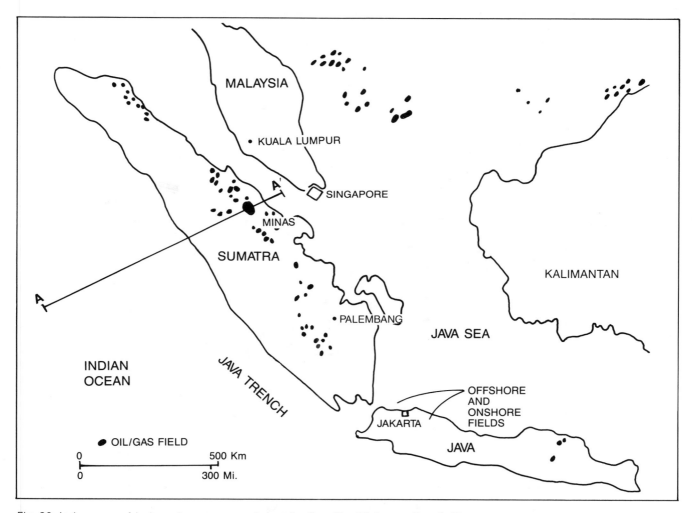

Fig. 30. Index map of Indonesian convergent margin. See Fig. 29 for section A-A'.

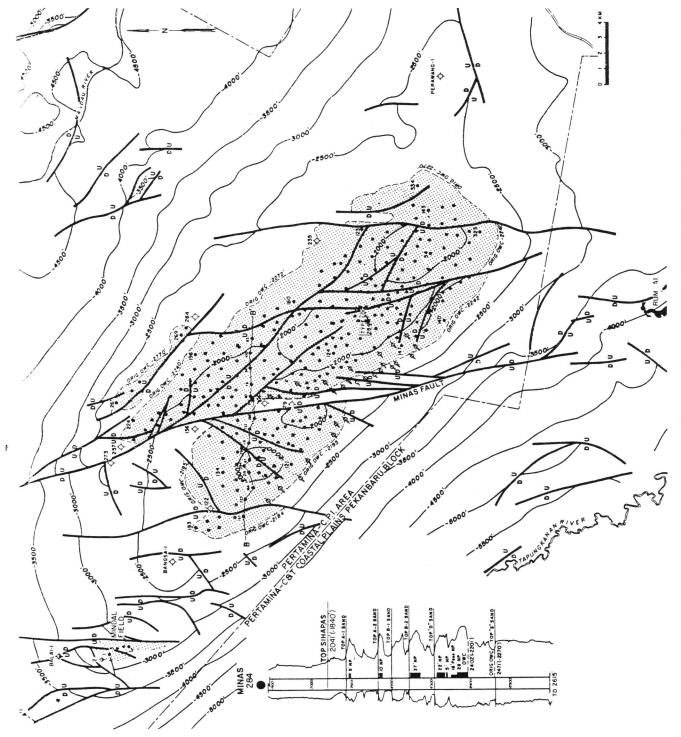

Fig. 31. Minas Field structure map. From Hasan, et al., 1977. Permission to publish by PT Caltex Pacific Indonesia.

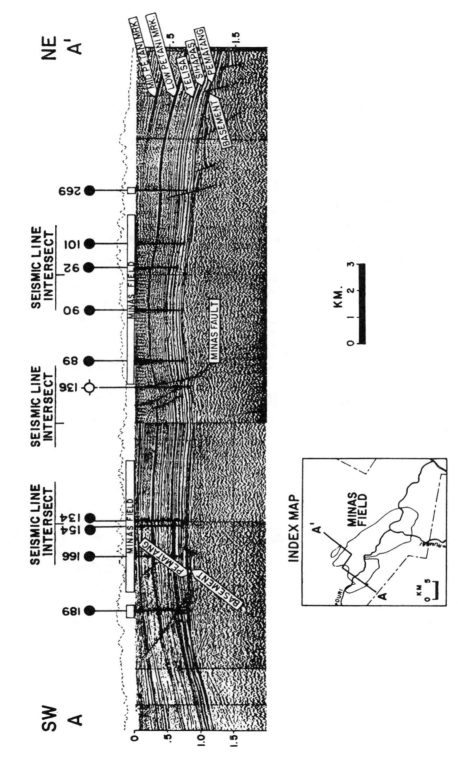

Fig. 32. Minas Field cross-section. See Figure 31 for location. From Hasan, et al., 1977. Permission to publish by PT Caltex Pacific Indonesia.

DOMINGUEZ OIL FIELD

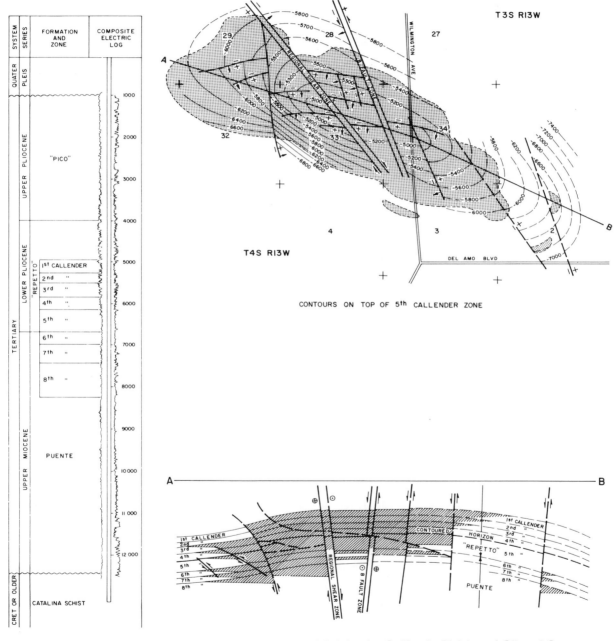

Fig. 33. Dominguez Oil Field, California. Permission to publish by the California Division of Oil and Gas.

2

Geologic Time, Historical Geology, and Stratigraphy

Geologic Time

For most of us, the concept of the enormity of geologic time is almost impossible to grasp in its abstraction. We are impatient when we must wait an hour for somebody who is late, and the thought of a ten-year jail term is frightening in its virtual finality. Holding one's breath for two minutes seems a substantial accomplishment, as does sitting on a flagpole for three months in search of some frivolous record. As expenditures of time, these events can seem tedious in the extreme when superimposed upon the rapid pace of our lives.

When such events are compared to geologic time, however, they lose their significance as important events upon the interval involved in the formation of the earth and its subsequent development. In everyday perceptions, geologic time is too great to conceive of because numerous consecutive lifetimes need to elapse before some of the most fundamental and rapidly occurring geologic events will have taken place.

Estimates of the age of the earth vary between four and seven billion years. Maximum ages of meteorites are about 4.6 billion years, which may suggest a similar age for the earth if all elements of the solar system and the universe formed at the same time. These ages are huge when compared to our concepts of time, indeed to our life spans, and require a different approach in the accommodation of such components of time. Accordingly, geologic time is often referred to in increments of millions, tens of millions, hundreds of millions, and billions of years; greater than any time increment we can ever experience.

Geologic time is represented by the presence of rock intervals in the geologic column or by the absence of equivalent rocks in correlative columns in adjacent or distant locales. It is also represented by periods of erosion, deformation, organic evolution, or floral and faunal extinction. Elapsed geologic time encompasses all of the events that affect the earth and, for the observer, the clues that represent those events. Petroleum forms, migrates and accumulates as direct results of those events represented by the clues to be interpreted as significant exploration data. Without the continuum there is no rationale for exploration.

Uniformity

Inasmuch as earth history extends to at least 4.5 billion years ago, the opportunity for change has been significant. For lack of a better approach and because of detailed observation, the principle of uniformity or uniformitarianism was developed in the late eighteenth century. This principle stated that internal and external processes affecting the earth today have been operating unchanged and at the same rates throughout the developmental history of the earth. This means that a historical geologic event preserved in the rock record can be identified, in terms of its elapsed time of development, with similar observable events taking place at the present time. So comparisons of rates of deposition, erosion, igneous emplacement, and structural development preserved in the geologic record are easily made with what we know about current similar processes. Because time criteria on these and other processes are submitting more and more to quantitative measurements, accuracies of time estimates are progressively increasing.

Relative Time

In the eighteenth and nineteenth centuries and until the middle part of the twentieth century, geologic time was based upon the occurrence of geologic events relative to each other. Dating by relative time required the development of a sequence of events that could be established on the basis of obvious consecutive criteria. In this procedure, the identification of a geologic continuum required that the events within the sequence be sufficiently identifiable and be widespread enough to have a real significance.

Within the concept of relative time and the establishment of a continuum, successions of various types have evolved and include depositional, erosional, intrusive, and faunal considerations. Combinations of several of these factors have facilitated relative age dating throughout the world.

Superposition. Superposition is fundamental to the study of layered rocks and means that in any normal sequence the oldest rocks, deposited first, are on the bottom and the youngest rocks, deposited last, are on the top.

Any sequence of sedimentary rocks is considered to have been deposited horizontally, or nearly so, and in order to reflect normal superposition must not be overturned by deformation. A well drilled through a normal sedimentary sequence will encounter the youngest rocks on the surface and progressively older rocks with greater depth. As erosion attacks terrain underlain by a normal rock sequence, successively older rocks will be exposed as younger rocks are removed.

Since sea level is known to vary, erosion may be curtailed at any time and be replaced by a period of deposition that covers the erosional surface. It is obvious that the gap between the oldest rocks exposed by the erosional surface and the rocks laid down by subsequent deposition is represented, in geologic time, by the period of erosion. Examination of a sequence of rocks containing such a gap would indicate that the section had undergone deposition, erosion, and subsequent deposition. If the amount of geologic time represented by the missing portion of the section had been established in adjacent areas, the history of the geologic section could be pieced together.

Succession of Fauna. As stratigraphic units are being deposited, fossils of plants and animals are being included with the sediments. As geologic time and deposition continue, evolution of the plants and animals changes some of the forms, eliminates others, and gives identity to the ages of the rocks being laid down. In this way, occurrences, sequences of occurrences, evolutionary development, and appearance and disappearance of fossil forms illustrate position in the rock record and therefore the ages of the strata involved.

Fossil successions and occurrences allow correlations, or comparisons, to be made between rock sections in different localities and thereby establish age similarities or disparities. Fossils can be useful in rock sections that change rock type from place to place but remain the same age. Comparison or correlation of strata at several localities illustrates the areal or regional behavior of stratigraphic sequences (Fig. 34).

Inclusion. Igneous rocks that intrude surrounding rocks are invariably younger than the rocks they intrude because they form later. During the process of intrusion, fragments of the older surrounding rocks become integrated into the intrusive mass and become inclusions.

Clastic rocks, such as sandstone and conglomerate, are dependent upon pre-existing older rocks for the sources of the particles of which they are composed. Particles and fragments that are derived from a recognizable source are older than the new rock into which they eventually become integrated.

In both of these cases, the pre-existing rock is older than either the intrusive body or the subsequently deposited sandstone or conglomerate. By establishing the sequence of occurrence of the pre-existing rocks and the subsequently formed rocks, the relative ages of the two can be determined.

Crosscutting Relations. Pre-existing rocks can be intruded by younger rocks or can be affected by later faulting and/or folding. It is evident that for a sequence of rocks to be affected by crosscutting events, the original sequence must be in existence before the crosscutting events occur. Scales of crosscutting features are variable, from regional in extent to those on the microscopic scale. Crosscutting features can be very useful in determining relative geologic ages (Fig. 35).

Physiographic Development. Landforms and the shape of the landscape are continuously subjected to constructive and destructive processes. Constructive processes augment the landscape and can consist of uplift, volcanic activity, and deposition of sediments. All of these either add to the surface of the land or enhance its topographic configuration.

Destructive processes consist primarily of erosional activity that degrades the landscape (Fig. 35). Cycles of erosion affect a number of the erosional processes that are brought to bear upon different aspects of the landscape. Since cycles of erosion follow recognized stages of development, land forms, streams, and shore lines can be identified according to how far they have developed in their particular cycles.

Landform and landscape modification is continuous

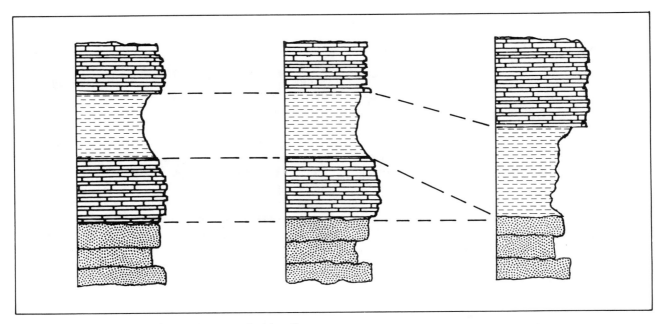

Fig. 34. Lithologic correlation between separated locations

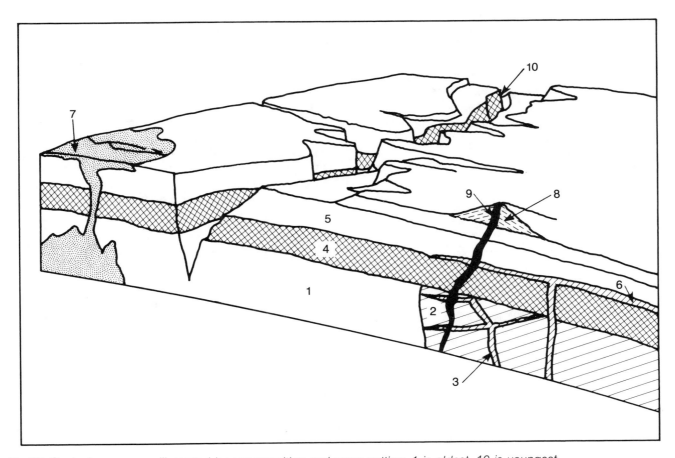

Fig. 35. Geologic sequence illustrated by superposition and cross-cutting. 1 is oldest, 10 is youngest.

and sequential. It therefore establishes a predictable continuity and can be helpful in determining certain aspects of relative geologic age dating.

Absolute Time

Until the mid-twentieth century, absolute geologic age dating techniques were nonexistent. All dating involved relative parameters that permitted estimates for absolute ages, which were subject to considerable error. The development of absolute age dating techniques, however, eliminated much of the error and now routinely confine geologic events to narrow time ranges.

Absolute geologic time is determined by measurements involving the half-lives of certain radioactive elements. Half-life is a measure of radioactive decay and is the time required for half of a certain element to be eliminated by the loss of its radioactive particles. If the half-life of an element is known and is compared to the amount of that element remaining in a deposit, the time elapsed since the mineral was originally crystallized can be calculated. This procedure is enhanced by the production of daughter products, the amounts of which are predicated upon the half-life of the original element and which are therefore indicative of the time elapsed since crystallization.

Some radioactive element isotopes decay very rapidly and have short half-lives. They are not useful for measuring long periods of geologic time. There are, however, other isotopes that have slow radioactive decay rates and correspondingly long half-lives. These elements are very suitable for measurements of large geologic time increments and are useful in reaching back into the early stages of earth history.

The process whereby radioactive elements decay is accomplished by the emission of particles from the nuclei of the atoms of which the element is composed. As the particles are released, the atom is converted to a different element, or daughter product, with a mass and atomic number that is smaller than that of the original radioactive element. The daughter product can be a different element altogether, or it can be an isotope of the original element. Uranium 238 is an example of a radioactive element that through decay produces a series of radioactive isotopes ending in the production of lead 206, a non-radioactive end product. Measurements derived from this particular decay process produce ages that are considered accurate to within less than 3% error.

Some radioactive elements can be used for dating young rocks as well as older rocks. The procedure using potassium-argon is one of these and is capable of dating rocks up to 1.5 billion years old (Fig. 36).

Dates of rocks obtained by these methods are called radiometric dates. Radiometric dates are determined from the time of crystallization of the radioactive element as it cools within its host magma. In metamorphic rocks, the time of formation occurs when new minerals are formed from pre-existing minerals during recrystallization in the metamorphism process.

A somewhat different radioactive age-dating process is based upon radiocarbon or the carbon isotope carbon-14, which has a half-life of 5690–5730 years. Nitrogen-14 is the resulting daughter product of decaying carbon-14 and as a diffusing gas cannot be measured. However, cosmic radiation of nitrogen-14 produces radioactive carbon-14, which becomes mixed with ordinary carbon-12 in carbon dioxide gas. In this form, it is absorbed by plants that are consumed by animals, resulting in a balance of carbon-14 as long as the organisms are alive. At death the balance is destroyed, and because replenishment has ceased, the carbon-14 in the organic tissues is lost through radio-

ISOTOPE				
Original	Daughter	Half-Life	Dating Range	Materials Dated
Carbon 14	Nitrogen 14	5700 ± yrs.	100-50,000 yrs.	Wood/plant material Animal material Sea water Subsurface water
Potassium 40	Argon 40 Calcium 40	1.3 billion yrs.	100,000-4.6 billion yrs.	Mica Amphibole
Uranium 235	Lead 207	710 million yrs.	10 million-4.6 billion yrs.	Uraninite Pitchblende Zircon
Uranium 238	Lead 206	4.5 billion yrs.		

Fig. 36. Isotopic geologic age dating

active decay at a predictably decreasing rate. The decreasing rate yields an amount of radioactivity commensurate to the amount of time elapsed since the death of the organism and provides the age of the specimen at the time of testing.

Carbon-14 dates have been compared to historically dated samples that are less than 5,000 years old and have correlated well. Because this method involves testing of organic samples, carbon-14 age dating is particularly useful in archeological studies. Wood samples taken from trees affected by glaciation, marine advances, and recent uplift give good information on when these events occurred in the relatively recent past.

Radiometric Time Scale

Radiometric age dating for sedimentary rocks is difficult since they often consist of fragments derived from several sources. The ages of the individual constituent grains can be determined, but these do not date the time of deposition of the entire rock. However, dating of igneous and metamorphic rocks within the sedimentary column is very satisfactorily done and when applied with relative age data can bracket periods of geologic time.

Igneous and metamorphic rocks that are closely associated with sedimentary rocks will most accurately establish the ages of the sediments. In this manner, lava flows and ash falls interbedded with fossiliferous sediments offer the most accurate age determinations to be found in the geologic column.

Intrusive igneous bodies that postdate the surrounding sediments will give radiometric dates that will be relative to the dates of deposition of the sediments. Fossils in the sediments, dated from other exposures, can be important in determining the time difference between the sediments and the intrusion. Exposure of the intrusion to erosion and subsequent deposition of fossiliferous beds over it result in more relative ages, as well as nearly absolute ages derived from the fossils. The result of these procedures is a geologic time scale based upon a continuous narrowing of the age brackets that limit segments of the time scale.

Geologic Time Scale

Development of the geologic time scale began with the work of early scholars in western Europe during the eighteenth century and involved the concept of relative time since radioactivity had not yet been discovered. Early workers described rocks and fossils accessible to them and slowly established sequences according to stratigraphic position and the fossils contained therein. As numerous workers continued their work in many areas of western Europe, it became evident that correlations between localities were possible, and regional concepts of stratigraphy and superposition emerged. Recognizable stratigraphic units proved to be traceable from place to place as additional information became available, and gaps in the vertical and lateral sequences were slowly filled. In this manner, standard sections evolved and eventually became the geologic column, and the geologic time scale was developed.

Eventually, conventional nomenclature was required to integrate information from many places in order to standardize the geologic column and to make it useful to subsequent workers. A system of descriptive terminology was applied to the various intervals throughout the years and evolved into the stratigraphy of today. As this process began and continued, it became evident that increments of geologic time, and the rocks represented by the increments, required identification in terms of their relative stratigraphic positions, the time segments they represented, and the rocks deposited during those time segments. So the concept of stratigraphic units and how they represent time and rocks emerged.

Historical geology or the study of geologic history required the identification of sedimentary rocks and their fossils as well as igneous rocks and their effects upon the sediments. Since these factors involved segments of geologic time, they had to be integrated with it, and time units, time-rock units, and rock units were used to describe time and equivalent rocks.

Early workers described the rocks and fossils at locations in western Europe, referring to the rock sequences as systems and naming them according to local geography. Eventually it became evident that systems were too limiting for some scales of reference or were not sufficiently limiting for others. Systems represented time-rock units or the actual rocks that were deposited during a specific time interval, which was called a period. Groups of periods were called eras, while periods were divided into epochs and ultimately into ages. Systems were divided into series and further into stages. Rock units irrespective of time considerations were laid down as groups, formations, and members and indicated some measure of stratigraphic and depositional consistency. The relationships between time units, time-rock units, and rock units are shown as follows.

Time Units	*Time-Rock Units*	*Rock Units*
Eon		
Era	Erathem	Group
Period	System	Formation
Epoch	Series	Member
Age	Stage	

Evolution of the geologic column has been marked by significant changes since the early eighteenth century when the study of geology began. For a long time, dates were established by relative occurrences of events that were finally assigned absolute or nearly absolute ages by means of radiometric dating. The latest geologic column will likely undergo additional change as dating efforts continue and as accuracy increases. The geologic column can be illustrated in several ways, as illustrated in Tables 1, 2, and 3.

If the age of the earth (between 7 and 4.5 billion years) is about 4.6 billion years, approximately 80% of earth history occurred during Precambrian time or before the beginning of the Paleozoic. Phanerozoic time, which includes the Paleozoic, Mesozoic, and Cenozoic, began 570 to 600 million years ago and is represented by rocks that can be correlated in many places around the world. Precambrian correlations, however, are substantially more difficult, and though Precambrian rocks represent most of earth history, unravelling their historical significance has not been without problems.

The passage of geologic time relative to the occurrence of mammals is noteworthy in that primitive mammals appeared upon the earth perhaps 200 million years ago and did not begin to flourish until less than 65 million years ago. In the context of elapsed earth history, Man appeared in less than a snap of the fingers, about two million years ago.

Historical Geology

Development of the earth and its processes of construction, destruction, and continuing change bear directly upon its features at any particular time. Important mountain-building episodes produce certain features that affect the crust and its internal and external shape. Erosion of the resulting mountain ranges changes them and produces different sets of features, which themselves are progressively modified by yet another realm of processes superimposed upon the continuum. Since the earth is constantly changing, certain of its features undergo continuous alteration. What we see and identify as familiar and stable is in reality a segment in the process of change that guarantees nothing remains the same.

Part of the expression of change is manifested in the processes that relate to the generation, migration, and accumulation of hydrocarbons. We have located and produced large quantities of oil and gas that accumulated through the life-bearing parts of geologic time. Consid-

erable time was required to concentrate the oil and gas into economically valuable accumulations. Considerably less time was required to produce and deplete many of the concentrations that took so long to form. Important in our ability to locate oil and gas has been the history of the accumulations and therefore the geological history of the earth during those times of formation and accumulation.

Inasmuch as geologic time does not stop and geologic processes continue, the formation of oil and gas has been occurring some place on the earth at all times since at least the early part of the Paleozoic era and probably the late Pre-Cambrian. Evaluation of geologic processes and their historical significance bear directly upon exploration for oil and gas. Not only are environmental conditions of erosion and deposition important, but so are their chronological relationships. Historical geology helps us find oil and gas.

Studying historical geology and the sequence of geologic events utilizes the integration of parameters from a variety of scientific disciplines. They are used to develop historical sequences that identify changes in environments of deposition, erosion, and deformation relating directly and indirectly to oil and gas considerations. Not only are sequences established but their relations to relative and absolute geologic time are evaluated and integrated into worldwide concepts of historical geology. In this way, geologic events from all over the world are placed in the sequence of the passage of time.

All geologic parameters are significant in the historical sequence of the earth. They change the earth in the ebb and flow of constructive and destructive processes. They do so by gradually evolving into each other and maintaining an orderly sequence, the occurrence of which is fragmentarily preserved in the geologic record. New geologic data continually improve the reliability of our concepts of historical geology and fill in the gaps in the geologic record.

Without integration into a time frame, geologic events cannot represent punctuations of geologic time or provide temporal values to durations of time segments. However, the geologic time scale is becoming more accurate as more data become available, and time considerations involving geologic processes have more significance. It is not impossible now to deal with geologic parameters in terms of how much time is required for them to occur.

The order of sequential geologic events is fundamental to the determination of geologic history. For several centuries during the early development of geology, geologic

Table 1
The Standard Geologic Column

Era	Period		Epoch	Atomic Time
Cenozoic	Quaternary		Holocene	
			Pleistocene	—2–3—
	Tertiary		Pliocene	—12—
			Miocene	—26—
			Oligocene	—37–38—
			Eocene	—53–54—
			Paleocene	—65—
Mesozoic	Cretaceous		Late / Early	—136—
	Jurassic		Late / Middle / Early	—190–195—
	Triassic		Late / Middle / Early	—225—
Paleozoic	Permian		Late / Early	—280—
	Carboniferous	Pennsylvanian	Late / Middle / Early	
	Systems	Mississippian	Late / Early	—345—
	Devonian		Late / Middle / Early	—395—
	Silurian		Late / Middle / Early	—430–440—
	Ordovician		Late / Middle / Early	—500—
	Cambrian		Late / Middle / Early	—600—
Precambrian				3600

*Estimated ages of time boundaries (millions of years)

From Hamblin, W. K., *The Earth's Dynamic Systems*, 2nd edition, 1978. Permission to publish by the Burgess Publishing Company.

Table 2
The Geologic Time Scale

Era	Approx. Age in Millions of Years (Radioactivity)	Period or System Period refers to a time measure; system refers to the rocks deposited during a period.
Cenozoic	7 26 37-38 53-54 65	Recent (Holocene) Pleistocene — Neogene Tertiary — Pliocene, Miocene — Oligocene, Eocene, Paleocene — Paleogene
Mesozoic	136 190-195 225	Cretaceous Jurassic Triassic
Paleozoic	280 310 345 395 430-440 500 570	Permian Pennsylvanian — Carboniferous Mississippian Devonian Silurian Ordovician Cambrian
Precambrian	700 3,400 4,000 4,500	First multi-celled organisms First one-celled organisms Approximate age of oldest rocks discovered Approximate age of meteorites

From Foster, 1979. Permission to publish by Charles E. Merrill Publishing Company.

events were placed in the geologic column on the basis of their relative occurrences; that is, a stratigraphic sequence was deposited, then folded, and subsequently eroded. All of these events related to each other because when placed in order they represented a sequence. With the discovery of radioactivity, these events were not only established in sequence, but the times of their occurrences were determined as well. Historical developments were correlated to the time of occurrence and the amount of time required for their occurrence. Geologic history became a time-related process and geologic events assumed new significance.

Stratigraphy

Strata, or layers of sedimentary rock, vary in distribution, thickness and character with changes in depositional environment and sediment supply (Fig. 37). Changes in characteristics of sedimentary rocks are important since they control parameters important to the discrimination of potential petroleum source and reservoir rocks. *Stratigraphy* is the study and classification of layered rocks, their depositional succession, and geographic distribution. In its name is specific allusion to study of the origin and character of layered sedimentary rocks.

Table 3
Geologic Column

Uniform Time Scale		Subdivisions Based on Strata/Time		Radiometric Dates (millions of years ago)	Outstanding Events	
		Systems/Periods	Series/Epochs		In Physical History	In Evolution of Living Things
Phanerozoic	Phanerozoic	Cenozoic — Quaternary	Recent or Holocene / Pleistocene	0 — 2?	Several glacial ages	Homo sapiens
		Tertiary	Pliocene	6	Colorado River begins	Later hominids
			Miocene	22		Primitive hominids / Grasses; grazing mammals
			Oligocene	36	Mountains and basins in Nevada	
			Eocene	58	Yellowstone Park volcanism	Primitive horses
			Paleocene	63	Rocky Mountains begin	Spreading of mammals / Dinosaurs extinct
		Mesozoic — Cretaceous	(Many)	145	Lower Mississippi River begins	Flowering plants / Climax of dinosaurs
		Jurassic		210		Birds
		Triassic		255	Atlantic Ocean begins	Conifers, cycads, primitive mammals / Dinosaurs
		Paleozoic — Permian		280	Appalachian Mountains climax	Mammal-like reptiles
		Pennsylvanian (Upper Carboniferous)		320		Coal forests, insects, amphibians, reptiles
		Mississippian (Lower Carboniferous)		360		
		Devonian		415		Amphibians
		Silurian		465		Land plants and land animals
		Ordovician		520	Appalachian Mountains begin	Primitive fishes
		Cambrian		580		Marine animals abundant
Precambrian		Precambrian (Mainly igneous and metamorphic rocks; no worldwide subdivisions.)		1,000 / 2,000 / 3,000	Oldest dated rocks	Primitive marine animals / Green algae
						Bacteria, blue green algae
~4,650		Birth of Planet Earth		~4,650		

From Flint and Skinner, 1974. Permission to publish by John Wiley and Sons.

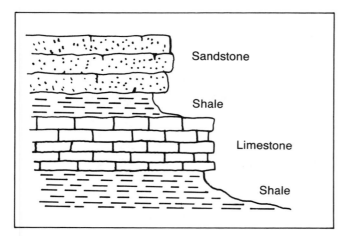

Fig. 37. Layered rock sequence illustrating strata of sandstone, shale, and limestone.

Sedimentary rocks are not the only rock type, however, since igneous and metamorphic rocks can be formed and emplaced before, after, or coincident with, them in the same or contiguous areas. Therefore, stratigraphic relations include the spatial and temporal formation or deposition of all rock types relative to each other.

Understanding rock units and sequences is paramount to classification of geographically and geologically significant stratigraphic increments that comprise paleographic and paleogeologic units in the geologic past. In this context the objective of stratigraphy is to describe and classify rock sequences and interpret them toward appreciating geologic events. Identification of rock sequences in the geologic record and in their depositional contexts defines the continuum of geologic history. It elucidates the four dimensional story necessary for complete prospect development and relates historically to deposition as well as generation, migration and accumulation of petroleum. A folded, conglomeratic, quartz-rich sandstone, for example, represents a high energy depositional environment followed by deformation. The sandstone is a coarse-grained reservoir that has the potential for containing petroleum in a structural trap.

Stratigraphic Methods: To define stratigraphic units it is necessary to study the compositional and physical parameters of rock constituents. If these factors are combined with faunal and floral considerations, the temporal and kinetic aspects of depositional environments can be determined. The addition of tectonic sequences, sediment source and transport, depositional systems, and life environments that are driven by time constructs the framework of geologic events in an historical continuum. The entire system establishes the geographic distribution of multiple rock sequences that preserve the story of geological history, which is determined by the stratigraphic method.

Each of the factors in the stratigraphic method is derived by detailed study of the observable parameters to construct the rock record in the context of geologic time. In this way geologic history is established from observation and classification of the rock sequences.

Rock Type or Lithology: Inasmuch as rocks are classified as igneous, sedimentary and metamorphic, their origins must be determined. The lithologies or rock types within the rock classifications illustrate the compositional and physical conditions of the formation.

Within the three classifications rocks are divided according to origin, composition, grain and crystal size and numerous other internal and external characteristics.

Stratified or layered sedimentary rocks contain features that demonstrate the spatial parameters of their depositional environments as they are manifest in the kinetic progress of time. By studying the parameters, the development of a channel sand, for example, can be observed as it occurs and extrapolated to similar events in the geologic record. The mechanics of depositional events can be temporally related to their development and the constructive or destructive sequence understood. Therefore, the characteristics of the channel sand illustrate stream-motivated clastic deposition of a particular current-related style that can be a reservoir rock under favorable conditions.

Fossils: Fossils are important in the understanding of depositional environments and in appreciating the increments of time involved. Many fauna and flora live within a narrow range of tolerable conditions and do so with a certain discernible life expectancy. Therefore, their environments of deposition and the time increments represented by their life expectancies contribute to understanding the entire sequence represented by them and many other similar and sequential scenarios.

Life forms appear, flourish, mix with new forms, evolve, decline, and eventually disappear. The environmental time significance of these events bears directly upon the depositional sequences and our understanding of them.

Evolution of the horse through the Cenozoic Era illustrates a change from a small multi-toed animal to the large, single hoofed form of today (Fig. 38). More than 50 million years and various documented physiological changes were required to modify the horse as it developed.

Rock Distribution: Distribution of rock types or facies is important to understanding the regional changes in depositional and faunal or floral environments. Identification of the changes establishes specific depositional en-

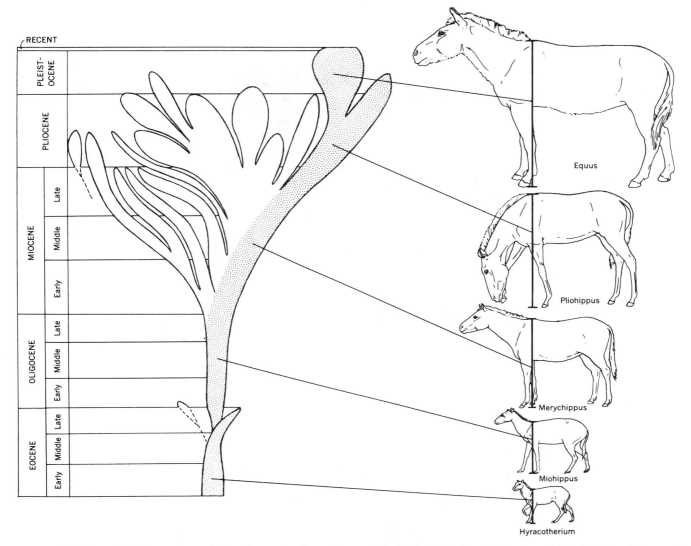

Fig. 38. Evolution of the horse through the Cenozoic. From Kay and Colbert, 1965. Permission to publish by John Wiley and Sons.

vironments in specific areas. In a delta, for example, (Fig. 39) distributary channels grade laterally into natural levees that become flood plain and swamp deposits away from the channels. The influence of marine deposition, manifest as channel mouth bars, lagoon sediments, and coastal sediments superimposed upon the delta illustrates the transition between terrestrial fluvial deposition and marine sedimentation. Similarly, an offshore barrier island grades seaward into thick, fine-grained shelf sediments and shoreward into thin, fine-grained lagoon deposits.

Distribution of various rock types demonstrates the limits

of the sedimentary and environmental conditions they represent. Lithofacies and isopach maps are among those that represent sediment distributions (Figs. 40 and 41).

Identification of the various environments permits discrimination of their areas of influence and establishes the dimensions of their distribution and geologic history. Facies maps based upon regional and local stratigraphic and sedimentologic analysis reveal the depositional development of the area and permits classification of the rocks.

Transgression and Regression: Inasmuch as sea level is routinely changing so do depositional conditions (Figs. 42 and 43). The sea encroaches upon the land and as it

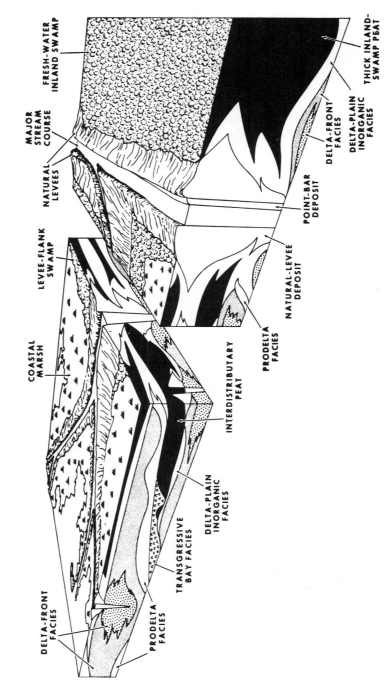

Fig. 39. Typical delta complex. From Frazier, 1967. Permission to publish by GCAGS.

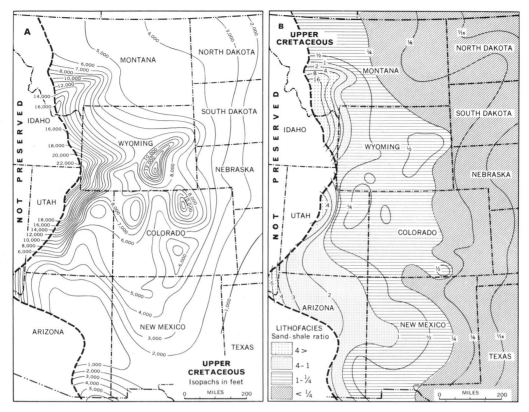

Fig. 40. Upper Cretaceous isopach and lithofacies maps. From Kay and Colbert, 1965. Permission to publish by John Wiley and Sons.

transgresses, conditions of deposition change: the water deepens and wave energy and sediment grain size diminish. As the sea retreats or regresses, depositional conditions change again and the water becomes shallower, and wave energy and sediment grain sizes increase. Perhaps emergence and erosion may occur. Water depths and energy conditions vary. Temperature, salinity, turbidity and faunal and floral conditions change. As all of these conditions change they are represented in the accompanying sedimentary deposits.

Periodic, consistent transgression and regression results in cyclic sedimentation that represents repetition of similar conditions. In many instances minimal changes in physical and/or chemical conditions produce cyclic stratigraphy, which can be repeated many times. Seasonal factors are manifest as cycles in the sediments of temperate latitude lakes, for instance, where summer and winter deposits are easily differentiated (Fig. 44). Layered, river floodplain deposits can be seasonally related to spring floods as well.

Paleogeography and Paleogeology: As depositional variations are correlated with regional factors, geographic considerations modify the sedimentation pattern. Geographic locations of a flood plain, delta, offshore barrier island, submarine fan or other feature are specified exactly and related to the time of their existence as the paleogeography (Fig. 45). Geologic parameters including rock types, facies changes, sediment thicknesses and depositional environment are geographically located, and the geology of the time is established as the paleogeology (Fig. 46).

Changes in paleogeography and paleogeology through time establish the geologic history and control its development. The temporal and spatial relationships become sequentially clear, and the influence of individual areas is integrated into understanding the entire regional scenario.

Criteria for stratigraphic study identify many rock

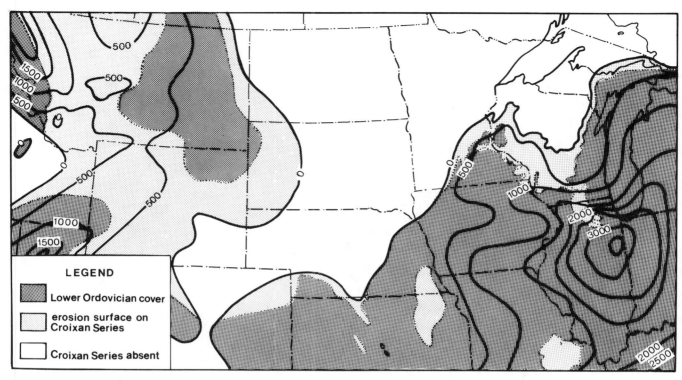

LEGEND

Lower Ordovician cover

erosion surface on Croixan Series

Croixan Series absent

Fig. 41. Isopach map of the Croixan (Upper Cambrian) Series. From Krumbein and Sloss, 1958. Permission to publish by W. H. Freeman and Co.

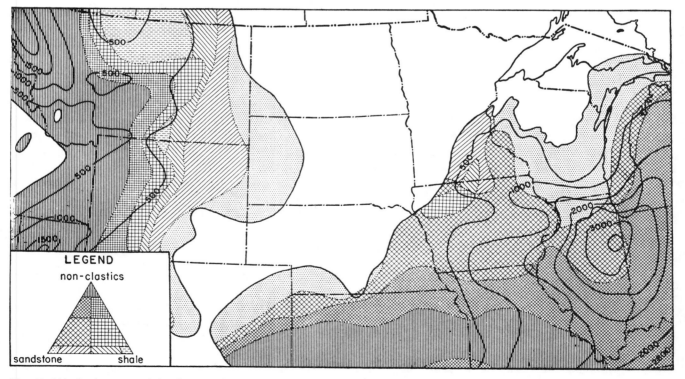

LEGEND

non-clastics

sandstone   shale

Fig. 42. Lithofacies map of the Croixan (Upper Cambrian) Series. From Kumbein and Sloss, 1958. Permission to publish by W. H. Freeman and Co.

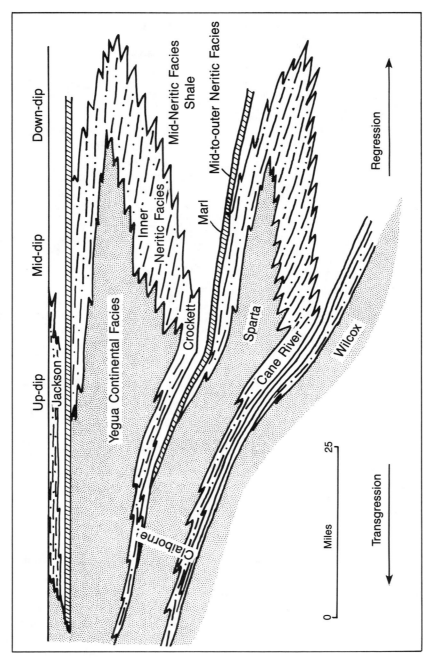

Fig. 43. Tertiary transgression and regression of the Gulf Coast of Texas. From Kay and Colbert, 1965. Permission to publish by John Wiley and Sons.

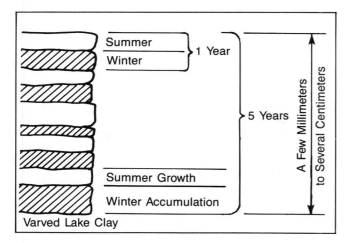

Fig. 44. Varved lake clay.

characteristics. They define the rocks, their distribution, sequences and how they illustrate geologic history and include:

1. Rocks
   a. Rock Types
      1. Igneous
      2. Sedimentary
      3. Metamorphic
   b. Sedimentary Depositional Regime
      1. Marine
      2. Deltaic
      3. Fluvial
      4. Lacustrine
      5. Glacial
      6. Desert
2. Stratigraphic Sequences and Correlation
   a. Lithology
   b. Stratification
   c. Superposition
   d. Fossils and Fossil Succession
   e. Facies Changes
   f. Depositional Cycles
   g. Unconformities
   h. Deformation
   i. Intrusion
   j. Metamorphism
3. Sedimentary Criteria
   a. Lithology
   b. Mineralogy
   c. Bedding
   d. Grain Size and Shape
   e. Grading

   f. Seasonal Bedding
   g. Shale, Ash, and Tuff
   h. Conglomerate
   i. Sedimentary Features
      1. Ripple Marks
      2. Crossbeds
      3. Mudcracks
      4. Borings
      5. Raindrops
      6. Channel Lenses
      7. Channel Cut and Fill
      8. Bottom Marks
      9. Soft Sediment Deformation
4. Deposition and Tectonics
   a. Deformation
   b. Source of Sediments
   c. Depositional Environments
   d. Sediment Distribution and Thickness
5. Depositional Environment
   a. Physical Parameters
   b. Rock Type
   c. Fossil Data
6. Paleogeography
   a. Sediment Distribution
   b. Depositional Environment Distribution
   c. Basins, Shelves, Land Areas
   d. Tectonic Features
7. Paleogeology
   a. Sediment Distribution
   b. Facies Changes
   c. Thickness Distributions
8. Geologic History
   a. Rock Types
   b. Deposition
   c. Deformation
   d. Fauna and Flora
   e. Integration

Paleontology

Fossil considerations in the stratigraphic record have been significant in assignment of ages to rock units and their lateral equivalents. The presence of fossils facilitated identification of rocks belonging to the same age when lithologic parameters changed significantly with geography. In early geologic endeavors, fossils with narrow vertical stratigraphic ranges (index fossils) represented the only means for realistic correlation and age assignment of rock sequences. As paleontologic methods became more sophisticated and accurate with increased geographic coverage and taxonomic (classification) com-

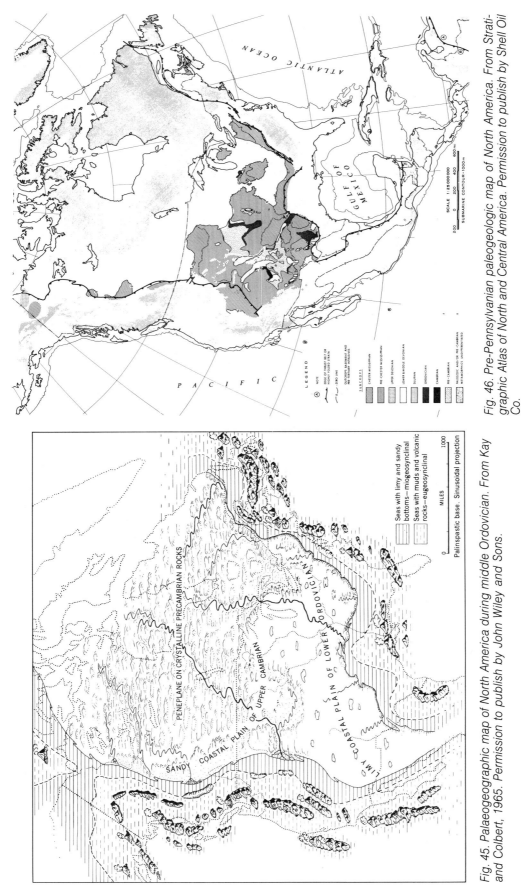

Fig. 46. Pre-Pennsylvanian paleogeologic map of North America. From Stratigraphic Atlas of North and Central America. Permission to publish by Shell Oil Co.

Fig. 45. Palaeogeographic map of North America during middle Ordovician. From Kay and Colbert, 1965. Permission to publish by John Wiley and Sons.

39

pleteness, geologic age dating improved. Subsequent application of radiometric parameters to floral and faunal methods provided absolute ages to a steadily widening sphere of accuracy.

Individual fossil occurrences are important age indicators and correlation devices for rock sequences from diverse areas. Like all organisms, those preserved as fossils change with time as they adjust to the influences of geologic time and environmental variations (Fig. 47). Evolution is the cause and result of these adjustments. It may be subtle over an extensive period or may be relatively abrupt. In either case the result is change to the extent that new forms emerge or old forms disappear. Recognition of the changes is paramount in the definition of age considerations, from the older to the younger. Fossil assemblages and progressive changes within them are as significant as individual forms and their changes.

Parameters for recognition of fossil changes through geologic time are manifold and include single fossil and fossil assemblage development:

a. Fossil characteristics: size, shape and physiology (Fig. 48)
b. Stratigraphic range of fossil (Fig. 49)
c. Fossil associations
d. Relative numerical proportions
e. Time of maximum occurrence (Fig. 50)
f. Fossil changes within fossil associations (Fig. 51)
g. Changes in numerical proportions within fossil associations and individual fossil changes
h. Times of fossil appearance, change and disappearance (Figs. 47 and 51)
i. Fossil characteristics at times of appearance, development and disappearance

Integration of single fossil and fossil assemblage development and radiometric data substantially improved the geologic time scale and provided absolute age bench marks for positive correlations and accurage geologic history.

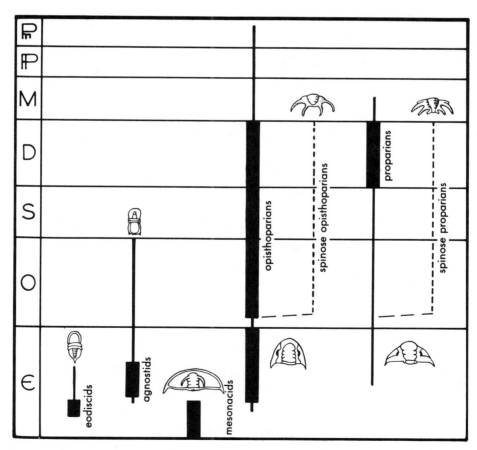

Fig. 47. Geologic range of trilobites. From Easton, 1960. Reprinted by permission.

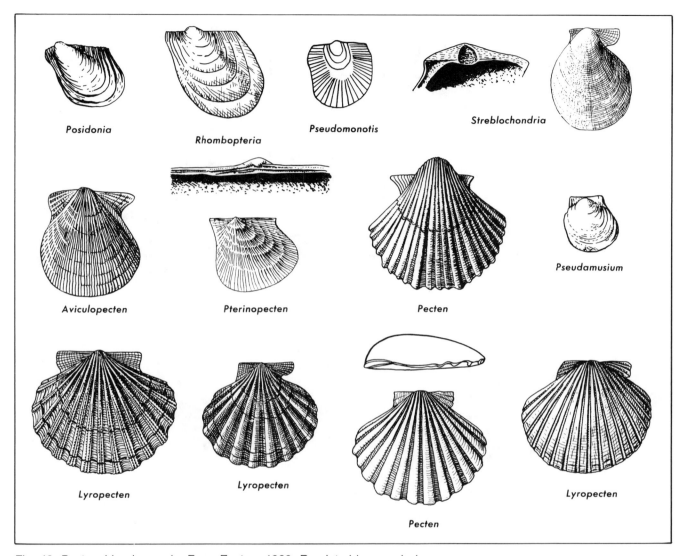

Fig. 48. Pectenoid pelecypods. From Easton, 1960. Reprinted by permission.

Classification of Geological Systems

Definition of discrete geological increments is accomplished by relating lithologic and rock units to geologic time. This delineates time-rock units, consisting of rock sequences that identify particular segments of geologic time. Regardless of the amount of time or section of rocks represented, they can be classified according to their mutually defined temporal and lithologic constraints as time-rock units of a particular dimension. Classification of geological increments in this manner permits construction of an orderly sequence of geologic time and the rock intervals related to the increments of time. Some of the factors involved in classification of time-rock units of geological systems include faunal, lithologic, historic, and deformational parameters.

1. In rocks where narrow-range index fossils or faunal assemblages occur, age determinations and relations to rock sequences can be made. Some rock sequences contain few or no fossils with no specific age significance and must be classified by other parameters. However, faunal characteristics are often important and can provide significant data to the classification process (Fig. 53).

Narrow-range faunal occurrences are best for establish-

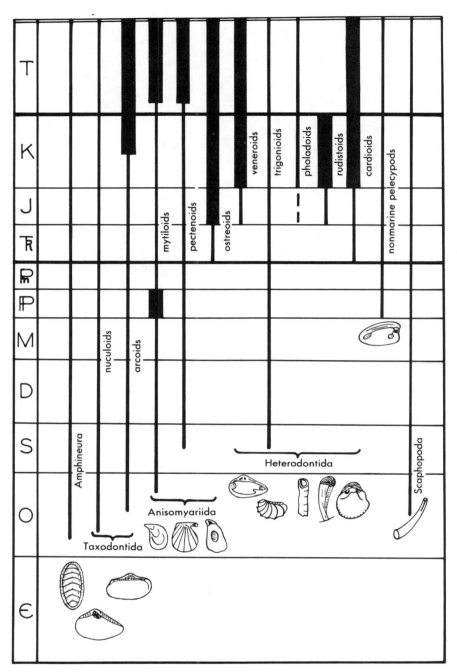

Fig. 49. Geologic ranges of pelecypods, chitons, and scaphopods. From Easton, 1960. Reprinted by permission.

ment of specific time segments. Wide faunal ranges encompass large segments of time and do not pinpoint specific ages. Groups of particular faunal assemblages, particularly within narrow stratigraphic ranges, are used in the same way as individual fossils. The presence or

absence of individuals within a faunal assemblage can indicate the beginning or end of a certain segment of geologic time and designate a time-rock unit.

Faunal evolution causes changes in fossil forms. Changes in fossils often occur at particular times in the

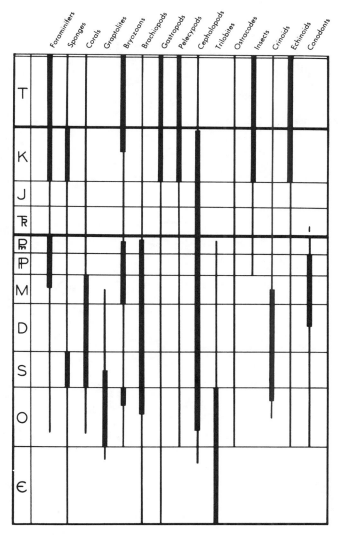

Fig. 50. Ranges of common fossil groups. From Easton, 1960. Reprinted by permission.

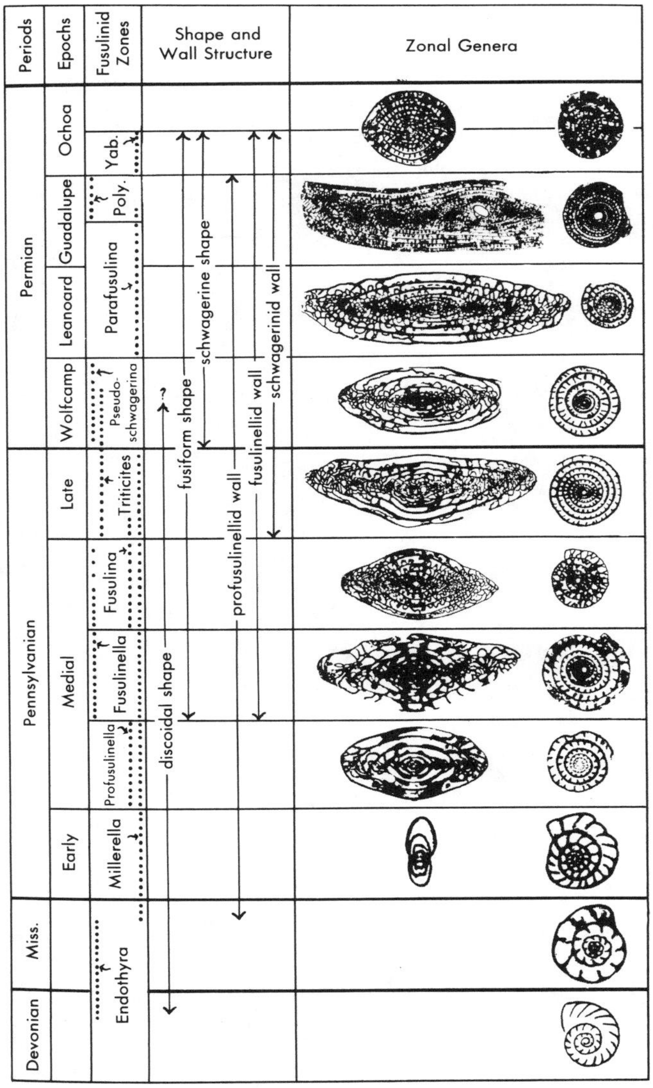

Fig. 51. Ranges of fusulinids. From Easton, 1960. Reprinted by permission.

evolution of the organisms and establish time segments by the introduction of newly evolved forms. Time-rock units can be classified by faunal successions as one organism characteristic of an older unit is replaced by another in a younger unit. As faunal forms evolve, they tend to become extinct as they are modified or die out completely. When organisms that occupy very narrow geologic ranges flourish and then abruptly die off, they are useful as time indicators.

2. In some cases, rock types are indicative of certain time-rock units and are important in their classification.

This is true of individual lithologies and can be equally true for recognizable sequences of rock types. Rock types often change laterally as depositional environments change from place to place (Fig. 54). This means that time-rock units classified at their type localities assume different characteristics elsewhere and are recognized by parameters other than the original classification factors. In the process of changing laterally, however, time-rock units assume the nomenclature of their local areas and their names change although their ages remain the same.

3. Rock units that represent time intervals are desig-

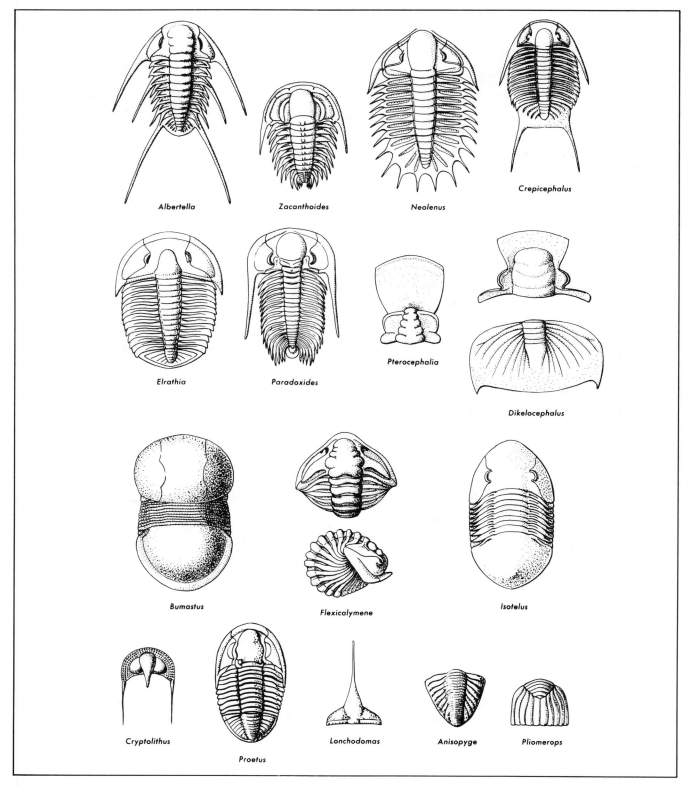

Fig. 52. Cambrian Opisthoparian trilobites. From Easton, 1960. Reprinted by permission.

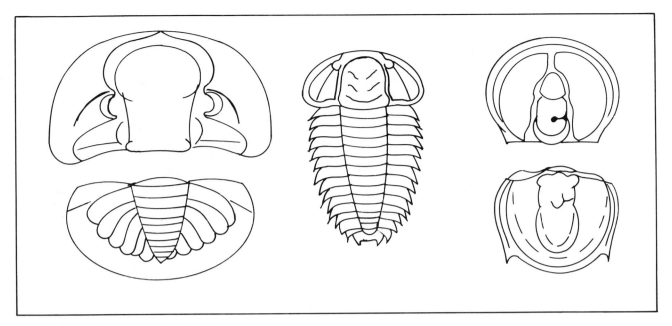

Fig. 53. Some Cambrian index fossils

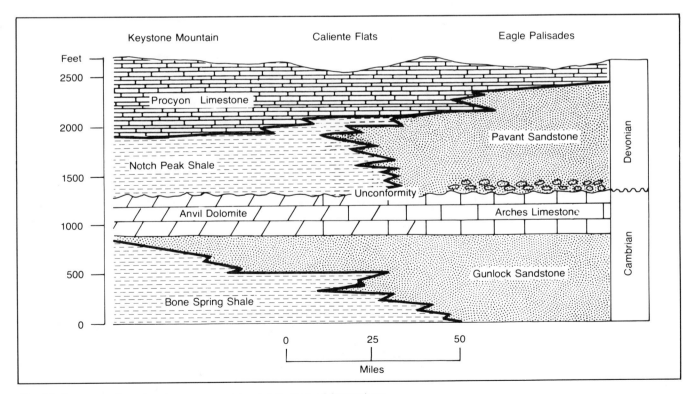

Fig. 54. Cross-section showing geographic distribution of formations

nated by geologic events that can mark the beginning or the end of the time segment when they were deposited (Fig. 55). Erosion, deformation, and intrusion are several geologic events that punctuate geologic time and define individual time-rock units.

4. Regional classifications involve identification of time-rock units between localities in order to maintain continuity and integrity of the geologic system. Regional identification is accomplished by correlation of sections of rock cropping out at various localities. Correlation utilizes rock thickness, lithology, trace elements, fossils, and depositional environments to establish lateral equivalency (Fig. 56).

5. Relative rock ages are now being given absolute values by radioactive dating methods. Absolute ages are correlated with faunal occurrences, which are subsequently used to establish absolute ages in rocks where radioactive procedures are not possible. Correlations based upon absolute ages are very significant in the time scale and permit comparisons of time-rock units of the same age that were deposited under different conditions.

6. Sedimentary rocks are deposited in sequences with the oldest beds on the bottom and the youngest beds on top. Superposition establishes the relative ages between the bottom beds and top beds but does not discriminate absolute ages (Figs. 57, 58). Vertical sequences and up-

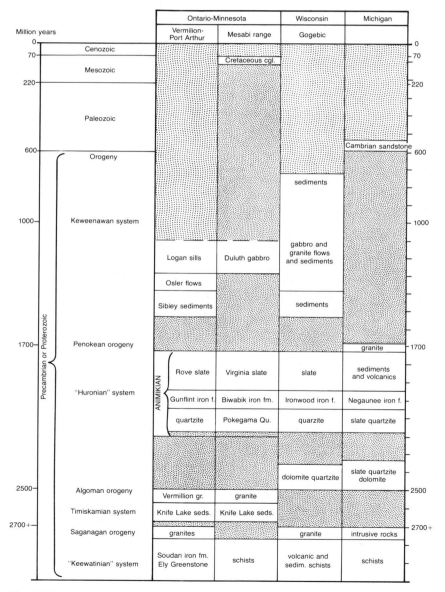

Fig. 55. Tectonic history of central North America. From Kay and Colbert, 1965. Permission to publish by John Wiley and Sons.

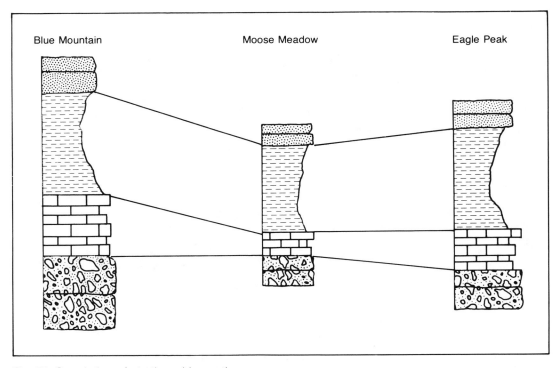

Fig. 56. Correlation of stratigraphic sections

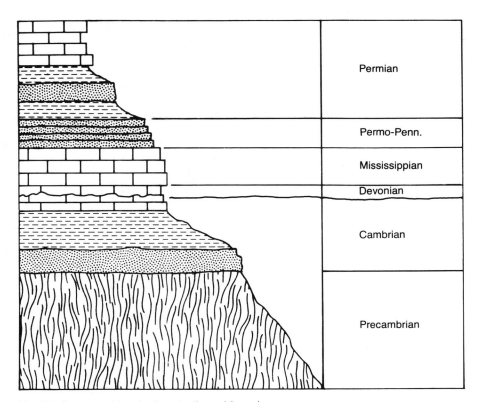

Fig. 57. Superposition in the stratigraphic column

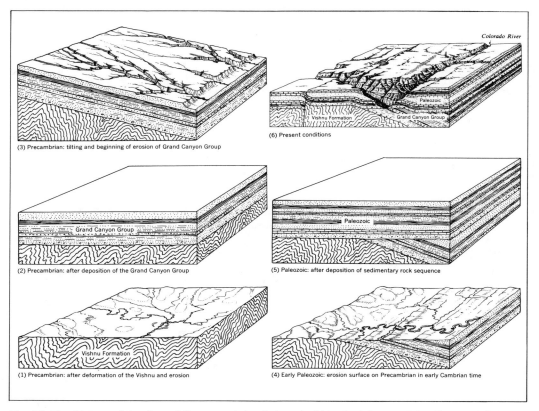

Fig. 58. The history of the Grand Canyon of the Colorado River in Arizona. From Kay and Colbert, 1965. Permission to publish by John Wiley and Sons.

ward facies changes are useful in the classification of time-rock units in localities where other information is not available.

7. Erosional periods often establish the top and bottom limits of time-rock units (Fig. 59). If the erosion maintains stratigraphic consistency, the resulting unconformity is particularly useful in establishing the beginning and/or end of a particular time-rock interval.

8. Rock sequences, in proper superposition, are sometimes the only criteria for time-rock classifications (Fig. 60). The sequences must be recognizable to be useful and must represent specific segments of geologic time. Some rock sections have no fossils, unconformities, or other defining features and must be classified according to lithologic and stratigraphic sequences.

Regional and Local History

Geologic history comprises consideration of all events that occurred throughout geologic time. Fundamental segments of geologic history involve how geologic events combined to produce a certain stratigraphic, faunal, and structural configuration, as well as the timing of the events.

History of deposition considers depositional or emplacement environment and factors related to the distribution of particular rock types. Igneous, sedimentary, and metamorphic environments are necessary to the history of deposition of emplacement.

The evolution of fauna and flora relate directly to depositional environments and geologic history. Their particular stages of evolution, succession, and extinction relate directly to when they occurred during geologic time. Therefore, they provide important information on depositional conditions as well as the time of deposition.

Deformation is often accompanied by deposition and therefore is suggestive of it by characteristics of depositional environment. Some deformation occurs after deposition and has no correlation with depositional conditions. In each case, however, deformation punctuates the geologic history and is representative of the constantly changing continuum.

Without *erosion*, there would be no sediments with which to supply depositional environments. Erosion can only occur, however, if there is some rock exposure to break down and transport. This must take place after deposition of rocks; otherwise there would be no erosion.

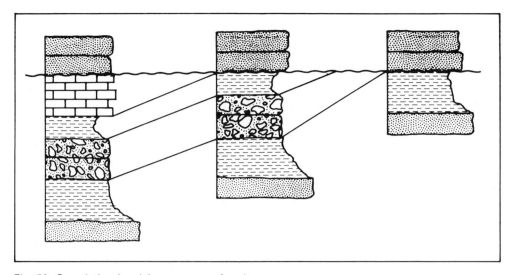

Fig. 59. Correlation involving an unconformity

So erosion becomes a part of continuing geologic processes and a part of geologic history.

Historical geology of any area involves the chronologic scheme of its development. It comprises deposition, evolution, deformation, and erosion set to a pattern of geologic time (Fig. 61).

History of Continents

Continental histories are a function of the same developmental factors involved in local or regional considerations. However, history of continents is normally expressed on a larger scale and integrates the histories of individual local or regional areas.

Deposition and erosion on the continental scale establishes the historical sequences of geological events. But continental history can also involve several different types of plate tectonic activity. This means that sequences of events will be different in various locations on a particular continent.

Geological processes affect all parts of the earth. Those processes that affect the crust affect the generation and accumulation of oil and gas and are important in how and what sequences they operate. Historical geology charts the continuum of geological processes and allows the explorationist to develop his prospecting parameters.

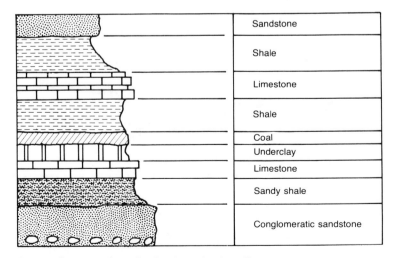

	Sandstone
	Shale
	Limestone
	Shale
	Coal
	Underclay
	Limestone
	Sandy shale
	Conglomeratic sandstone

Fig. 60. Section of cyclically deposited sediments

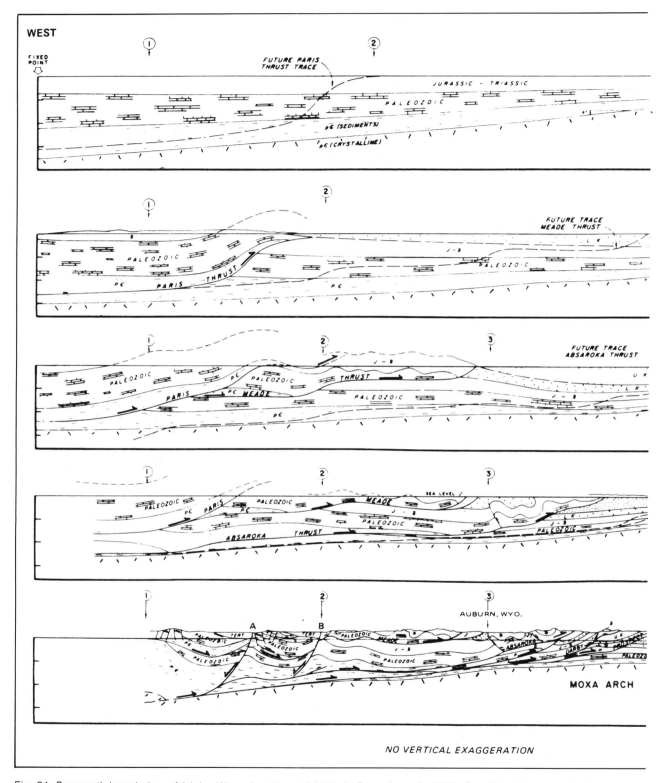

Fig. 61. Sequential evolution of Idaho-Wyoming thrust-fold belt. From Lowell, 1985. Reprinted by permission.

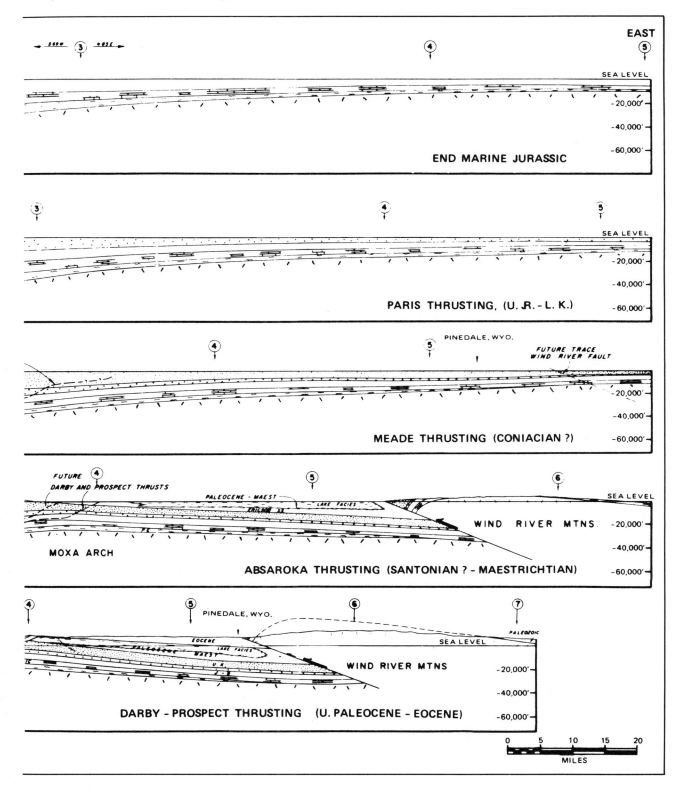

EAST

SEA LEVEL

-20,000'

-40,000'

-60,000'

END MARINE JURASSIC

SEA LEVEL

-20,000'

-40,000'

-60,000'

PARIS THRUSTING, (U. JR. - L. K.)

PINEDALE, WYO.

FUTURE TRACE
WIND RIVER FAULT

-20,000'

-40,000'

-60,000'

MEADE THRUSTING (CONIACIAN ?)

FUTURE
DARBY AND PROSPECT THRUSTS

SEA LEVEL

PALEOCENE - MAEST

LAKE FACIES

WIND RIVER MTNS. -20,000'

-40,000'

MOXA ARCH

ABSAROKA THRUSTING (SANTONIAN ? - MAESTRICHTIAN) -60,000'

PINEDALE, WYO.

PALEOZOIC

EOCENE

PALEOCENE

LAKE FACIES

MAEST

SEA LEVEL

U. K.

WIND RIVER MTNS

-20,000'

-40,000'

DARBY - PROSPECT THRUSTING (U. PALEOCENE - EOCENE) -60,000'

0   5   10   15   20

MILES

3 Minerals and Rocks

Minerals

Minerals are the fundamental building blocks of rock materials in the earth. They are defined as naturally occurring inorganic substances with a definite chemical composition and specific crystal structure. Over 2,000 minerals have been identified in the earth. There are over 100 elements in the crust, which consists almost entirely of eight elements. The remaining elements account for less than 1% of the crust, which therefore has a very simple composition.

Average composition of the continental crust includes larger amounts of the most common minerals and much smaller percentages of the rare constituents (Table 4). The first eight elements (Table 5) are the most common and constitute the basic ingredients for most continental crustal rocks, which are predominantly silicates and to a lesser extent oxides. Even less common are sulfides, chlorides, carbonates, sulfates, and phosphates.

Inasmuch as minerals are defined by composition and by crystal structure, these two parameters must be consistent to result in the predictable constancy of each mineral. Crystal structure is controlled by composition that directly determines which atoms of which elements will be distributed in which arrangement. With consistency in composition and crystal structure, certain physical characteristics of each mineral will be established within certain limits. However, because some minerals can occur in several polymorphs or physical forms, some of the physical properties expected for one particular polymorph will not apply to another form. At the same time, some physical characteristics will have a sufficiently wide range to be nondiagnostic by themselves but will be indicative when combined with other features.

Not all minerals have a distinctive taste, for example. However, it would be difficult to improperly identify rock salt on the basis of its taste.

Physical Properties

Physical properties of minerals are often the only guidelines for field identification of specimens that may occur combined with other minerals or in very small amounts in a particular sample. A hand lens is often useful in identification of these specimens, which will display enough of several physical properties to be sufficiently diagnostic.

Physical properties of minerals can permit identification of hand specimens without involving the time consuming and often difficult laboratory procedures required for detailed chemical analyses (Table 6).

1. Minerals grow into consistent crystal shapes, which are controlled by the chemical composition of the mineral (Fig. 62). There are six crystal systems that include all minerals. Mineral specimens that are preserved as well-developed crystals can be much more easily identified than those that are not.

Some mineral compounds can form more than one mineral with entirely different crystal structures. These compounds occur as polymorphs, which are quite different in appearance. Several common minerals occur as polymorphs (Table 7). The study of mineral crystal forms is called crystallography, which is a highly refined geological specialty.

2. How a mineral resists scratching is the measure of its hardness. A mineral will resist scratching or demonstrate its hardness relative to the substance used to scratch

Table 4
Average Composition of Continental Rocks

Symbol	Name	Weight per cent	Symbol	Name	Weight per cent	Symbol	Name	Weight per cent
O	Oxygen	46.4	Nd	Neodymium	0.0028	Mo	Molybdenum	0.00015
Si	Silicon	28.15	La	Lanthanum	0.0025	Ho	Holmium	0.00015
Al	Aluminum	8.23	Co	Cobalt	0.0025	Eu	Europium	0.00012
Fe	Iron	5.63	Sc	Scandium	0.0022	Tb	Terbium	0.00011
Ca	Calcium	4.15	N	Nitrogen	0.0020	Lu	Lutetium	0.00008
Na	Sodium	2.36	Li	Lithium	0.0020	I	Iodine	0.00005
Mg	Magnesium	2.33	Nb	Niobium	0.0020	Tl	Thallium	0.000045
K	Potassium	2.09	Ga	Gallium	0.0015	Tm	Thulium	0.000025
Ti	Titanium	0.57	Pb	Lead	0.00125	Sb	Antimony	0.00002
H	Hydrogen	0.14	B	Boron	0.0010	Cd	Cadmium	0.00002
P	Phosphorus	0.105	Th	Thorium	0.00096	Bi	Bismuth	0.000017
Mn	Manganese	0.095	Sm	Samarium	0.00073	In	Indium	0.0000100
F	Fluorine	0.0625	Gd	Gadolinium	0.00073	Hg	Mercury	0.000008
Ba	Barium	0.0425	Pr	Praseodymium	0.00065	Ag	Silver	0.000007
Sr	Strontium	0.0375	Dy	Dysprosium	0.00052	Se	Selenium	0.000005
S	Sulphur	0.026	Yb	Ytterbium	0.0003	A(r)	Argon	0.000004
C	Carbon	0.020	Hf	Hafnium	0.003	Pd	Palladium	0.000001
Zr	Zirconium	0.0165	Cs	Cesium	0.003	Pt	Platinum	0.000001
V	Vanadium	0.0135	Er	Erbium	0.0028	Te	Tellurium	0.000001
Cl	Chlorine	0.013	Be	Beryllium	0.00028	Ru	Ruthenium	0.000001
Cr	Chrome	0.010	U	Uranium	0.00027	Rh	Rhodium	0.0000005
Rb	Rubidium	0.009	Br	Bromine	0.00025	Os	Osmium	0.0000005
Ni	Nickel	0.0075	Ta	Tantalum	0.0002	Au	Gold	0.0000004
Zn	Zinc	0.0070	Sn	Tin	0.0002	He	Helium	0.0000003
Ce	Cerium	0.0067	As	Arsenic	0.00018	Re	Rhenium	0.0000001
Cu	Copper	0.0055	Ge	Germanium	0.00015	Ir	Iridium	0.000000
Y	Yttrium	0.0033	W	Tungsten	0.00015			

From Foster, 1979. Permission to publish by Charles E. Merrill Publishing Company.

Table 5
Main Elements in the Continental Crust

Element	Weight per cent	Atom per cent	Volume per cent
Oxygen	46.40	62.17	94.05
Silicon	28.15	21.51	.88
Aluminum	8.23	6.54	.48
Iron	5.63	2.16	.48
Calcium	4.15	2.22	1.19
Sodium	2.36	2.20	1.11
Magnesium	2.33	2.05	.32
Potassium	2.09	1.15	1.49
Totals	99.34	100.0	100.00

From Foster, 1979. Permission to publish by Charles E. Merrill Publishing Company.

it or the substance that the mineral can scratch. A hardness scale involving ten minerals from the softest to the hardest was developed by German mineralogist Friedrich Mohs during the nineteenth century. His scale, known as Mohs' hardness scale (Table 8), is the standard by which mineral hardnesses are determined. The hardness scale is not linear since the variation in hardness between successive minerals may be many times the hardness variation between other successive minerals.

Only fresh mineral surfaces should be tested for hardness. Weathered mineral surfaces will result in hardness measurements that can be quite different from the hardness of the fresh unweathered mineral.

3. In numerous cases, color is a physical property of limited value in mineral identification. Quartz can be of nearly any color because of slight mineral impurities. Specimens of quartz have been collected that are clear,

Table 6
Physical Characteristics of Some Minerals and Rocks

	Comp.	Xt. Sys.	Sp.G.	H	Color	Streak	Cleavage	Luster
				MINERALS				
Graphite	C	Hexagonal	2.1–2.2	1–2	Dark Gray	Dark Gray	Basal	Metallic
Chalcopyrite	$CuFeS_2$	Tetragonal	4.1–4.3	3.5–4	Brass Yellow	Green Black	Indistinct	Metallic
Galena	PbS	Isometric	7.4–7.6	2.5–2.75	Gray	Gray	Cubic	Metallic
Pyrite	FeS_2	Isometric	4.9–5.1	6–6.5	Brass Yellow	Green Black	Indistinct	Metallic
Hematite	Fe_2O_3	Hexagonal	4.9–5.3	5.5–6.5	Earthy-Silvery	Red-Brown	Indistinct	NonMetallic/ Metallic
Magnetite	Fe_3O_4	Isometric	5.17	5.5–6.5	Black	Black	Indistinct	Metallic-Dull
Limonite	$2Fe_2O_3 \cdot 3H_2O$	None	3.6–4	5–5.5	Brown, Yellow	Brown, Yellow	Yellow Brown	Silky
Sphalerite	ZnS	Isometric	3.9–4.1	3.5–4	Yellow, Brown Black	Brown to Yellow	Dodecahedral	Resinous
Corundum	Al_2O_3	Hexagonal	3.95–4.1	9	Blue, Red, Brown	—	Basal	Vitreous
Fluorite	CaF_2	Isometric	3–3.25	4	White, Purple	White	Pyramidal	Vitreous
Calcite	$CaCO_3$	Hexagonal	2.7	3	White	White	Rhombic	Vitreous
Gypsum	$CaSO_4 \cdot 2H_2O$	Monoclinic	2.3	1.5–2	White, Pink	White	1 Good + 2	Pearly
Halite	NaCl	Isometric	2.1–2.6	2.5	White	White	Cubic	Vitreous
Quartz	SiO_2	Hexagonal	2.65	7	White, Clear Variety	—	None	Vitreous
Milky Quartz	SiO_2	Hexagonal	2.65	7	White	—	None	Vitreous
Flint	SiO_2	Hexagonal	2.65	7	Gray, Brown	—	None	Subvitreous
Microcline	$KAlSi_3O_8$	Triclinic	2.54–2.57	6–6.5	White, Buff, Pink, Green	Colorless	1 Good + 2	Vitreous
Labradorite	$NaCaAlSi_3O_8$	Triclinic	2.6–2.62	6–6.5	Iridescent, Blue	Colorless	1 Good + 2	Vitreous
Pyroxene	$CaMgFeAlSiO_6$	Monoclinic	3.2–3.6	5–6	Brown, Black, Green	Colorless	1	Vitreous, Dull
Amphibole	$CaMgFeAlNaSiO_{12}$	Monoclinic	2.9–3.4	5–6	Black, Green	Colorless	1 Good + 2	Vitreous
Muscovite	$H_2KAl_3(SiO_4)_3$	Monoclinic	2.76–3	2–2.25	Colorless	Colorless	Basal	Vitreous
Biotite	$H_2K(MgFe)_3Al(SiO_4)_3$	Monoclinic	2.7–3.1	2.5–3	Black, Green	Colorless	Basal	Pearly
Talc	$H_2Mg_3(SiO_3)_4$	Orthorhombic	2.7–2.8	1–1.5	Green, White	White	Basal	Pearly
Serpentine	$H_4Mg_3Si_2O_9$	Monoclinic	2.5–2.65	2.5–4	Green	White	Indistinct	Greasy
Garnet	$Fe_3Al_2(SiO_4)_3$	Isometric	3.15–4.3	6.5–7.5	Red, Brown, Yellow	—	Indistinct	Vitreous

	Orgin	Texture	Porous	Permeable	Reservoir	Source	Remarks
			ROCKS				
Obsidian	Volcanic	Glassy	No	No	No	No	Volcanic Glass
Pumice	Volcanic	Bubbly	Yes	No	No	No	Frothy Glass
Granite	Plutonic	Coarse Crystals	No	No	No	No	Identifiable Crystals
Basalt	Volcanic	Fine Crystals	No	No	No	No	Unidentifiable Crystals
Conglomerate	Clastic	Granular	Yes	Yes	Yes	No	Good Reservoir
Sandstone	Clastic	Granular	Yes	Yes	Yes	No	Good Reservoir
Shale	Clastic	Very Fine Grains	Yes	No	No	Yes	Good Source
Limestone	Organic/ Clastic	Lime Mud/Shells/ Crystals	Yes	Yes	Yes	At Times	Reservoir/Source
Dolomitic Limestone	Clastic/ Precipitate	Granular/ Crystalline	Yes	Yes	Yes	At Times	Good/Source
Gneiss	Metamorphic	Banded	No	No	No	No	Banded Minerals
Schist	Metamorphic	Foliated	No	No	No	No	Foliated Minerals
Quartzite	Metamorphic	Massive	No	No	No	No	Tightly Cemented
Slate	Metamorphic	Foliated	No	No	No	No	Hard/Laminated
Marble	Metamorphic	Massive/Banded	No	No	No	No	Recrystallized

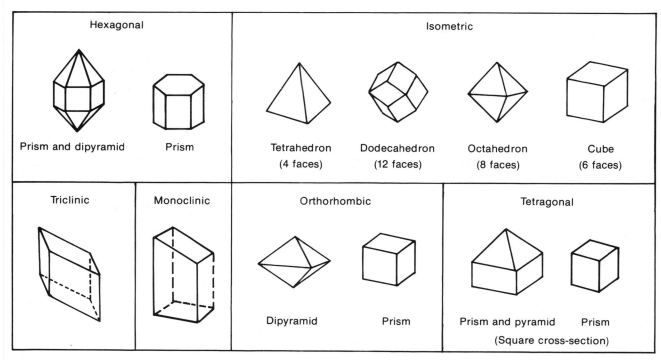

Fig. 62. Forms of the six crystal systems

cloudy, white, pink, purple, red, black, smoky, orange, and brown to name but several. Color is not usually sufficient by itself to identify a mineral; however, in combination with other physical properties it can assist in mineral identification.

4. The streak of a mineral is its color in the powdered form. Streak is obtained by drawing the mineral across an unglazed porcelain plate. In many minerals the streak is more diagnostic than color and is often very different from the mineral color. Minerals that are harder than the porcelain plate produce no streak.

5. Minerals have different weights relative to each other. Some are relatively heavy and others are light. Specific gravity of a mineral is its weight relative to the weight of an equal volume of water. It is obtained by dividing the weight of the mineral in air by the difference between its weight in air and its weight in water. Most mineral specimens are difficult to measure for specific gravity because they are usually associated with other minerals. However, very heavy or very light minerals are most easily recognized by specific gravity.

6. How a mineral reflects light determines its type of

Table 7
Polymorphs of Some Common Minerals

Composition	Minerals
SiO_2	Quartz Chalcedony
C	Diamond Graphite
$CaCO_3$	Aragonite Calcite
FeS_2	Marcasite Pyrite

Table 8
Hardness Scale

Mineral	Hardness (Increasing Downward)	Reference Materials
Talc	1	
Gypsum	2	
Calcite	3	Fingernail
Fluorite	4	Copper Penny
Apatite	5	
Orthoclase	6	Pocket Knife
Quartz	7	Glass
Topaz	8	
Corundum	9	
Diamond	10	

luster. Most minerals are considered to have metallic or nonmetallic luster, with a few having a submetallic luster.

Metallic	Submetallic	Nonmetallic
Bright		Adamantine (bright)
Dull		Vitreous (glassy)
		Pearly
		Waxy
		Silky
		Greasy
		Resinous
		Earthy (dull)

Numerous terms are used to describe luster, and beyond metallic, submetallic and nonmetallic there is no standard scale currently in use.

7. How a mineral parts along weaknesses in its crystal lattice is referred to as cleavage (Fig. 63). Some minerals such as quartz are tightly bonded in their crystalline structure and produce no cleavage. Other minerals like mica may have one direction of cleavage while calcite has three directions. Cleavage is often very diagnostic as a mineral identification method since weaknesses in the crystal lattice will always be consistent in the same mineral. Some minerals with more than one cleavage direction may have one well-developed cleavage and one or more less well-developed cleavages.

8. Random breakage of a mineral along no particular orientation is referred to as its fracture. Some types of fracture are particularly diagnostic and others are entirely not diagnostic. Quartz and some other similar silicate minerals break along characteristic curved surfaces that look like slightly concave, circular seashells. These are called conchoidal fractures and can also be found along the chipped edges of window glass.

9. Some minerals have very distinctive taste qualities that facilitate their immediate identification. Rock salt (halite) and sylvite are two minerals with distinctive tastes.

10. A few minerals respond to common chemical reagents that can be applied to hand specimens. Limestone (calcite) responds readily to hydrochloric acid. Dolomite responds weakly to hot hydrochloric acid but reacts more vigorously when it is powdered. Other minerals respond to other chemicals that are more appropriately administered in the laboratory.

11. Radioactive minerals emit particles that activate various types of detectors and counters. This characteristic is important in the identification of some radioactive minerals and in exploration for them.

Rocks

Rocks are important to the petroleum industry because they provide source beds and reservoirs for hydrocarbons. They consist of aggregates of minerals (Table 9) in various proportions and are identified by their origin and composition. Differentiation of rock types is on the basis of their origin, which can be igneous, metamorphic, or sedimentary. Since the earth is considered to have derived from a molten source, all rock materials on the earth came from an original igneous beginning.

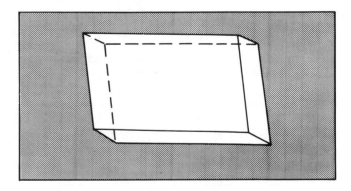

Fig. 63. Rhombohedral calcite cleavage

Table 9
Rock Forming Minerals

Silicates			Oxides	Sulfides	Carbonates	Sulphates	Phosphates
Garnet Olivine Pyroxene Amphibole	}	Iron and Magnesium Rich	Magnetite Hematite Limonite Corundum Quartz	Galena Sphalerite Pyrite	Dolomite Calcite	Gypsum Anhydrite Barite	Apatite
Feldspar Clay Mica	}	No Iron and Magnesium					

Hydrocarbons are usually considered to be associated with sedimentary rocks. Source beds for hydrocarbons are always sedimentary rocks. However, igneous, metamorphic, and sedimentary rocks produce hydrocarbons as reservoirs at various places around the world. Reservoir rocks need porosity, permeability, and access to source beds to be productive. It is not essential that they be exclusively sedimentary to meet these criteria.

Igneous Rocks

Igneous rocks are those that have resulted from the cooling and crystallization of magma. During the cooling process, minerals in the magma crystallize according to their compositions and sequentially according to their temperatures of crystallization. In the ideal slow-cooling situation, minerals in the magma have sufficient time to differentiate themselves from the other minerals and form large identifiable crystals. As minerals cool and crystallize, the remaining minerals are left to cool and differentiate according to their compositions and cooling temperatures, eventually leaving end-stage minerals as the last to form crystals. This process is called magmatic differentiation, which will proceed to whatever stage is allowed by the mineralogy of the magma and the length of the cooling process. Since the cooling process controls magmatic differentiation, igneous rocks are classified according to composition, origin, and related physical and chemical considerations, which include the following.

Origin. Igneous rocks are formed beneath the surface of the earth or on the surface. If they are emitted upon the surface they are extrusive rocks and are said to be volcanic or have a volcanic origin. Volcanic rocks cool rapidly and therefore crystallize rapidly since there is little time for crystal growth and differentiation of the mineral constituents. Rapid cooling causes the development of small crystals that are typical of aphanitic texture normally associated with volcanic rocks. Some volcanic rocks are ejected very rapidly, cool very quickly, and develop a glassy texture with no crystal structure.

Rocks that form beneath the surface are called intrusive or plutonic rocks. They cool slowly, differentiate their minerals, and form large crystals that result in phaneritic texture in which mineral crystals can be seen without a microscope.

Some igneous rocks originate as slow-cooling plutonic rocks in which minerals with high cooling temperatures crystallize and form large crystals. In some cases, the molten material that forms in this manner will be exposed to rapid cooling by erosion or contact with water, resulting in rapid crystallization and small crystals from the remaining melt. Two crystal sizes form from this activity. Large early crystals, or phenocrysts, surrounded by a late-stage finely crystalline matrix, or groundmass, produce porphyritic texture.

Some volcanic rocks are composed of fragments of ash and early-formed crystalline material. Explosive eruptions eject these fragments, which fall upon the surrounding landscape and accumulate. These deposits have a pyroclastic texture because they consist of "fire fragments." Tuff and ash flow tuff are examples of pyroclastic rocks.

Composition. Igneous rocks vary in chemical composition because of differences in the constituents of the original melt and stages of differentiation of the melt. Mineral constituents of igneous rocks provide another criterion for classification and are compositionally designated by their quartz content, type of feldspar, and type of ferromagnesian minerals. These criteria provide classification according to the reaction series described by Bowen (Table 10).

High-temperature igneous rocks consist of a mixture of olivine and pyroxene called peridotite. Peridotite is rarely found at the earth's surface or in continental crust. It is a phaneritic rock that comes from the mantle and occurs along the Mid-Atlantic Ridge and in some localities in the Alps.

Gabbro is a phaneritic rock consisting of olivine, pyroxene, and calcium plagioclase (feldspar) and forms at a lower temperature than does peridotite. The aphanitic equivalent of gabbro is its extrusive form, basalt.

Intermediate temperature rocks consist of amphibole, biotite, and a plagioclase that contains more sodium than calcium. The intrusive form of this rock is called diorite and has andesite as its aphanitic equivalent. Diorite sometimes contains small amounts of quartz and potassium feldspar.

Low-temperature igneous rocks are represented by granite as the intrusive form. Granite contains quartz, muscovite, and potassium feldspar (orthoclase). Rhyolite is the extrusive (aphanitic) equivalent of granite.

To illustrate the various classifications of igneous rocks, granite is selected to represent the criteria used. Granite is plutonic (intrusive), cooled slowly, cooled late, and contains large identifiable crystals. Quartz, muscovite, and potassium feldspar are its primary constituents, which indicate low temperature and late-stage differentiation. Large crystals in granite indicate that it cooled slowly.

Igneous Rocks and Petroleum. Igneous rocks have no

Table 10
Bowen's Reaction Series

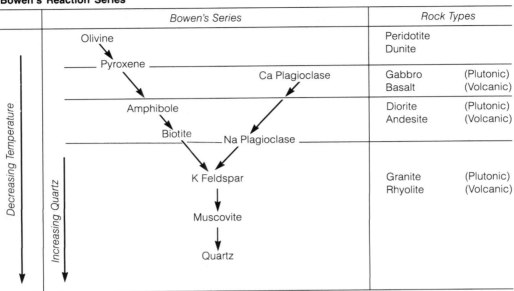

source potential. Fractured igneous rocks or porous and permeable weathered igneous rocks can and do provide adequate reservoirs in many locations including California, Nevada and Utah. The volcanic Garrett Ranch Formation (Fig. 64), which formed as a cooling incandescent cloud of airborne extrusive material, covered much of central Nevada less than 50 million years ago. It is an important oil reservoir in the Eagle Springs and Trap Springs fields of east-central Nevada. Oil produced from the Garrett Ranch formation was generated in underlying lake sediments and older marine beds and migrated into the reservoir.

Sedimentary Rocks

Rocks formed from materials derived from pre-existing sources that are transported and deposited are called sedimentary rocks. They may result from the accumulation of particles weathered from rock exposures, transported and ultimately deposited, may accumulate by precipitation from solution, or may result from buildup formed from skeletons or life processes of marine animals and plants. Sedimentary rocks can be classified according to environment of deposition, rock type, or by origin.

Rocks classified according to environment will include those classified by rock type and by origin. This means that an exclusive classification by environment is not realistic because it is not unique. Classifications by rock type and origin are useful only in the identification of a rock in absolute terms. Since sedimentary rocks come from many sources that can produce nearly identical lithologies, a combination of classification criteria is often useful. An example of this is a marine near-shore sandstone versus a near-shore sandstone deposited in a lake. The sandstone in each case is a clastic rock. However, the composition of the sandstone grains in each case is not determined and would need identification to complete the classification.

Classification by Environment of Deposition. Many environments of deposition are possible in the formation of sedimentary rocks. They include: marine, swamp, lacustrine (lake), fluvial (river), eolian (wind), estuarine, glacial, and lagoonal among others. Combinations of these environments are possible, for example, where a desert lake and sand dunes may come together.

Classification by Rock Type or Lithology. Since there are numerous types of sedimentary rocks, there can be many classifications based upon lithology. There are many types of limestone, but if a limestone is deposited as a reef, it is a reef limestone that can be further classified according to the reef-building organisms and whether or not it has been altered by the rock-forming or diagenetic process. A number of rock types include: limestone, feldspathic sandstone, sandy siltstone, chert, etc. All of these rock types came from a particular environment of deposition that further identifies them.

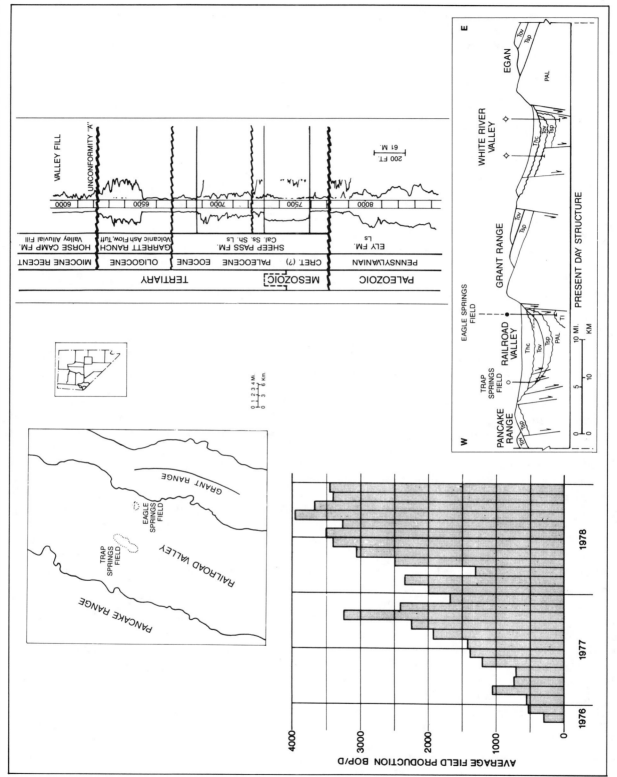

Fig. 64. Trap Spring Field map and sections illustrating Garrett Ranch igneous rocks as reservoirs. From Dolly, 1979. Permission to publish by RMAG.

Classification by Origin. Sedimentary rocks come from clastic sources, a variety of precipitates, and from organically related processes. Here again the origin is further classified by the rock type and the environment of deposition. In this manner, the rock is completely classified and identified by all significant parameters. An example of this classification is a quartzose deltaic channel sandstone. The sandstone was laid down as a clastic rock in a deltaic channel under marginal marine conditions and is composed of quartz grains. A complete description of the rock would include bedding characteristics, grain sizes, grain shapes, trace minerals, and modifiers where applicable to fully cover the background and present features of the deposit.

Clastic rocks are composed of clasts, which are fragments or grains of pre-existing rocks. The fragments are derived from the pre-existing rock, transported, deposited, and hardened, or lithified, into a new rock. Mechanical weathering is usually the manner in which the clastic particles are derived from the original rock. However, chemical weathering of susceptible materials surrounding resistant grains in the original rock produces rock and mineral fragments as well.

Clastic sedimentary rocks can consist of virtually any material as long as they are made of fragments of pre-existing rocks. It is on the basis of the size of the fragments that make up clastic rocks by which they are classified (Table 11).

Conglomerate is a clastic sedimentary rock composed of gravel of boulders, cobbles, and/or pebbles cemented together. Boulders are clastic fragments larger than 256 mm in diameter. Cobbles range from 64 to 256 mm, and pebbles from 2 to 64 mm. Boulders, cobbles, and pebbles in conglomerates are usually rounded by abrasion and are laid down in stratified units. The large conglomerate constituents are surrounded by a matrix of much finer sand and mud and are cemented together to form the hard rock. Porosity and permeability in conglomerate are variable with grain shape and amount of matrix materials. Conglomerates are often good reservoirs for hydrocarbons.

Sandstone fragments vary in size from 1/16 mm to 2 mm. Sandstone is composed of these sand-size particles that have been compacted and cemented together. Sorting of sandstone particles ranges from good to poor in a rock that can be designated as coarse, medium, or fine within the size range for sandstone. Since sandstone fragments can be of virtually any composition, the sandstone rock itself can be designated according to its mineral constituents or its combinations of constituents (Table 12). Examples are quartz sandstone, which consists of resistant quartz grains, feldspathic or arkosic sandstone with more than 20% feldspar grains, graywacke, which consists of poorly-sorted grains with abundant feldspar, and a clay matrix, some of which may be altered to chlorite.

Clean, well-sorted sandstones can be tightly or loosely cemented and therefore can have a wide range of porosity and permeability. Many Gulf Coast sandstones are clean, well-rounded, and weakly cemented. They provide good hydrocarbon reservoirs. Certain Ordovician sandstones of the mid-continent area are some of the best-sorted, best-rounded, clean sandstones developed anywhere in the world. They are loosely cemented and are excellent reservoirs in a number of mid-continent oil and gas fields. Sandstone bodies are formed in a large variety of environments and acquire a variety of shapes and characteristics. Because sandstone is deposited virtually everywhere, it is the rock type that occurs most often as a reservoir rock.

Sandstone reservoirs produce large volumes of petroleum in the North Sea, China, Soviet Union, and North and South America. The upper Paleozoic Gardner barrier bar is a productive sandstone in central Texas (Fig. 65).

Petroleum is produced from several lithologically dissimilar zones in the Prudhoe Bay Field of northern Alaska (Fig. 66). Principal production comes from the coarse clastics of the Sadlerochit Group (Fig. 67).

Sandstone can consist of grains of virtually any composition, and it follows that a sandstone of limestone grains is not only a possibility but a common reality.

Table 11
Clastic Sedimentary Rocks

Sediment	Size	Rock
Gravel	> 2 mm	Conglomerate
Sand	2–¹⁄₁₆ mm	Sandstone
Silt	¹⁄₁₆–¹⁄₂₅₆ mm	Siltstone
Clay	< ¹⁄₂₅₆ mm	Shale

Table 12
Mineral Composition of Some Common Sandstones

Sandstone Classification 2 mm–1/16 mm		
Particle Composition	<15% Clay Matrix	>15% Clay Matrix
Rock Fragments	Sandstone (Rock Fragment)	Argillaceous Sandstone (Rock Fragment)
Feldspar	Feldspathic: >30% Sandstone (Arkosic)	Argillaceous Feldspathic Sandstone
Quartz	Quartz Sandstone	Argillaceous Quartz Sandstone

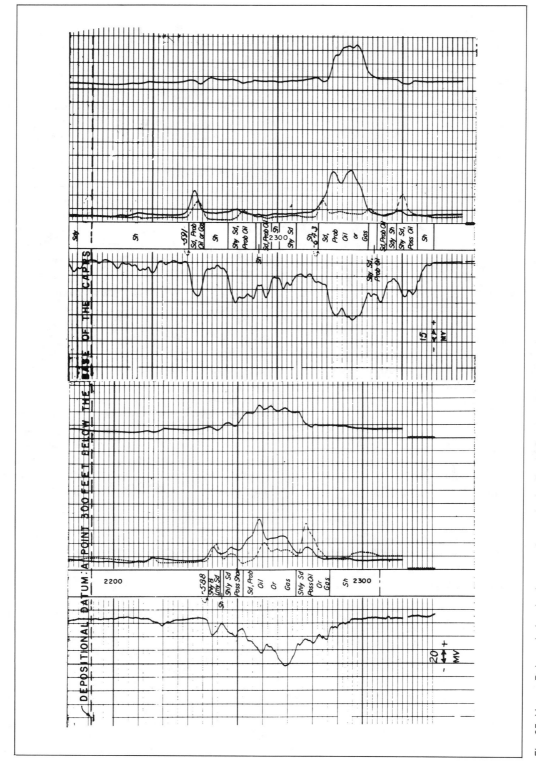

Fig. 65. Upper Paleozoic barrier bar sandstone in the Davis-Gardner Field, Coleman County, Texas. From Rothrock, 1972. Permission to publish by AAPG.

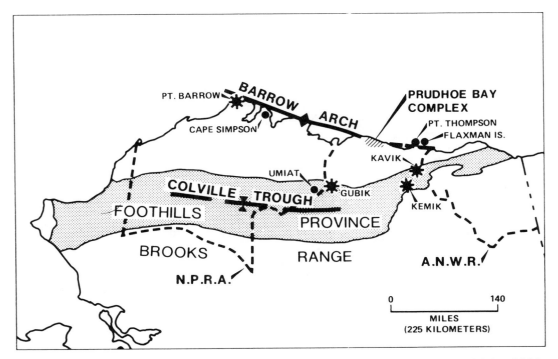

Fig. 66. Index map, Northern Alaska. From Jamison, et al., 1980. Permission to publish by AAPG.

Fragments derived from a pre-existing limestone deposit, transported and deposited, form a sandstone composed of limestone fragments. This particular rock is called a calcarenite, or calcareous sandstone.

Mud and fine-grained materials that are coarse enough to feel gritty on the teeth are the components of siltstone. Siltstone grain sizes range from 1/16 to 1/256 mm. Rocks of this classification have limited porosity and permeability because of their potentially high clay content and small grain sizes. They have limited potential as reservoir beds.

Clay and mud with particles of less than 1/256 mm diameter combine to form shale that is smooth and not gritty on the teeth. Shale is usually finely laminated and fissile, which allows it to split easily along its bedding surfaces. Rocks of similar composition that are not fissile are called mudstone, which has a blocky character.

To say that shale is the product of mechanical accumulation of small clastic fragments in a manner similar to the formation of sandstone is not completely accurate. Shale materials are often stable end-stage weathering products derived from unstable clay-forming minerals. Because they are stable they no longer respond to chemical breakdown. Clay minerals can be carried in chemically stable suspensions and colloids and will be deposited by chemical or electrolytic differences in the waters in the depositional areas. In this way the mode of clay mineral transport is different from the simple mechanical process of moving chemically inert grains from origin to place of deposition.

Shale, with high-preserved organic content, is considered to represent the most favorable of potential source rocks. After their formation waters and fluids are expelled, shales provide sealers for porous reservoir rocks and are instrumental in forming hydrocarbon traps. They are not generally considered to be reservoirs, however, although fractured shales can act in that capacity.

Organic-rich shale provides the petroleum source and subsequent seal for the highly productive Tertiary Zelten reef limestone of the Zelten Field in east-central Libya. Upper Paleozoic shale provides source and seal to numerous fields in northeastern Oklahoma as well.

Sedimentary rocks not formed by derivation from preexisting rocks include those that are inorganically and organically precipitated from water and those that form by accumulation of organic material or organically derived material (Table 13). Rocks that are inorganically precipitated are deposited from a sea water or lake water solution that achieves saturation and causes the deposition of the material in solution (Table 14). Limestone formed by precipitation from solution is deposited as lime

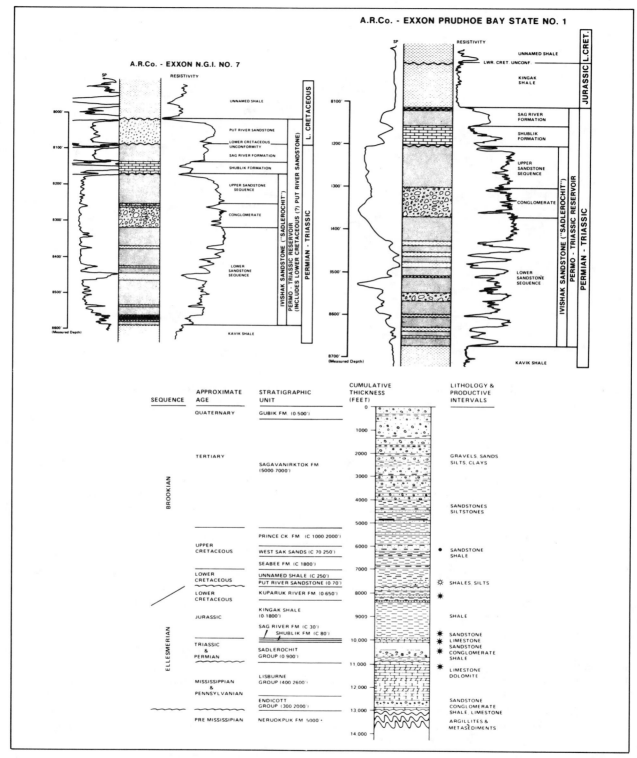

Fig. 67. Permo-triassic reservoir at Prudhoe Bay Field, Alaska. From Jamison, et al., 1980. Permission to publish by AAPG.

Table 13
Nonclastic Sedimentary Rocks

Rock	Physico-Chemical	Organic	Secondary
Coal		X	
Rock Salt	X		
Gypsum	X		
Chert	X	X	X
Limestone	X	X	
Dolomite	X Rare		X

mud and is typically dense and lacking in significant porosity. Chert is formed by the precipitation of silica from sea water. Unweathered chert has no porosity or permeability and is not a potential reservoir. Evaporite deposits result from increase in saturation of a solution as water is evaporated. The mineral constituents that are left behind precipitate out of the solution as their concentrations increase. Common evaporite deposits consist of rock salt, gypsum, potassium, and magnesium salts. Limestone and dolomite are potential evaporite deposits since they are sea water constituents. However, dolomite as an evaporite deposit is very rare.

Organically precipitated rocks are those that come out of solution in response to life processes of marine plants. Marine flora extract carbon dioxide from the sea water and cause the precipitation of calcium carbonate, which remains in solution as long as there is sufficient carbon dioxide dissolved in the water.

Marine fauna extract calcium carbonate and some phosphate minerals from sea water as part of the metabolic processes involved in forming their shells and exoskeletons. Reefs and banks of fossils and fossil fragments, including shells and corals, form by accumulation of these skeletal materials (Table 14). Rocks formed in this manner often have extensive porosity and permeability and are excellent reservoirs.

A variety of carbonate reservoirs produces petroleum in Mexico, Indonesia, the North Sea and the Gulf of Mexico. The Empire-Abo Field of southeastern New Mexico (Fig. 68) is an additional example.

Small spherical concretions of calcium carbonate develop in special marine environments where limestone precipitation and ocean waves and currents occur in a shallow water environment. Wave and current action continuously move the small concretions as calcium carbonate concentrically precipitates on a nucleus consisting of a small shell fragment or sand grain. The result of this precipitation under wave and current conditions is the formation of individual ooliths, which are presently being formed and deposited in the shallow waters of the Bahama Islands. Beds of oolitic limestone are often very porous and can be excellent reservoir rocks.

Sedimentary rocks provide petroleum source beds and reservoirs. Their classification, based upon origin, environment of deposition, and grain size, covers their wide range of occurrence but does not provide commentary on features acquired during their transportation and deposition.

Sedimentary Features. Deposition of sedimentary rocks includes transportation from the source area, reworking of sediments, settling rates, current and wave effects, and drying or desiccation. Common sedimentary features are those that, because of their frequent occurrence, are intrinsic to the concept of sedimentary rocks and have bearing on the type of origin involved in their deposition. Many sedimentary features have some relation to currents in rivers, along shorelines, and in shallow marine areas where sediments are deposited, reworked, redistributed,

Table 14
Depositional Texture of Carbonates

DEPOSITIONAL TEXTURE RECOGNIZABLE					DEPOSITIONAL TEXTURE NOT RECOGNIZABLE
Original Components Not Bound Together During Deposition				Original components were bound together during deposition ... as shown by intergrown skeletal matter, lamination contrary to gravity, or sediment-floored cavities that are roofed over by organic or questionably organic matter and are too large to be interstices.	*Crystalline Carbonate*
Contains mud (particles of clay and fine silt size)			Lacks mud and is grain-supported		(Subdivide according to classifications designed to bear on physical texture or diagenesis.)
Mud-supported		Grain-supported			
Less than 10 percent grains	More than 10 percent grains				
Mudstone	*Wackestone*	*Packstone*	*Grainstone*	*Boundstone*	

From Dunham, 1962. Permission to publish by AAPG.

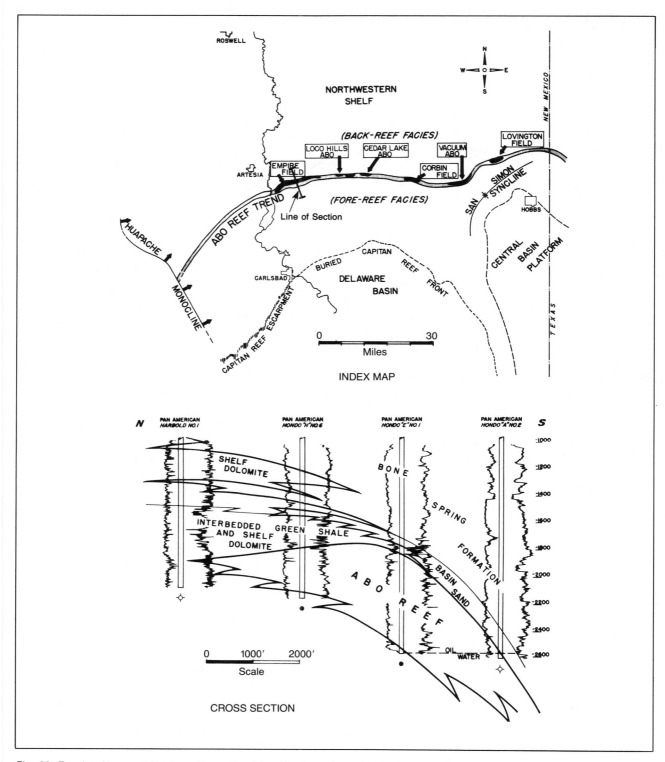

Fig. 68. Empire-Abo reef field, southeastern New Mexico. From LeMay, 1972. Permission to publish by AAPG.

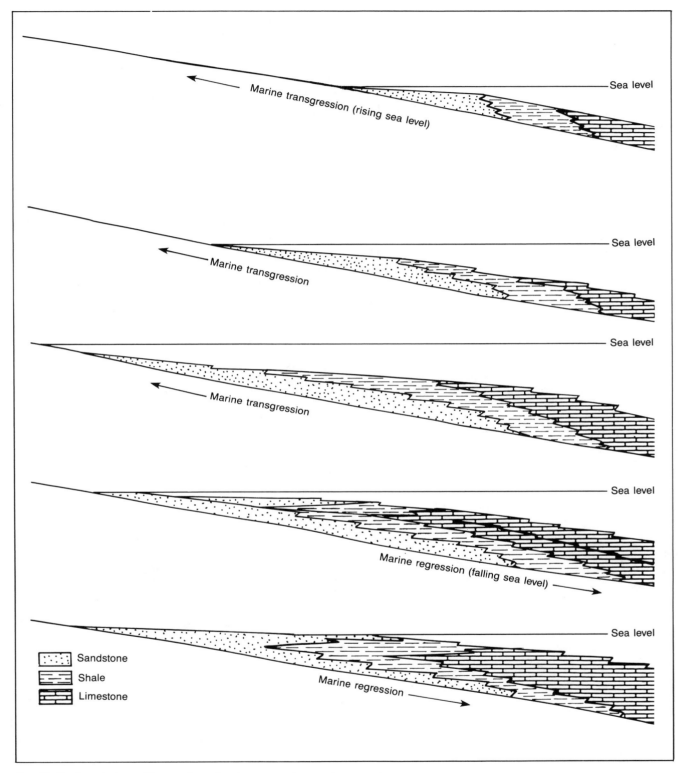

Fig. 69. Sediment deposition and marine transgression and regression.

and ultimately redeposited. In these areas, the types of sedimentary features reflect the environment of deposition as well as the dynamic conditions involved.

Stratification, or bedding, is the layering that is typical of sedimentary rocks (Fig. 69). It is essentially an indication of changes in derivation of the source material, transportation and deposition. Obvious stratification occurs when layers of different rock types succeed each other or are interbedded. It is easy to differentiate between successive layers of sandstone, shale, and limestone. Within a single rock type, however, bedding can also be obvious as variations in bed thickness and weathering characteristics evident in the outcrop.

Current-induced bedding is called crossbedding. It consists of wind or water current-formed dunes or ripples, the surfaces of which are not horizontal and are therefore across normal horizontal bedding. Crossbeds are asymmetrical and develop with their steeper slopes oriented in the downstream direction (Fig. 70). They can be less than an inch high or greater than several hundred feet high. The largest crossbeds can often be found in subaerial sand dunes though large dunes are associated with marine deposition as well.

A vertical, progressive, upward decrease in grain size in a sedimentary unit produces graded bedding (Fig. 71).

It occurs when a current bearing a sediment load suddenly slows down and loses its suspended materials, which settle out with the progressive reduction of its ability to transport them. Heaviest particles settle first and are followed by successively smaller particles. Graded bedding can often be repeated in a sedimentary section as depositional currents subject to velocity variations vary their load-carrying capacities.

Symmetrical ripple marks indicate that the sediments on which they are developed were subjected to wave action. Because there is no current involved, symmetrical or oscillation ripples indicate only that the sediments have been redistributed by wave processes. Asymmetrical ripples indicate current effects. They are steepest on their downstream sides and are therefore indicative of current direction.

Subaqueous sediments that are occasionally exposed to drying conditions often develop mud cracks. Mud cracks usually indicate a shallow water environment.

Sedimentary rocks include a variety of additional sedimentary features, some of which indirectly indicate the type of depositional environment. They include animal tracks, worm borings, raindrop imprints, and imprints of plant leaves and stems.

The following sandstone is an example of a sedimen-

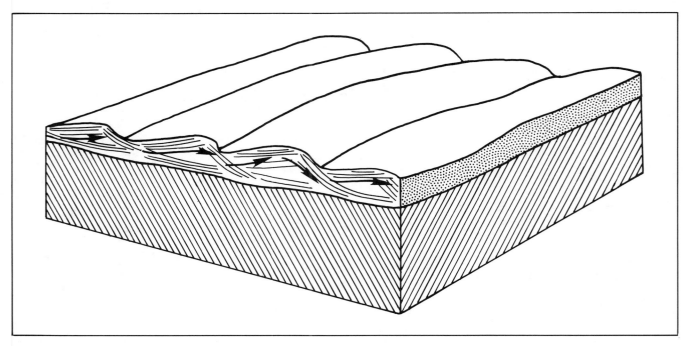

Fig. 70. Current-induced ripples and crossbedding

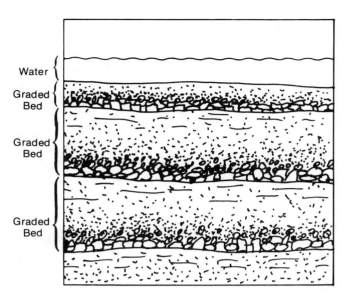

Fig. 71. Graded bedding

tary rock with various classification criteria and sedimentary features: sandstone, quartzose, medium-grained, well-sorted, well-rounded, frosted, with minor limonite cement, medium-bedded (1–3 ft), oscillation ripple marks on bedding surfaces, with some asphaltic stain on some grains. Interpretation of the environment of deposition of this sandstone is as follows:

1. Clastic sedimentary rock consisting of medium quartz grains in the mid-range of sandstone grain sizes, which vary between $\frac{1}{16}$ mm and 2 mm. Grains loosely cemented by limonite (hydrous iron oxide).
2. Grains are well-sorted, well-rounded, and frosted, indicating considerable transportation and substantial current and wave action.
3. The sandstone is medium-bedded with beds between 1 and 3 feet in thickness, indicating very uniform marine depositional conditions.
4. Oscillation ripple marks are indicative of wave action.
5. Medium grain size and minor limonite cement mean good reservoir characteristics that are compatible with the asphaltic oil stain.
6. This sandstone was probably laid down as a beach or barrier bar deposit and has good potential as a reservoir.

Sedimentary Rocks and Petroleum. All petroleum source rocks are sedimentary. Black shale, rich in preserved organic materials, is considered to have the best

source-bed characteristics. Some carbonate rocks have excellent source-bed potential. Recent data suggest that evaporites from highly saline environments may act as source beds.

Petroleum reservoir rocks include many types of sedimentary rocks. Sedimentary rocks containing intrinsic porosity and permeability such as sandstone, conglomerate, and reef limestone are considered good reservoirs. Sedimentary rocks with limited porosity can become good reservoirs if fractured. Sedimentary rocks can and do act as reservoir seals.

Metamorphic Rocks

Pressure, heat, directed pressure, and catalytic action related to chemical agents are important factors in changing pre-existing rocks into rocks with different textures and compositions. The process which involves these agents is called metamorphism and results in metamorphic rocks. Crystal structure, bedding, texture, and other characteristics are changed by metamorphism, which occurs locally or regionally depending upon conditions.

Metamorphic agents act upon rock in its solid phase. There is no return to a liquid phase for any of the minerals in a rock subjected to metamorphism. Water, which is present in the pore spaces of most rocks, can be an important constituent in the metamorphism process. It permits exchanges of ions and atoms to form new crystalline structures and promotes alteration of textures into new forms. In this process, however, water is the only liquid involved since rock materials remain solid.

Pressure. Most pressure that affects rocks is due to the weight of overlying rocks and is called overburden pressure (Fig. 72). Overburden pressure increases at an approximate rate of one pound per square inch for each foot of depth. At 5,000 feet, for example, the overburden pressure would be 5,000 pounds per square inch. Changes due to overburden pressure must occur at great depths and even under those conditions must involve others of the possible agents of metamorphism. Pressure derived from overburden is equally directed, is uniform on all sides, and represents an essentially closed pressure system.

Pressure resulting from a directed source is known as stress (Fig. 72). Stress occurs when pressure is unequally directed, and deformation or strain within the stress system is possible. Directed pressure results in dynamic metamorphism, which causes the formation of oriented minerals in the newly formed metamorphic rocks. It is also responsible for shear, which consists of failure of the

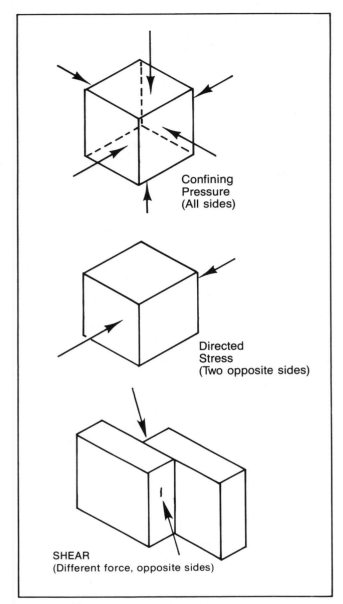

Fig. 72. Pressure (confining), stress (directed), shear (failure)

tal collisions and related tectonic events are important sources of directed pressure.

Chemical Agents. Chemical changes during metamorphism usually occur without addition or removal of mineral material to the rock undergoing change within the closed system. The process involves recombination of mineral constituents that are already present and results in more stable materials compatible with the new pressure and temperature regime. Fluids derived from the mineral materials already present provide the medium for transportation of atoms through existing pore spaces and promote the development of new crystalline materials. No materials have been added or removed from the system while new minerals are being formed.

Chemical changes can also involve the addition or subtraction of mineral materials in an open metamorphic system. Changes in the composition and mass of the rocks occur. Chemically charged fluids, including water, are introduced by magmatic intrusion, for example, and distributed through the rock. Chemical reactions take place, and new rocks are formed by enrichment or by elimination of some mineral constituents.

Types of Metamorphism. One classification of metamorphic rocks is based upon origin and includes static or regional, dynamic, contact, and pneumatolytic metamorphism.

Overburden pressure produces static metamorphism. Rocks which are statically metamorphosed are affected by the lithostatic and geothermal gradients, and those which are appreciably altered must have been very deeply buried. Regional metamorphism produces rocks, with recrystallized oriented mineral grains, which are said to be foliated.

Tectonic stress or directed pressure is responsible for dynamic metamorphism, which alters rock by breaking and crushing along zones of shear. Little recrystallization occurs during dynamic metamorphism.

Intrusive igneous bodies cause thermal metamorphism when they penetrate a pre-existing rock section. Batholiths, dikes, and sills are several types of intrusive bodies that produce contact metamorphic rocks, which are known as hornfels.

Fluids and solutions introduced into a rock sequence from an external source produce pneumatolytic metamorphic rocks. Pneumatolytic agents may be derived from hot intrusive bodies but need not be hot to be effective.

Textures of Metamorphic Rocks. Metamorphic rock textures consist of those that are layered or foliated and those that are massive or non-foliated.

Recrystallization of mineral constituents in metamor-

rock materials and is manifest along preferentially oriented zones of movement. In metamorphic rocks, shear is distributed along movement surfaces, which can be so numerous and closely spaced that deformation appears to be the result of flowage.

Directed pressure requires that stress originate and be delivered along a primary axis of compression. Continen-

phic rocks under regional metamorphism causes the development of oriented mineral grains that grow in the direction of least stress and result in foliation of the rock. As metamorphism is prolonged and recrystallization continues, recognizable mineral grains are aligned along the foliation and can grow into large mineral crystals.

Foliated or layered metamorphic textures are classified according to the grade of metamorphism, which ranges from low to high (Tables 15 and 16). Low grade metamorphism results in minimal change of the original rock and high grade metamorphism in significant change (Fig. 73). Foliation is divided into three categories, slate, schist, and gneiss.

Slaty cleavage is characteristic of low metamorphic grade and includes fine mineral grains oriented along closely spaced planes perpendicular to the direction of metamorphic stress (Fig. 74).

Table 15
Metamorphism of Clay Minerals and Silica

Sediments	Low Grade	Medium Grade	High Grade
Clay → Shale → Slate → Phyllite → Schist → Gneiss → Granite			

Table 16
Metamorphism of Basalt

	Low Grade	Medium Grade	High Grade
Basalt	⟶ Greenschist	⟶	Amphibole Schist

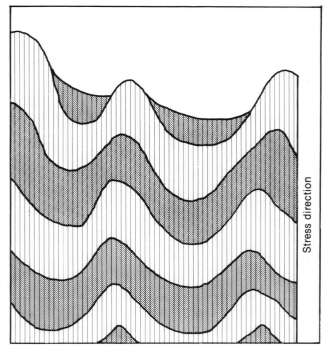

Fig. 74. Slaty cleavage in metamorphic rock

Fig. 73. Metamorphic texture and foliation

Medium-grade metamorphism results in schistosity, which contains larger grains than slaty cleavage and consists of platy minerals such as talc, chlorite, and mica. Other mineral crystals such as garnet may be found aligned along the well-developed foliation of the platy minerals.

Alternating layers of variable mineral composition result from high-grade metamorphism that produces gneiss and gneissic layering. Mineral constituents have large recognizable crystals that have had time to grow under sustained metamorphic conditions.

Slate, schist, and gneiss are foliated metamorphic rocks. Schist and gneiss are identified by their mineral constituents and are called chlorite schist, or mica schist, and granite gneiss, or amphibolite.

Metamorphic rock texture that contains no foliation or mineral grain orientation is massive or nonfoliated. Massive rocks can comprise a single mineral constituent like marble and quartzite or hornfels, which results from alteration produced by contact metamorphism.

Metamorphic Rocks and Petroleum. Metamorphic rocks (Fig. 75) have no petroleum source potential. Fractured and weathered porous and permeable metamorphic rocks are locally significantly productive (Table 17). The La

Original Rock		Low Grade	Medium Grade	High Grade
Shale	*Foliated*	Slate	Schist	Gneiss
Sandstone			Schist	Gneiss
Basalt		Greenschist	Amphibolite	Amphibolite
Gabbro			Amphibolite	Amphibolite
Granite			Schist	Gneiss
Limestone		Calcite Schist	Schistose-	Schistose
Dolomite			Marble	Marble
Limestone } Dolomite }	*Massive*	Marble	Marble	Marble
Quartz-Sandstone			Quartzite	Quartzite
Peridotite		Serpentinite	Pyroxenite	Pyroxenite

Fig. 75. Development of common metamorphic rocks

Paz field, Venezuela (Fig. 76), the El Segundo field, California (Fig. 77), and portions of the Central Kansas uplift (Fig. 78) are all productive from metamorphic reservoirs. Unfractured and unweathered metamorphic rocks have no permeability and can provide effective reservoir seals.

Rock Cycle

The molten origin of the earth indicates that igneous rocks are the fundamental and original rock type prior to any chemical and physical degradation. Development of the hydrosphere and atmosphere resulted in the break-down of igneous rocks into sedimentary rocks. Continuation of dynamic earth processes created metamorphic rocks from igneous, sedimentary, and other metamophric rocks and in turn were source materials for other sedimentary rocks.

New igneous, sedimentary, and metamorphic rocks are continually being formed from each other by currently active earth processes. This indicates a continuous cycle of rock production, alteration, and consumption to produce new rock types. The changing of rock types comprises the rock cycle (Fig. 79), which demonstrates that virtually any rock type can result from alteration of a pre-

Table 17
Basement Reservoirs

Name of Field	Location of Field	Reservoir Rock	Nature of Porosity	Initial Daily Production (per well)
La Paz	Venezuela	Metamorphic and igneous rocks	Fractures	Average 3,600 bbl; maximum 11,500 bbl
Mara	Venezuela	Metamorphic and igneous rocks	Fractures	Average 2,700 bbl; maximum 17,000 bbl
Orth	Kansas, USA	Precambrian quartzite	Fractures	Average 20 bbl; maximum 939 bbl
Ringwald	Kansas, USA	Precambrian quartzite	Fractures	Production low, 190 bbl
Silica	Kansas, USA	Precambrian quartzite	Fractures	100 bbl
Beaver	Kansas, USA	Precambrian quartzite	Fractures	55 bbl

Table 17 (Continued)
Basement Reservoirs

Name of Field	Location of Field	Reservoir Rock	Nature of Porosity	Initial Daily Production (per well)
Trapp	Kansas, USA	Precambrian quartzite	Fractures	173 bbl
Eveleigh	Kansas, USA	Precambrian quartzite	Fractures	434 bbl
Kraft-Prusa	Kansas, USA	Precambrian quartzite	Fractures	108 bbl
Kraft-Prusa	Kansas, USA	Arbuckle dolomite	Fractures, fissures, caverns	Up to 3,000 bbl
Hall-Gurney	Kansas, USA	Precambrian granite	Joints, fractures	Average 355 bbl
Gorham	Kansas, USA	Precambrian granite	Joints, fractures	306 bbl
Apco	Texas, USA	Ordovician dolomite	Fractures solution cavities	60 to 2,416 bbl
Embar	Texas, USA	Ordovician dolomite	Intercrystalline porosity, fractures and solution cavities	2,275 bbl
Edison	California, USA	Schist	Fractures	1,000 to 2,000 bbl
Santa Maria	California, USA	Knoxville sandstone	Fractures and intergranular	200–400 bbl; maximum 2,500 bbl
El Segundo	California, USA	Schist	Fractures	Maximum 4,563 bbl
Wilmington	California, USA	Schist	Fractures	1,200 to 2,000 bbl
Morrow County	On east flank of Cincinnati arch, Ohio, USA	Upper Copper Ridge dolomite (Upper Cambrian)	Fractures, caverns	200 bbl
Amal	Libya	Paleozoic Amal quartzose sandstone	Fractures	From hundreds of bbl to 1,000 bbl
Augila field	Libya	Weathered granite and rhyolite	Fracturing weathering, and solution cavities	1,500–7,627 bbl
Hassi Messaoud	Algeria	Cambrian sandstone	Primary porosity	30 wells at 6,000 bbl/day per well
Shaim	USSR	Paleozoic metamorphic rocks	Fractures	39 bbl
Yaerxia	China	Caledonian metamorphic rocks	Fractures	1,050 bbl
Yihezhuang	China	Ordovician limestone	Fractures and caverns	1,400 to 6,548 bbl
Renqiu	China	Sinian dolomites and Cambro-Ordovician limestones	Fractures and caverns	7,000 to 33,200 bbl
Xinglongtai	China	Archean granite and Mesozoic volcanic rocks	Fractures	210 to 756 bbl

(From Chung, 1982). Permission to publish by AAPG.

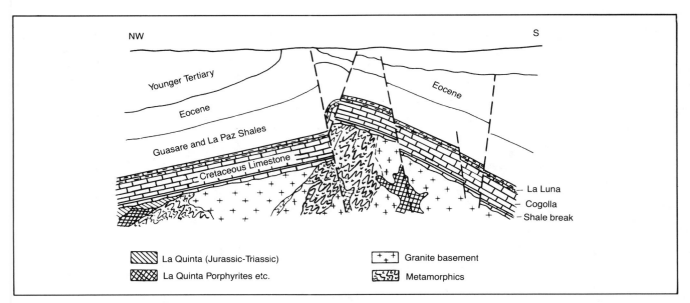

Fig. 76. Cross section of La Paz field, western Venezuela, metamorphic reservoir. After Smith, 1956. Permission to publish by AAPG.

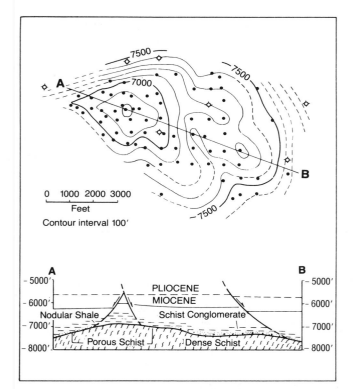

Fig. 77. El Segundo oil field, California, metamorphic reservoir. After Reese, 1943. Permission to publish by the California Department of Conservation, Division of Oil and Gas.

existing rock type whether by remelting, metamorphism, or erosion.

Rock Cycle and Petroleum

Sedimentary rocks are source and reservoir beds, and are important components of the rock cycle. They are the products of breakdown, transportation, and deposition of all three rock types and are developed in many different environments, some of which generate and accumulate petroleum.

The rock cycle forms, changes and destroys rocks as a continuum. Petroleum experiences a similar kinetic history during which time it is being formed, changed and destroyed with its host rocks. At any instant oil is being generated in some rocks, changing in others, migrating in still others and being eliminated by geologic processes. Oil is found where physical, chemical and temporal conditions are appropriate; where source beds are properly mature, migration has occurred, and the trap has integrity.

Igneous and metamorphic rocks do not generate petroleum but under proper circumstances can provide reservoirs and seals for petroleum accumulations. They may be additionally significant because igneous activity and metamorphism are not compatible with petroleum generation and accumulation and when present, destroy petroleum.

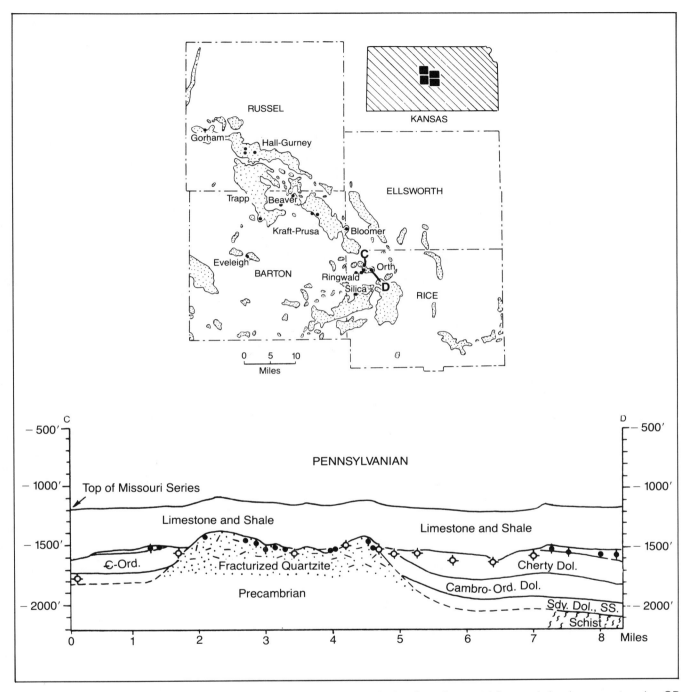

Fig. 78. Basement reservoirs of Central Kansas uplift and wells producing from fractured Precambrian basement rocks. CD is line of section shown below. Lower diagram: Section along line CD of figure above. After Walters, 1953. Permission to publish by AAPG.

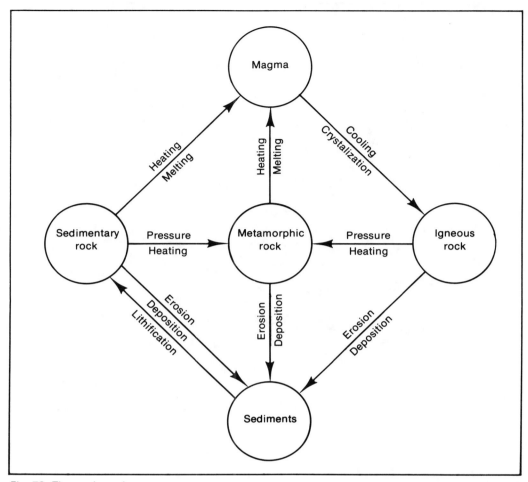

Fig. 79. The rock cycle

4

Weathering, Erosion, and Deposition

Weathering

Rock exposures undergo constant change when subjected to the elements. Wind, rain, temperature, and chemical agents combine to degrade the landscape. In the process, they produce sedimentary materials that are transported and subsequently deposited to form new rocks.

Weathering is the mechanism by which rocks are broken up and supplied to the agents that transport the fragments to their new depositional sites. The mechanism of breakdown is accomplished by mechanical, or physical, and chemical means, which often work together and are called mechanical and chemical weathering (Fig. 80).

Mechanical Weathering (Disintegration)

When rocks are physically reduced to fragments, mechanical weathering has taken place. The original rock has been broken down by physical means with no chemical changes.

1. Water that percolates through cracks in rock exposures freezes during cold weather and breaks fragments of rock away from the outcrop (Fig. 81). A daily freeze-thaw cycle causes considerable rock breakage during the course of a year and results in abundant mechanically derived rock material.

2. Erosion often removes rock material from underlying rocks and removes the weight of the overburden from them. Rocks that have been deeply buried under considerable overburden expand as the pressure is removed and break along layers parallel to the rock surface. Rock layers formed by sheeting in this manner are then susceptible to additional weathering and continued breakdown.

3. Rocks exposed to diurnal temperature changes over extended periods expand and contract many times and are subject to fatigue, which continuously produces rock fragments. Exfoliation results from mechanical and chemical activity which produces rounded, weathered shapes.

4. Roots of plants, growth of vines, and contact with moving tree branches can contribute to the mechanical weathering process. Lichens on rock surfaces can alter rock compositions, which can result in susceptibility to weathering.

Animals do relatively little weathering. However, game trails used by large animals and burrows formed by digging animals contribute to the overall process of mechanical weathering. Man, by virtue of his altering the landscape with earth-moving machines and off-road vehicles, causes considerable weathering in many localities.

Chemical Weathering

Chemical alteration and breakdown of rock exposures is accomplished by contact with chemical agents in the atmosphere, soil, flowing streams, rivers, and runoff. Water is a significant agent in chemical weathering, which is most active where the climate is warm and moist. Water places chemical agents in contact with rock outcrops and removes rock fragments keeping fresh surfaces exposed.

Most rock-forming minerals have some solubility in water. Chemical agents in aqueous solution often increase selected mineral solubilities. Water is an important solvent in most localities, and chemical weathering is a

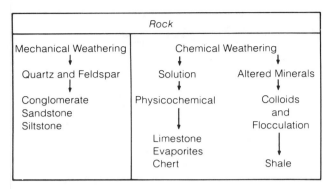

Fig. 80. Weathering and its products

function of the amount of rainfall and the chemical composition of the rock outcrops.

Limestone, which is soluble in water, is more soluble in water containing carbon dioxide. Rock salt and gypsum are also examples of water soluble materials.

Minerals that react to atmospheric oxygen form oxides that are often mobile and susceptible to other weathering agents. In this way, oxidation contributes to chemical weathering and rock breakdown.

Chemical combination of water and rock minerals breaks the minerals down and creates new minerals. Hydrolysis is an important chemical weathering process that affects a number of common rock-forming minerals. Potassium feldspar is broken down by hydrolysis as follows:

$$2KAlSi_3O_8 + H_2CO_3 + H_2O \rightarrow Al_2Si_2O_5(OH)_4 + K_2CO_3 + 4SiO_2$$

| Potassium Feldspar | Carbonic Acid | Water | Clay | Potassium Carbonate | Silica |

Potassium carbonate is readily soluble in this process. Clay minerals are much less soluble, and silica is virtually insoluble.

Weathering Products

The process of weathering can produce products of wide-ranging benefit. Some of these products can be economically important.

Weathering of bedrock results in the formation of soil where the weathering products remain essentially in place (Fig. 82). The soil profile represents the upper part of the regolith, which is the weathered transition between it and the underlying bedrock. Soil is an important weathering

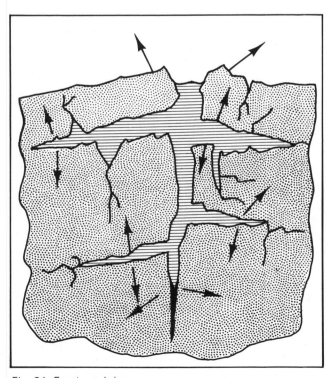

Fig. 81. Frost wedging

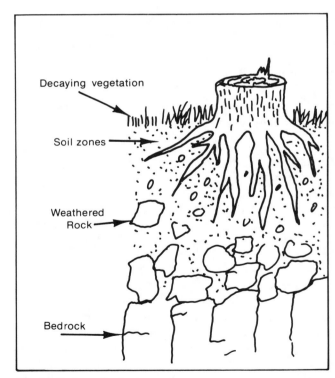

Fig. 82. Soil profile horizons

product because it is essential for plant life and the agricultural industry.

Weathering of exposed bedrock surfaces can improve porosity and permeability. Solution of carbonate terranes causes voids to develop as fossils are dissolved and cracks are widened. Improved reservoir characteristics are formed in rocks that, if not exposed to weathering, would have limited hydrocarbon accumulation potential.

Ore bodies of various kinds are enriched by weathering, which concentrates the important minerals by converting their original compounds to those with higher desirable mineral contents. Aluminum, iron, nickel, and lead are a few metals that, by alteration of their originally formed compounds, become commercially extractable.

Mineral components of bedrock exposures can be dissolved, carried in solution, and redeposited elsewhere. Some solubles, including rock salt and calcium carbonate, can form new deposits in which higher concentrations of the original minerals are laid down.

Rock fragments, derived from weathering outcrops, are transported and deposited as sediment. They ultimately become sedimentary rocks, which have significance to the petroleum industry as source and reservoir beds.

Factors Affecting Weathering

Inasmuch as weathering consists of a combination of variable rates, intensity and types of weathering are variable. However, there are general factors that are fundamental to weathering.

Some rock types are easily chemically weathered because their mineral constituents are susceptible to chemical decomposition by reagents that are common where they crop out. Limestone, rock salt, and anhydrite are easily chemically weathered and break down quickly in humid environments.

Other rock types that are chemically resistant may be strongly fractured and will respond to mechanical weathering. Fractured granite and sandstone are resistant to chemical weathering but yield to mechanical effects.

A humid climate is conducive to high rates of chemical weathering. Rocks susceptible to chemical weathering are significantly less weathered in a dry climate than under humid conditions (Fig. 83). For this reason, limestone, anhydrite, and salt can form important outcrops in an arid environment where they may underlie ridges and high topography. In a humid climate these same rocks may be exposed only in valleys that are topographically low.

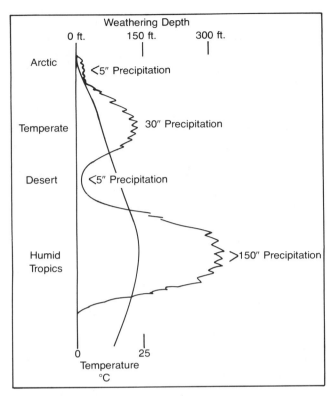

Fig. 83. Weathering relative to latitude, temperature, and precipitation

Unless weathering products are removed as they form, they will cover what solid rock remains and protect it from weathering. Steep slopes, however, promote removal of weathered rock material by rainfall and gravity, leaving fresh rock exposed to continued weathering. Weathered material accumulates on gentle slopes and flat surfaces, slowing the weathering process.

No weathering process is accomplished rapidly. The time required to weather a variety of rock types under widely varying conditions is accordingly variable. However, the longer the weathering process persists, the more thoroughly the affected rocks will be weathered.

Weathering and the Rock Cycle

Rocks of all kinds are continually being formed by igneous, sedimentary, and metamorphic processes. They are continually undergoing chemical and mechanical weathering, which breaks them down into sedimentary components that are transported to sites of deposition where they become sedimentary rocks. The new sedimentary rocks are susceptible to additional rock formation processes after which more weathering occurs. In this way the rock-forming and weathering cycles are continuously operational in the construction and destruction of rock materials.

Erosion

Rock materials are moved during weathering by various processes and transported to a site of deposition. Removal of rock materials is called erosion, which is responsible for the degradation of the landscape. Rock materials are removed during erosion and transported to where they are deposited. Along the way, the rock materials are further broken down, rounded, and sorted to the extent of their transportation.

Erosion occurs in several ways. All of them are involved in reducing the landscape to equilibrium where erosion and deposition are offsetting processes. The process of erosion is important to the petroleum industry because rock materials are derived, transported, concentrated, and deposited as source beds, reservoir rocks, and reservoir seals.

Inasmuch as erosion occurs in many ways, the products of erosion are numerous as well. Some erosional deposits have petroleum significance, and so do the processes that produce them. However, not all erosional processes are equally significant, but because petroleum occurs in rocks produced by virtually every sedimentary environment, they deserve some consideration.

Downslope Movement

Gravity forces all weathered rock material to move downhill with or without the aid of flowing water. Except in very arid areas, downslope movement is routinely affected by the presence of water, which may act as a lubricant for downwardly mobile rock and soil masses. In fact, because most desert areas receive some sporadic or very occasional rainfall at some time, included water can be a contributing downslope factor in those very dry areas as well.

Mass Wasting. Downslope movement of rock and soil materials without the aid of running water constitutes mass wasting. However, because most geographic areas receive some rainfall, a totally dry mass-wasting environment is unlikely. In any event, mass-wasting does not include removal and transportation of rock material by streams and rivers, although water-saturated rock and soil masses respond to gravity and move downhill.

Fragments of weathered and broken rock drop off cliffs and very steep slopes and accumulate as talus at the bases of slopes (Table 18). Larger fragments eventually accumulate at the bottom of the slope, and progressively smaller particles are distributed at successively higher levels up the slope. Talus slopes develop slowly over extended periods of time. Talus in arid or semi-arid regions often forms impressive cones of loose material that are important topographic features on mountain slopes. Humid climate talus slopes are often deeply weathered and/or covered by soil and vegetation.

Rock fall deposits have limited lateral extent. However, they are often very porous and permeable. If preserved in the stratigraphic column, they have reservoir potential but because of lack of preserved organic material have no source potential.

Rock fragments that slide rapidly in mass to accumulate at the bottom of a steep slope comprise a rock or debris slide (Table 18). Rock slides occur in heaps and irregular masses of poorly sorted and angular rubble. Individual rock slides occur rapidly but can accumulate with other slides at the bottom of a slope over extended periods of time. Like talus deposits, rock slides have good porosity and permeability and can act as reservoirs. Source bed deposition is not characteristic of rock or debris slides.

Rapid downhill movement of overburden results in a landslide or avalanche (Fig. 84). Soil, trees, and accumulated materials that comprise the overburden are affected by landslides that can cover large areas. Landslides can dam rivers to form lakes or can cover buildings and towns located along mountain bases (Fig. 85). Landslide or avalanche deposits are very heterogeneous and poorly sorted. They are not considered to have appreciable oil or gas potential.

Table 18
Mass Wasting Processes

Process	Characteristics
Rock and Debris Fall	Talus slope formation by accumulation of fallen rock
Rock and Debris Slide	Rubble accumulation at foot of slope
Slump	Rotation of blocks separating from slope or cliff
Debris Flow	Rapid downslope debris movement along concave surface in slope
Mud Flow	Water-borne debris accumulating and spreading out at base of slope

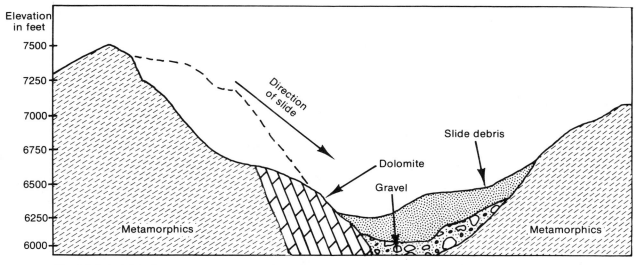

Fig. 84. Landslide in response to an earthquake

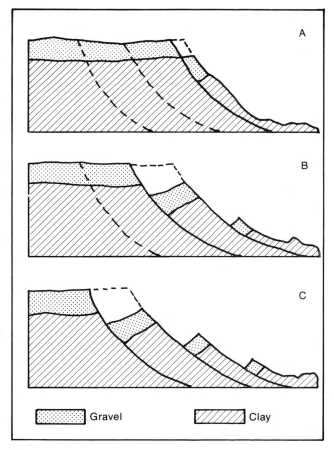

Fig. 85. Development of a slide

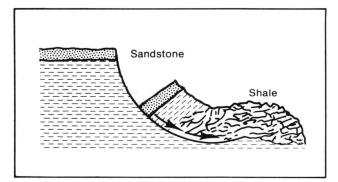

Fig. 86. Slumping and rotation

Mass movement occurring where competent materials overlie incompetent materials results in slump or movement along curved surfaces that tend to flatten at depth (Fig. 86). Slump blocks are disturbed little by their movement. However, some deformation can occur at the leading edge or toe of the slump mass.

Slump deposits have potential as oil and gas reservoirs and sources according to their intrinsic properties, which are not a function of the downslope mechanism. However, slump or "down to the coast" faulting is an important structural hydrocarbon trapping mechanism in the United States Gulf Coast, the Niger Delta and the Orinoco Delta.

Slow downslope movement of overburden and bedrock is known as creep (Fig. 87). Bedrock along the zone of

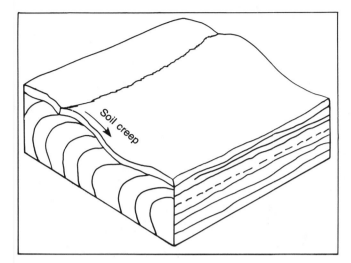

Fig. 87. Soil creep and bedrock effects

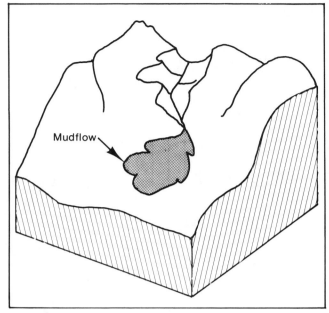

Fig. 88. Mudflow

movement may be deformed downslope, and trees grow-ing in the moving overburden may be deflected down-slope from the vertical. Creep deposits result from the weathering of bedrock and have little source or reservoir potential (Table 19).

As water content increases in materials that are creep-ing downslope, they move faster and eventually gain momentum as a rapidly moving nudflow (Fig. 88). Mud-flows result from heavy rainfall mixing with creeping overburden and sudden turbulent sliding downhill. Strong and vigorous mudflows occurring during flash floods can move large boulders, overturn vehicles, and destroy buildings. Mudflow deposits have little petroleum poten-tial as source beds and limited potential as reservoirs.

When frozen soil melts from the top down during spring thaws, it often contains water and flows downslope as a mudflow. Solifluction causes appreciable damage by

erosion in permafrost areas. Deposits formed by solifluc-tion have petroleum potential similar to that of mudflows.

Downslope movement and mass wasting are respon-sible for the relocation of large amounts of weathered rock material, soil, talus, and various forms of overbur-den. Some of the products of these processes have limited potential as petroleum reservoirs and almost no potential as source beds. Rocks formed from these processes and preserved in the stratigraphic column can have enough porosity and permeability to act as reservoirs if placed adjacent to source beds. However, subaerial deposits of this type have little or no opportunity to accumulate and subsequently preserve organic material necessary to gen-erate petroleum.

Hydrologic Cycle. Weathered rock materials derived from outcrops on cliffs or mountainsides move by down-slope movement to accumulate in talus slopes or slides and come to rest. However, rainfall upon the landscape also moves rock material and concentrates it in streams, where it is transported and deposited. In order for rain to fall and streams to collect the runoff, water must be avail-able to the atmosphere, from the evaporation of water from seas, lakes, streams, and springs, which when trans-ported by winds and convection finds its way to precipi-tation sites (Fig. 89).

The processes of evaporation, convection, advection, precipitation, runoff, transport and re-evaporation are the

Table 19
Causes of Regolith Creep

Intermittent Wetting	—Clay minerals expand and contract as they absorb water and desiccate. Vol-ume changes result.
Ground Water Solution	—Ground water dissolves rock and pro-duces porosity, which is infilled from above.
Snow Movement	—Snow accumulates and moves down-slope, moving the upper surface of the slope with it.
Plants	—Plant roots wedge soil downslope, form-ing porosity, which is filled from above.
Frost Heave	—Frost moves the regolith by expansion and contraction.

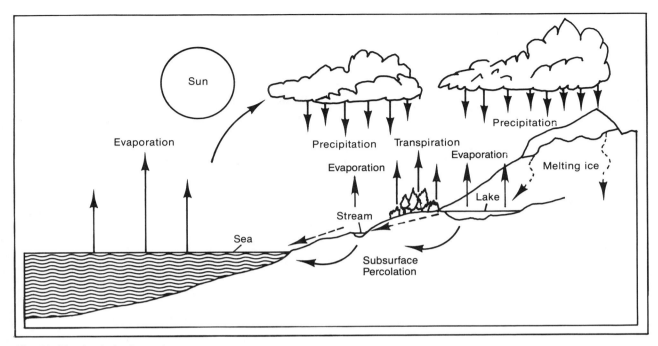

Fig. 89. The hydrologic cycle

mechanisms of the hydrologic cycle. Continuity of the cycle creates an endless source of erosive and transportive power under which rock outcrops are slowly but inexorably broken down, removed, concentrated, or ultimately redeposited.

As climates change so do the hydrologic cycle characteristics of different geographic areas. Humid climates have very active hydrologic cycles. Arid climates have considerably less active cycles. In the end, however, rock materials on the land portions of the earth's surface find their way to the sea for redeposition in response to the hydrologic cycle. Only continued crustal deformation precludes the permanent removal of all continental rocks to the ocean basins, as erosion and uplift work endlessly against each other and against the rock cycle, which is persistently forming new rocks. The hydrologic cycle is important because it helps in the formation of source and reservoir rocks significant to petroleum.

Stream Erosion

A rushing stream hurries down the mountain carrying sand and rolling boulders, abrading the bedrock, and cutting down its valley. It helps to slowly wear down the mountain, removing the derived rock materials to another place for deposition.

Running water is an important eroding agent because it carries abrasive particles that degrade the bedrock. It also removes weathered and eroded rock materials, revealing fresh rock surfaces to subsequent abrasion.

Stream Patterns. Stream erosion is a function of a number of factors, including climate, runoff, and vegetation. Geologic factors are also significant because structure and rock type underlying the landscape have direct effects on erosion rates and drainage patterns.

Streams that drain areas underlain by uniform bedrock develop a dendritic drainage pattern, which resembles the arrangement of veins in a leaf (Fig. 90). Bedrock under-

Fig. 90. Dendritic drainage pattern

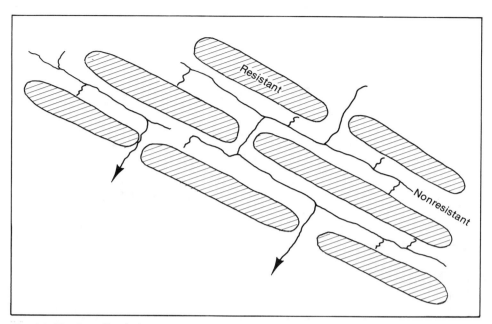

Fig. 91. Block trellis drainage pattern

lying the dendritic pattern can be relatively flat-lying, uniform in rock type, or both.

Areas underlain by dipping beds of alternating hard and soft layers will develop linear drainage patterns with streams occupying valleys in the soft layers in a block trellis arrangement (Fig. 91). A similar pattern will develop where parallel folding and/or faulting are prominent bedock characteristics.

Streams developed on terrain that includes domes and basins flow in a radial pattern (Fig. 92). Drainage is radial, away from the apex of a dome, and toward the center of a basin.

Gradient. Since water must flow downhill, a stream valley must be higher at one end than at the other. The difference in elevation is the gradient of the stream and is represented by the drop in elevation related to the horizontal distance. Headwaters of streams characteristically have a steep gradient that becomes progressively shallower as the stream approaches its mouth.

A section from the headwaters of a stream to its mouth is its *longitudinal profile* (Fig. 93). It is a representation of the progressively changing stream gradient and from headwaters to mouth assumes a shape that is concave upward.

Velocity. Stream velocity is the speed at which the water moves downstream. Inasmuch as stream channels are neither symmetrical nor straight, water velocities vary within individual streams (Fig. 94).

Fig. 92. Radial drainage pattern

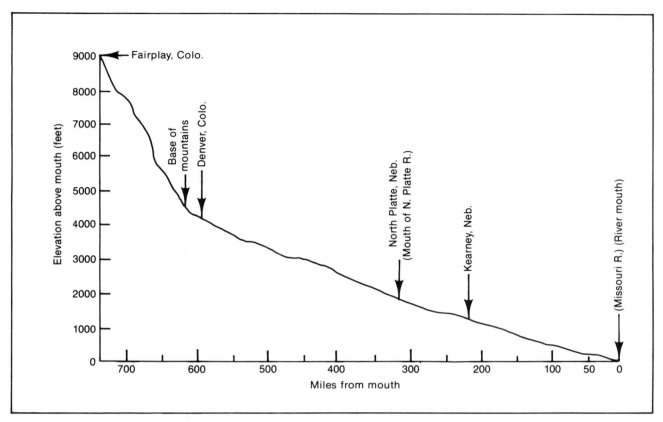

Fig. 93. *Longitudinal profile of the Platte River, Colorado and Nebraska*

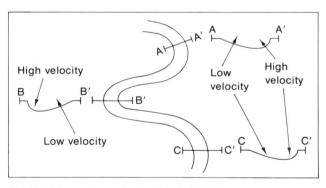

Fig. 94. *River channel velocity distribution*

Maximum stream velocity occurs in the main channel and represents the greatest water flow. Velocity decreases away from the main channel and in some streams can be virtually negligible in eddies and backwater areas. Water velocity has a direct relation upon how much sediment a stream can carry.

Bed Load. Water flow in a stream channel has the ability to move sedimentary particles. Material moved by sliding and rolling along the channel bottom comprises the bed load (Fig. 95). A rapidly moving stream can move large rocks and boulders in its bed load, which is being abraded as well as abrading the channel during its progress downstream.

Materials carried by a stream in suspension above the bed load comprise its *suspended load* (Fig. 95). These materials consist of smaller particles and can be carried greater distances to where they settle slowly as water velocity decreases. Grain sizes in a suspended load are also a function of the velocity of a stream and its ability to carry solid materials.

Chemical weathering and dissolving of bedrock by a stream contribute the *dissolved load* to the stream. It is carried in solution and has no physical effect upon the ability of the stream to flow or carry suspended loads or bed loads.

Base Level. As a stream evolves, it cuts downward toward a plane of equilibrium known as base level (Fig. 96). A stream flowing into the sea will cut down to sea

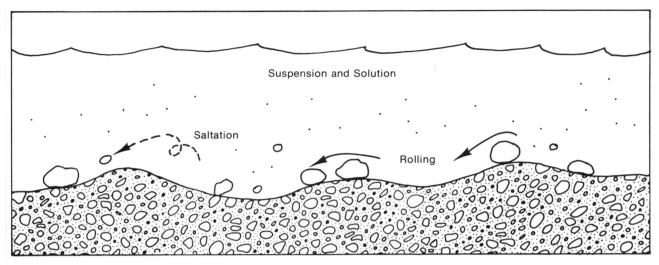

Fig. 95. Particle and solution loads in a stream

level, which is base level for the stream. A stream flow-ing into a lake will cut down to the level of the lake, which is the base level for that stream. As the stream approaches its base level, its ability to cut downward is increasingly limited as topography that creates stream velocity is reduced.

Balance between stream erosion and deposition is achieved as base level is approached and the stream is *at grade* or becomes a graded stream. Gradient is sufficient to move the erosion products of the stream and its water-shed. A graded stream is essentially at equilibrium, which represents an average of erosion versus deposition in various places along its longitudinal profile and at various times throughout the year.

Cycles of Erosion. If stream erosion had been allowed to proceed normally throughout geologic time, there would be no mountains or topography, and all continental

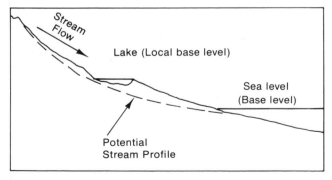

Fig. 96. Local and ultimate base levels

material would have long since been removed to the sea. During the process of erosion, streams and land forms change. If allowed to proceed undisturbed, the erosional process would degrade the land toward the establishment of a state of equilibrium (Fig. 97). Crustal activity pre-cludes the development of equilibrium and regenerates the erosional process by interruption of cycles already in operation. So with constant erosion operating against pe-riodically constant crustal movement, erosional cycles are constantly being adjusted, rejuvenated, and superim-posed upon each other.

Early stream development occurs during the youthful stage of the stream, which is a product of crustal uplift and the destruction of previously established quilibrium (Fig. 97). Youthful streams cut down rapidly and form V-shaped valleys entrenched in the newly uplifted flat surface. Floodplains are narrow or non-existent, and stream channels are nearly straight. Erosional and carry-ing capacities of youthful streams are great because their gradients are steep and water velocities high.

As youthful streams cut down, they approach base level and reduce the rate at which they erode. Deposition begins, a floodplain with bounding bluffs is developed, and the channel meanders across the floodplain as gra-dient and water velocity are reduced. Topography in the mature stage (Fig. 97) is manifested as rounded hills integrated into a well-developed drainage system.

Slow moving water in widely meandering streams is characteristic of old age drainage. Deposition is the pre-dominant mode as there is little or no erosion. Old age floodplains are very wide and tend to blend together along

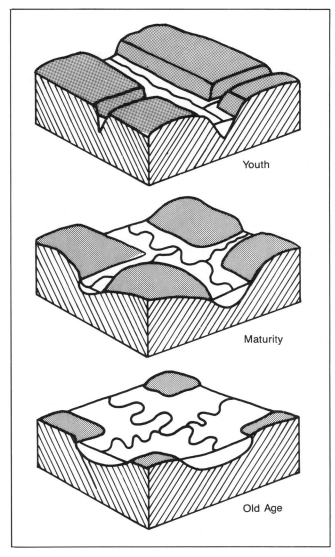

Fig. 97. Drainage and topographic cycles of erosion

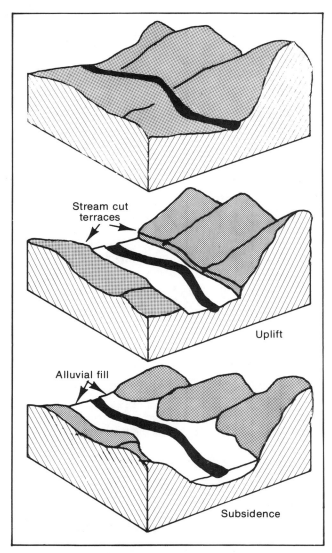

Fig. 98. Uplift and subsidence and stream erosion

very low, widely separated divides (Fig. 97). The onset of old age indicates the approach of the stream to base level and the establishment of the stream at grade. Because the development of the old age stage occurs so late in stream progression, the time required usually exceeds the interval between crustal movements. This means that the old age stage is seldom reached because streams are subject to prior rejuvenation.

Inasmuch as stream progression is interrupted by crustal movements, erosion cycles are necessarily superimposed by new erosion cycles (Fig. 98). In this manner,

a cycle of erosion having progressed to a certain stage can have its features occurring with those from a younger one. Cycles of erosion offer significant commentary upon the history of an area and therefore are useful as indicators of geologic events.

Stream erosion and transportation are important considerations in the movement and ultimate deposition of sedimentary rocks. Without the erosion and concentration of rock materials, sedimentary deposits would not exist. Without sedimentary deposits, there could be no petroleum source and reservoir beds.

Deposition

After rock materials are eroded and transported, they are deposited. Some time after deposition they are reworked, removed, and redeposited elsewhere. Ultimately, however, they are permanently deposited where they will eventually be retransformed into rocks.

Deposition of sediments occurs in many different ways, one of which is deposition by streams. Stream deposition is fundamental to the progression of the stream cycles of erosion. Environments of stream deposition are important because they provide source and reservoir rocks for petroleum.

Floodplain Deposits

Progression of stream development and the gradual shift from erosion to deposition results in the formation of a floodplain (Fig. 99). As a floodplain develops, the stream meanders between the bounding bluffs, depositing gravel, sand, and mud within the channel and, during periods of flooding, over adjacent areas outside of the channel. Some of these deposits are porous and permeable, offering excellent reservoir potential. Other deposits are much finer-grained and, when deposited with

organic matter that is preserved by burial, can act as petroleum source beds. River deposits preserved in the stratigraphic column are high potential exploration prospects in the mid-continent areas of Kansas, Nebraska, and Oklahoma.

Rivers and streams are often in flood, at which time they rise over their channel banks and spread out upon their floodplains. Increased water volume and velocity during flood means greater carrying capacity and an increased sediment load. As the river rises over its banks and spills over the floodplain, its velocity immediately diminishes and reduces its carrying capacity. Coarse materials settle immediately and are deposited as linear ridges or *natural levees* (Fig. 99) adjacent to the channel. Natural levee deposits that build up each year during river flood stage are normally poorly sorted but can contain sufficient porosity and permeability to be prospective as reservoirs (Fig. 100).

As a river in flood rises above its banks and spills over to deposit natural levees, additional sediment-carrying water extends farther from the channel toward the margins of the *floodplain* (Fig. 99). The flood waters move more and more slowly with distance from the channel, allowing finer and finer particles to settle. Fine-grained floodplain deposits can include organic material, which

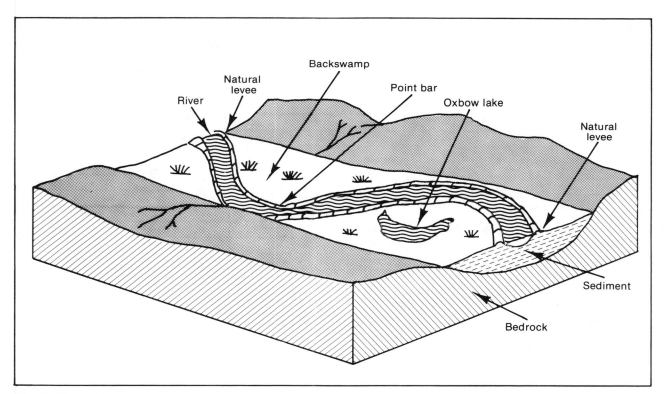

Fig. 99. Floodplain features

can be preserved by burial caused by subsequent flooding. Organic material accumulated this way and subsequently buried by younger sediments can be converted to petroleum and expelled from the fine-grained floodplain source beds into coarse-grained natural levee and channel deposits (Fig. 100).

Since channel sands occupy the channel portion of the river flood-plain system they grade laterally into fine-grained flood-plain deposits. The channel sands become thinner and eventually disappear with distance from the center of the channel (Fig. 101).

Sediments deposited in the *channel* (Fig. 99) of a stream are coarse and, though moderately well sorted, are the best sorted of floodplain deposits. Stream channels are subject to fluctuations in velocity, which results in differences in deposition and sorting. However, channel deposits have good porosity and permeability, which increase toward the base of the deposit.

Water velocity in stream channels varies, and when it is reduced causes settling of the heaviest and largest components first (Fig. 102). Continuing reduction of velocity causes progressively finer materials to be deposited over the coarse beds. Typical electric log response to channel sands indicates an increase in spontaneous potential (SP) in the bottom of a channel sand, which can occasionally imply increased permeability (Fig. 103). Shallow resistivity increases downward in a channel sand because of mud filtrate invasion and illustrates maximum contrast with underlying shale at the base of the sand. However, if the sand contains oil underlain by salt water, resistivity is higher along the oil leg and is reduced to inside the shale line in the underlying salt water column.

Since rivers tend to flow along curved paths, velocities in their channels increase along the outside of the bends and decrease along the inside. River sand accumulates on the inside of the bends in response to slower velocities, and point bars are formed as a particular type of channel deposit (Fig. 99).

When a river is supplied with more debris than it can carry by its tributaries, water flow is impeded. The channel divides to flow around the obstructions and forms a *braided channel* (Fig. 104). Braided channel deposits are often very coarse, contain boulders and cobbles, and are poorly sorted.

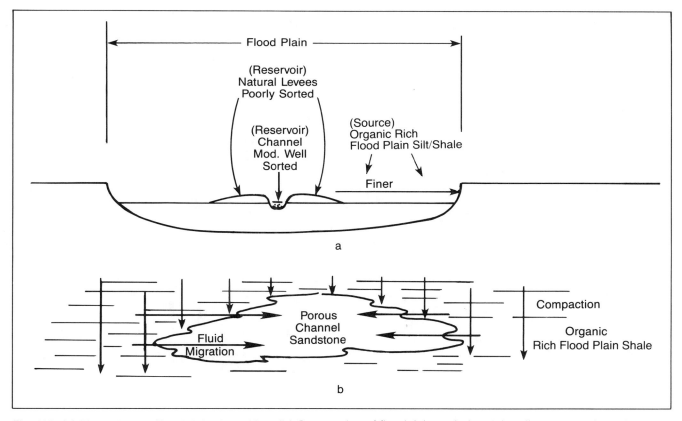

Fig. 100. (a) River-channel/floodplain deposition. (b) Compaction of floodplain and channel sediments causing migration of fluids from shale into sandstone.

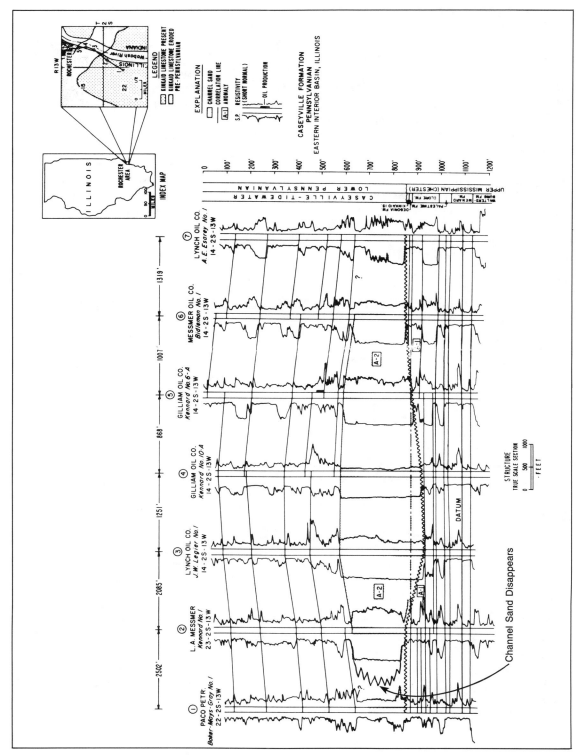

Fig. 101. Buried basal Pennsylvanian channel sandstone near Rochester, Illinois. From Rees, 1972. Permission to publish by AAPG.

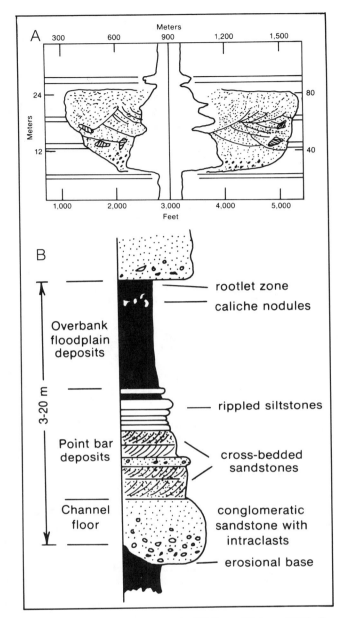

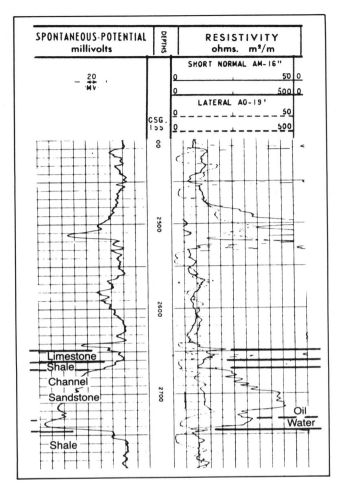

Fig. 103. *Electric log response of channel sandstone*

Fig. 102. Channel sand deposit. (A) From Dickey, 1979. Permission to publish by PennWell Publishing Company. (B) From Sedimentary Rocks, 3rd ed. by F. J. Pettijohn. Copyright 1975 by Francis J. Pettijohn. Reprinted by permission of Harper & Row, Publishers, Inc.

The curving and undulating pattern of a river channel within its floodplain results from turbulent deflections of the current around piled-up deposits, resistant bedrock, slight bends in the channel, or any other obstacles. As the river bends around the obstacles, water velocities change, deposition occurs in the slow parts of the channel, and erosion occurs in the fast portions. Sinuosity of the channel is increased as the river slows or speeds up in different places. This results in the development of *meanders* (Fig. 105) that direct the river channel from side to side through the floodplain (Fig. 106).

With continued undercutting where velocity is high and point bar development where it is low, the river channel becomes progressively more sinuous. Eventually, points of undercutting meet across the narrow portion of the meander, and a channel or chute that cuts off the meander is formed (Fig. 107). The meander is isolated and becomes an oxbow lake (Fig. 108).

Quiet sedimentation affected by occasional flooding is typical of oxbow lakes where organic deposits can accumulate over deposits of the old river channel.

Water trapped by floods in low floodplain areas behind natural levees often forms *swamps* (Fig. 99). These areas

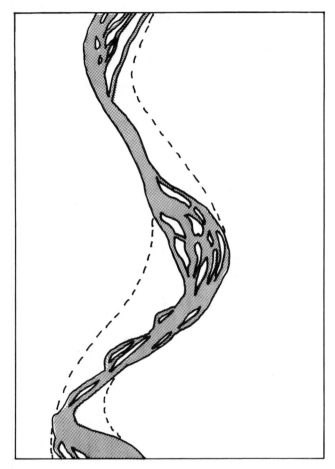

Fig. 104. Braided river channels

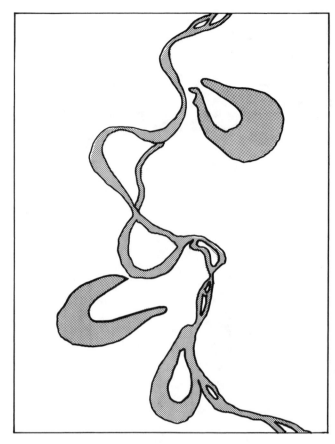

Fig. 105. Floodplain features showing oxbow lakes and meanders

are also affected by small tributaries that are unable to flow over the natural levees and drain into the swamps. Organic materials accumulated in the fine-grained swamp sediments and distributed liberally over the floodplain have good petroleum generating potential.

Any time that terrain is uplifted and streams are rejuvenated downward, erosion resumes. Erosion in an established river system with floodplain deposits results in down-cutting through those deposits, which are left as *terraces* (Fig. 109) at higher levels than the eroding channel. Periods of rejuvenation are reflected in several levels of terraces.

Deposition of a stream is a function of its channel and its floodplain, which are intrinsically related. A stream erodes and deposits sediments, developing its channel and floodplain and forming potential source and reservoir sediments as it does so. Not all river systems have the proper combinations of channel and floodplain deposits and organic material to be prospective petroleum gener-

ating areas. However, when the factors combine favorably, river and stream systems and their sediments can be very prospective of petroleum (Fig. 110).

Delta Deposits

When a sediment-carrying stream enters a lake or the sea, its velocity decreases and some of the coarse sediment settles to form a delta (Fig. 111). The remaining finer sediment is transported farther from shore to build the delta seaward (Fig. 112). Turbidity currents contribute graded sediments to the distal slope and its transition with the ocean bottom. Deltaic sediments are deposited in a marginal marine environment and represent the transition between terrestrial and marine conditions (Fig. 113). Where stream or river sediments are confined to several delta channels, a birdfoot delta is built out into the sea, provided that ocean waves and currents do not redistribute the sediments. The Mississippi delta in the Gulf of

Erosion
Deposition
Deposition
Erosion

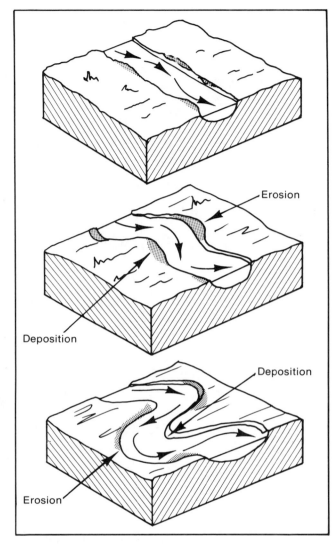

Fig. 106. Meander development

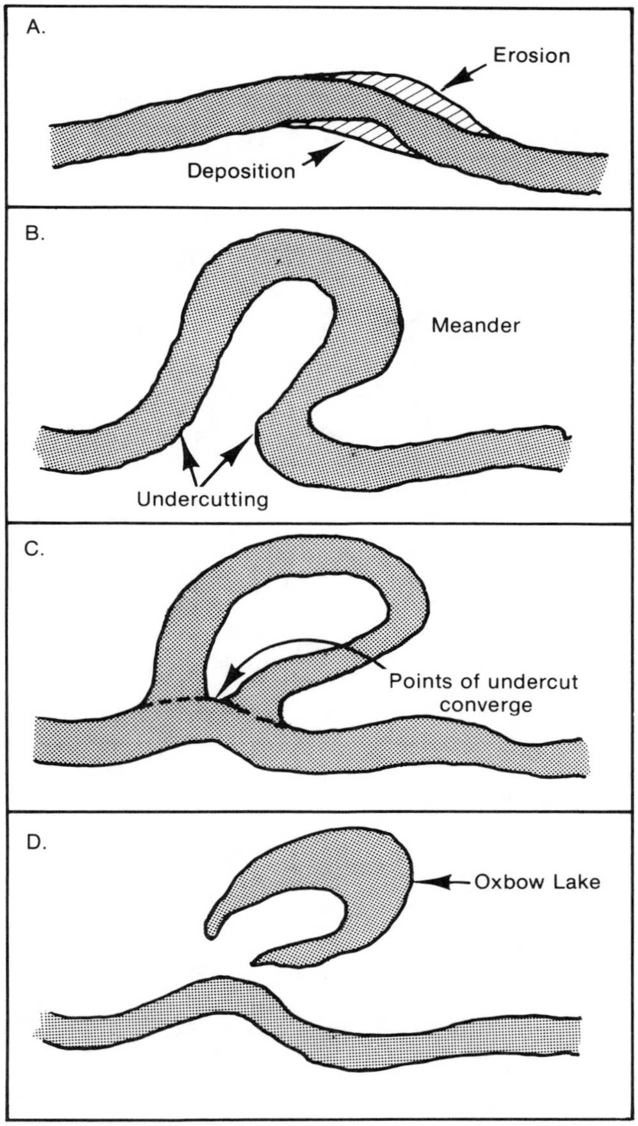

A.
Erosion
Deposition
B.
Meander
Undercutting
C.
Points of undercut converge
D.
Oxbow Lake

Fig. 107. Oxbow lake development

Mexico is a birdfoot delta where large amounts of sediments are being deposited. The large rivers of the world transport significant amounts of sediment to the sea and in many instances form deltas (Table 20).

Ocean currents and waves redistribute sediments of the Nile delta in Egypt and prevent it from building seaward. The Nile River breaks up into numerous channels far inland at Cairo and brings sediments to the Mediterranean Sea across a wide front, rather than a single channel, further preventing the development of a birdfoot delta.

Delta deposits are typically crossbedded (Fig. 114) and consist of three related but distinct bedding types (Fig. 115).

When streams and rivers reach the sea and slow down, the coarsest materials are deposited first in thick sloping layers called foreset beds (Fig. 116). Foreset beds grade seaward into fine-grained bottomset beds, which spread out, often as turbidites, to cover wide areas of the bottom.

Continued sedimentation causes coarse foreset beds to build seaward and overlap the bottomset beds. As the delta builds seaward, the stream carrying the sediment

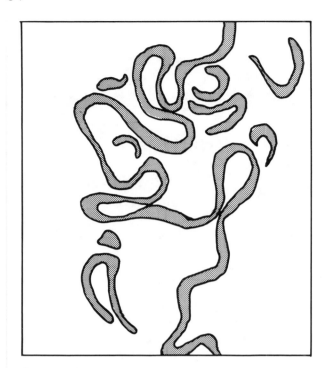

Fig. 108. Meanders and oxbow lakes in Canada

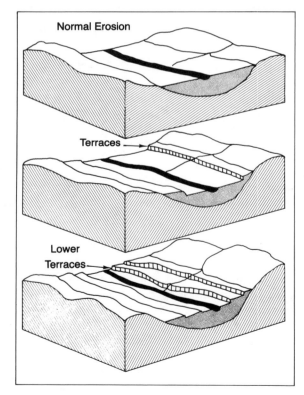

Fig. 109. Stream terrace development

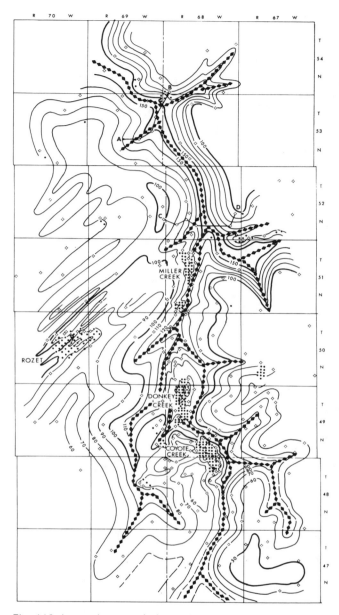

Fig. 110. Isopach map of channel sandstone. Contour interval, 10 ft. From Busch, 1974. Permission to publish by AAPG.

extends farther to seaward as it travels over the foreset and bottomset sediments. Channel and inter-channel deposits of the stream deposited horizontally over the building delta comprise the topset beds. When the stream is in flood, it spills over its banks on the delta and deposits floodplain sediments, which become part of the topset system.

Tributaries coalesce in the upper reaches of a stream to

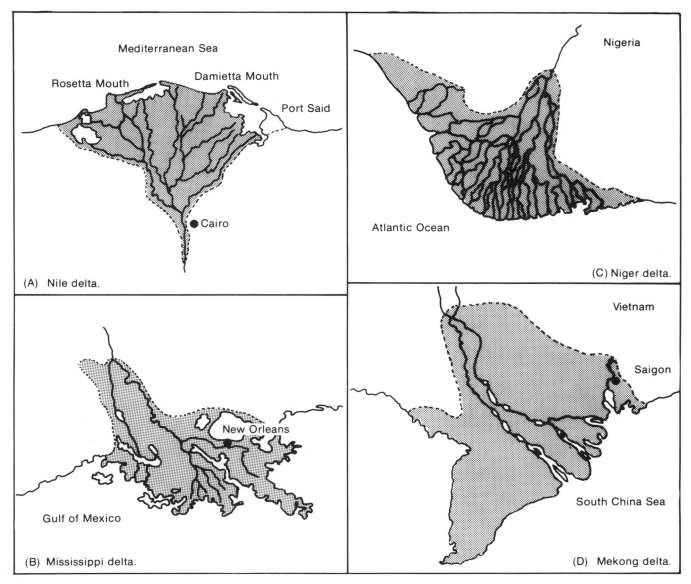

Fig. 111. Examples of modern deltas

form the main channel, which collects additional tributaries on its way to the sea. When the stream reaches the sea and builds a delta, the main channel breaks up into several or many *distributary channels* (Fig. 115), which can be simultaneously or individually operational. Distributary channels are typically like continental stream channels and produce similar sedimentary deposits, which are often coarse, porous, and permeable with good reservoir potential (Fig. 117).

Individual distributary channels are instrumental in concentrating sediments in specific areas of a delta and form subdelta lobes as long as they are active, particularly on birdfoot deltas. As depositional conditions change, an active distributary channel can become inactive, terminating deposition at its subdelta lobe. When this happens, another distributary channel will become active and deposit in a subdelta lobe along another part of the main delta (Fig. 118). In this manner, a delta can distribute itself within the marginal marine environment as it responds to variations in depositional conditions, runoff, and sediment load.

Flooding of a river across a delta results in deposition

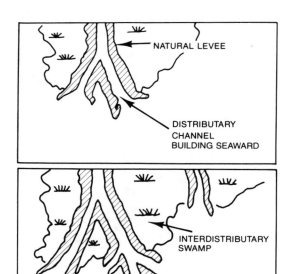

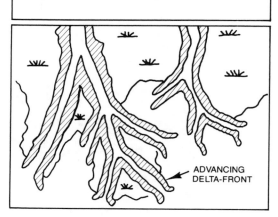

Fig. 112. Delta development

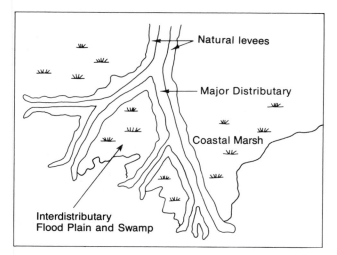

Fig. 113. Delta features

Table 20
Water Discharge and Sediment Suspension

River (location)	Yearly Discharge of Water (km³)	Mean Concentration of Suspended Sediment (mg/l)	Characteristics of Mouth of River
Rhine (Netherlands)	68.5	6.6	Estuary
St. Lawrence (Canada)	304.0	11.8	Estuary
Columbia (USA)	174.0	20.7	Estuary
Yenisey (USSR)	610.0	21.6	Estuary
Lena (USSR)	511.0	30.1	Estuary
Mackenzie (Canada)	15.0	34.1	Small delta
Ob (USSR)	400.0	39.5	Estuary
Congo (The Congo)	1,350.0	47.9	Estuary
Amur (USSR)	350.0	71.1	Neither
Amazon (Brazil)	3,187.5	156.2	Estuary
Orinoco (Venezuela)	442.0	195.7	Delta
Zambesi (Zaire)	500.0	200.0	Neither
Rio de la Plata (Uruguay-Argentina)	600.0	215.0	Delta in estuary
Niger (Nigeria)	293.0	228.7	Delta
Danube (Romania)	199.0	339.2	Delta
Po (Italy)	48.7	369.6	Delta
Mekong (Vietnam)	387.0	438.2	Delta
Yukon (Alaska, USA)	185.0	475.7	Delta
Tigris-Euphrates (Iran-Iraq)	210.0	500.0	Delta
Rhône (France)	52.7	597.7	Delta
Irrawaddy (Burma)	428.0	698.6	Delta
Mississippi (Louisiana, USA)	600.0	833.3	Delta
Nile (Egypt)	70.0	1,578.5	Delta
Orange (Southwest Africa)	91.0	1,681.3	Neither
Ganges-Brahmaputra (India-Bangladesh)	1,210.0	1,799.3	Estuary
Indus (India-Pakistan)	175.0	2,488.0	Delta
Yangtze (China)	690.0	2,734.6	Delta
Colorado (California, USA)	20.3	6,666.1	Delta
Yellow (China)	126.0	14,975.4	Delta

of *natural levees* in the same way they are formed on a floodplain upstream. The coarsest particles settle along the margins of the distributary channels and form the narrow ridges typical of natural levees. These deposits have reservoir potential only slightly less favorable than distributary channel deposits.

Upper surfaces of deltas lie at or very near sea level, particularly along their seaward margins. They are subject to periodic tidal flooding as well as river flooding.

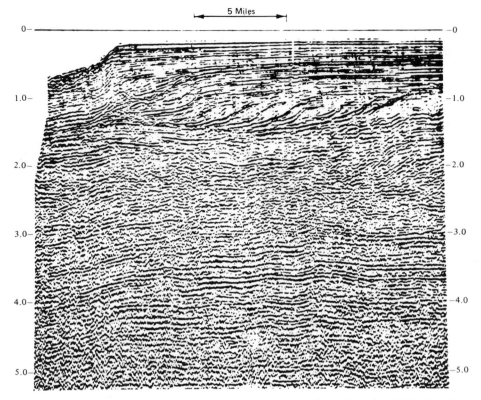

Fig. 114. Seismic section of deltaic deposition patterns. From Dobrin, 1977. Permission to publish by AAPG.

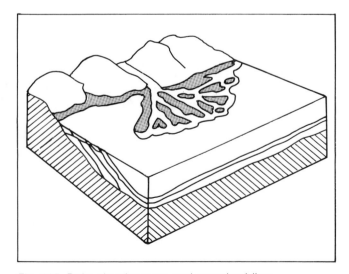

Fig. 115. Delta development and crossbedding

Vegetation growing in the inter-channel areas is occasionally inundated and buried by the flooding that carries fine-grained sediments, which are often rich in organic matter. As delta sediments accumulate, the thickness of vegetation and organic-rich sediments increases and is preserved by burial in the inter-channel *swamp areas* (Fig. 113). Inter-channel swamp or marsh deposits are highly favorable for generation of petroleum and are excellent source beds. Abundant production in the Gulf of Mexico comes from channel and fan reservoirs into which inter-channel organic-rich shales expelled their hydrocarbons.

Natural levees along distributary channels occasionally rupture and allow part of the stream to spill out over the adjacent swamp (Fig. 119). A reduction in velocity results in deposition of a *crevasse splay* or small delta adjacent to the crevasse or rupture in the levee (Fig. 120). Crevasse splays build in the direction of water flow and

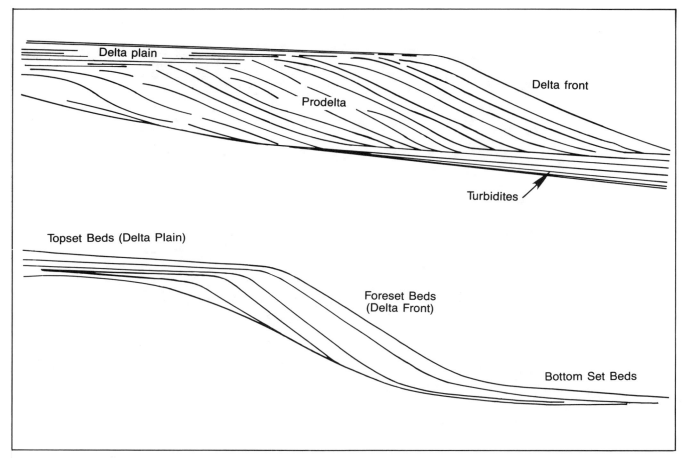

Fig. 116. Model of prograded fluvial-dominated delta and associated turbidite sequence.

form their own small distributary patterns, which are also subject to development of smaller splay systems.

Stream Deposits and Petroleum

Stream deposition includes channel and floodplain sediments, which are closely interrelated. The coarsest deposits are characteristic of channel deposition, and are subject to highest stream velocities. Floodplain deposits are variable from coarse-grained, poorly sorted natural levees to fine-grained swamp deposits.

Channel deposits include channel sands and point bars (Figs. 121, 122). These deposits are coarse and permeable at the base and become finer and less permeable upward. Their electric log response assumes the shape of an inverted V or pyramid (Fig. 123), on the self potential and resistivity curves (Fig. 124). Channel deposits have good reservoir potential.

Floodplain deposits include coarse, poorly sorted natural levee sediments, fine sand, shale, and swamp sediments. Natural levee deposits have some reservoir potential. Fine-grained organic-rich floodplain deposits and swamp sediments have good source potential.

Channel deposits are important petroleum reservoirs in the Norwegian Statfjord oil field in the North Sea (Figs. 125 and 126). Deltaic deposits also produce in the Statfjord field, which contains about three billion barrels of oil.

Delta Deposits and Petroleum

Delta sediments contain abundant reservoir and source beds (Fig. 127). Reservoirs include distributary channel, distributary front, and natural levee sediments (Fig. 128). Deltaic distributary channel, distributary front, and cut and fill sandstones produce about 46,000 barrels daily from 26 wells in the Bekapai Field in East Kalimantan,

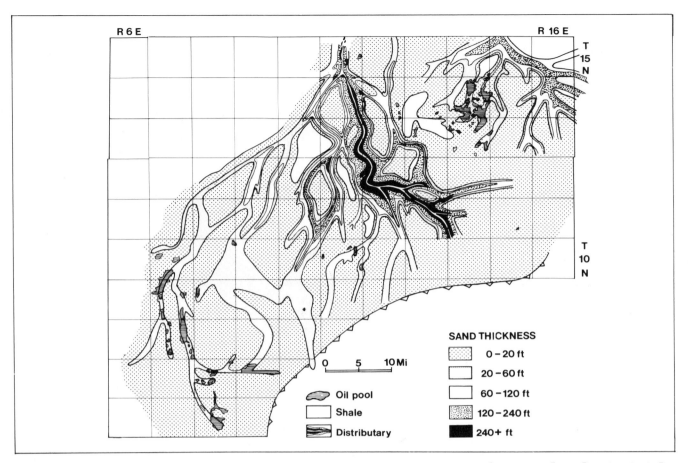

Fig. 117. Isopach map of Booch channel sandstone, greater Seminole district, eastern Oklahoma. From Busch, 1974. Permission to publish by AAPG.

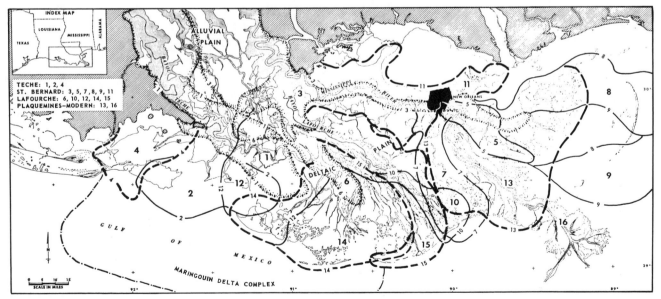

Fig. 118. Mississippi River delta lobes of the past 6,000 years. From Frazier, 1967. Permission to publish by the Gulf Coast Association of Geological Societies.

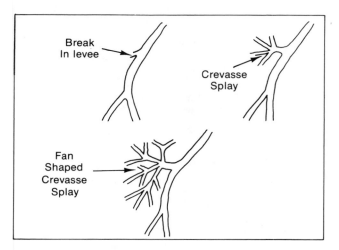

Fig. 119. Crevasse splay development

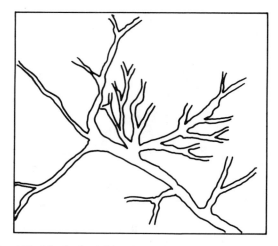

Fig. 120. Mississippi River crevasse splay (1838–1903)

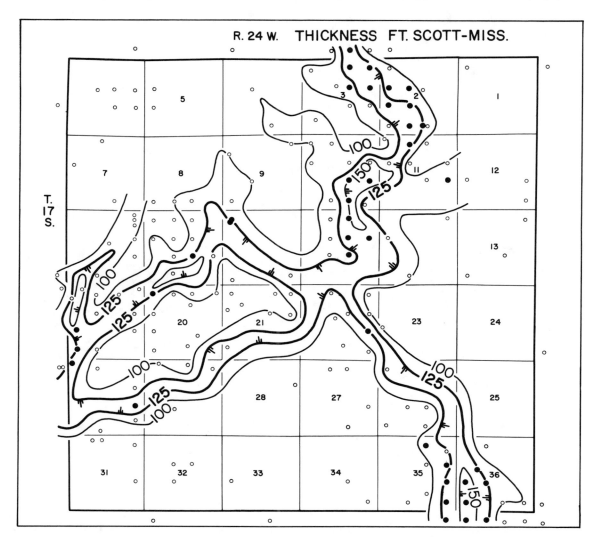

Fig. 121. Pennsylvanian channel reservoir, Kansas. From Walters, Gutru, and James, 1979. Permission to publish by the Tulsa Geological Society.

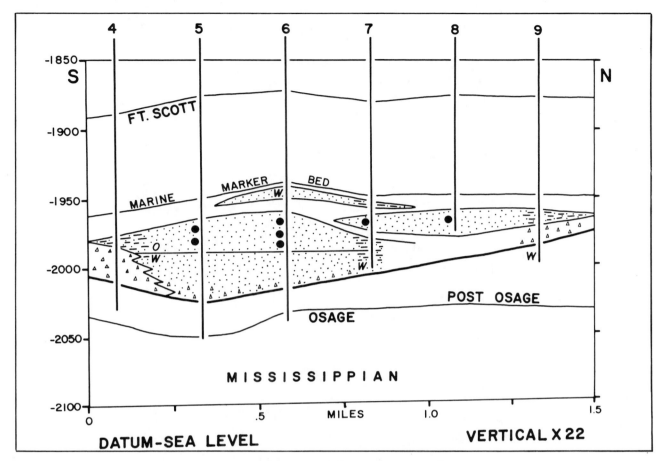

Fig. 122. Northeast-Southwest cross-section across Figure 121. From Walters, Gutru, and James, 1979. Permission to publish by the Tulsa Geological Society.

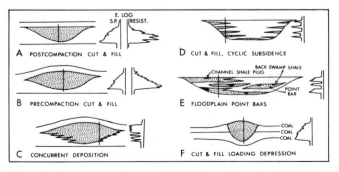

Fig. 123. Six types of channel fill. From Busch, 1974. Permission to publish by AAPG.

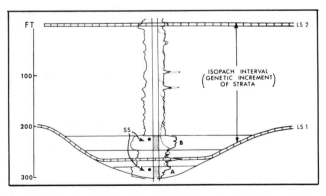

Fig. 124. Channel fill profile. From Busch, 1974. Permission to publish by AAPG.

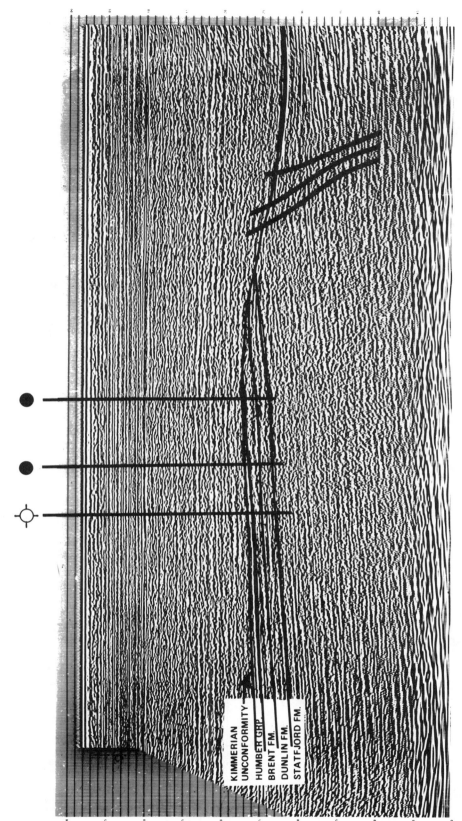

KIMMERIAN
UNCONFORMITY
HUMBER GRP
BRENT FM.
DUNLIN FM.
STATFJORD FM.

Fig. 125. Seismic line across Statfjord Field. From Kirk, 1980. Permission to publish by AAPG.

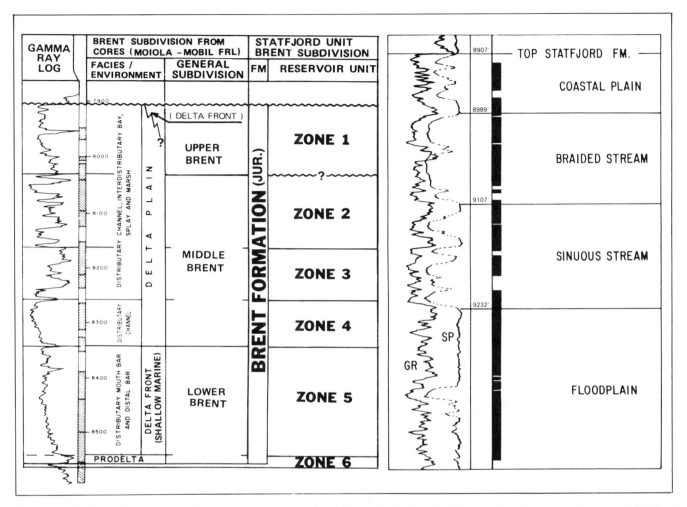

Fig. 126. Left: Brent Formation deltaic environment, Statfjord Field. Right: Statfjord Formation stream environment, Statfjord Field. From Kirk, 1980. Permission to publish by AAPG.

Indonesia (Fig. 129). Most of the sandstones have 25–35% porosity and permeability exceeding one darcy. They are of limited lateral extent and typical of deltaic deposition. Intra-channel shale and floodplain sediments and swamp deposits that contain organic material are good source beds.

Distributary channels contain porous and permeable rocks analogous to continental channel deposits. Their reservoir potential is very good and is indicated by their electric log response, which is similar to that of stream channels.

Intra-channel deposits include floodplain and swamp sediments, which have excellent petroleum source potential. Growth-faulted deltaic sediments have calculated oil reserves in excess of 645 million barrels in the Nembe Field on the Niger delta, Nigeria (Fig. 130 and 131).

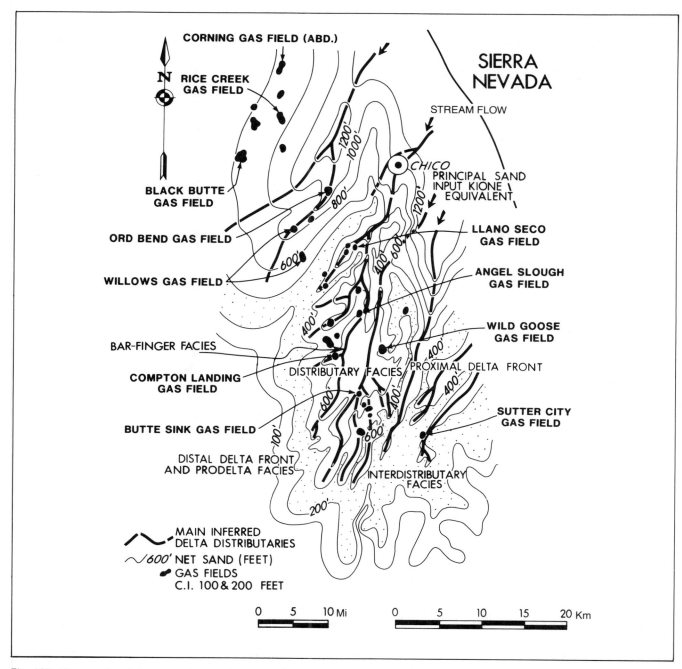

Fig. 127. *Kione delta plain producing and prospective trends, Sacramento Valley, California. From Garcia, 1981. Permission to publish by AAPG.*

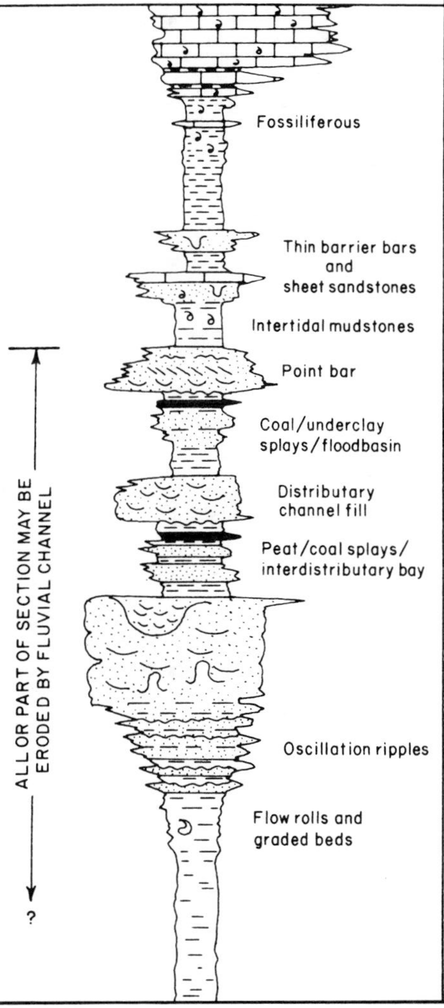

ENVIRONMENTS/FACIES — **IDEALIZED LOG PATTERN AND LITHOLOGY** — **DEPOSITIONAL PHASES** — **DESCRIPTION**

System	Subdivision	Environment	Facies	Depositional Phase	Description
SHELF SYSTEM	SUBMARINE	SHALLOW MARINE	OPEN-MARINE LIMESTONE	MARINE TRANSGRESSION / SUBMARINE AGGRADATION	Commonly mixed biomicrites, fusulines near base, grades upward into algal limestone, well bedded, very fossiliferous, persistent, grades downdip into shelf-wide limestones, grades updip into brackish shales and littoral sandstones.
			TRANSGRESSIVE SHALE	MARINE TRANSGRESSION / SUBMARINE AGGRADATION	Shale becomes more calcareous and fossiliferous upward, assemblage becomes less restricted, highly burrowed. In northern and eastern Mid-Continent, phosphatic black shale common at base.
DELTA SYSTEM	SUBAERIAL	SHOALS: WAVES AND TIDES	BARRIER BAR, STORM BERMS, SHEET SAND	DELTA DESTRUCTION / SUBMARINE AGGRADATION	Local barrier-bar sandstone: thin, coarsening upward, commonly fringe abandoned delta. Sheet sandstone: widespread, coarsening upward, burrowed, oscillation ripples on top. Storm berm: local, shelly bars composed of broken shells. Intertidal mudstone: laminated, red/olive.
		UPPER DELTA PLAIN / MID- AND LOWER DELTA PLAIN	POINT BAR; DISTRIBUTARY CHANNEL-FILL; CREVASSE SPLAYS; FLOODBASIN/INTERDISTRIBUTARY BAY; MARSH/SWAMP PEAT	DELTA CONSTRUCTION / SUBAERIAL AGGRADATION	Point-bar sandstone: fining upward from conglomerate lag to silty levees, upward change from large trough-filled crossbeds to tabular crossbeds and uppermost ripple crossbeds. Distributary channel-fill sandstone: fine- to medium-grained, trough-filled crossbeds, local clay, clast conglomerate: abundant fossil wood. Crevasse splay sandstone: coarsening upward, trough and ripple crossbeds, commonly burrowed at top. Floodbasin/interdistributary mudstone: burrowed, marine fossils, grade updip to non-marine, silty near splays. Coal/peat: rooted, overlie underclay (soil).
	SUBMARINE	DELTA FRONT	BAR CREST	DELTA CONSTRUCTION / PROGRADATION	Well-sorted, fine- to medium-grained sandstone, plane beds (high flow regime) common, channel erosion increases updip, distal channel fill plane-bedded, some contemporaneous tensional faults.
			CHANNEL-MOUTH BAR	DELTA CONSTRUCTION / PROGRADATION	Fine- to medium-grained sandstone, trough-filled crossbeds common, commonly contorted bedding, local shale or sand diapirs in elongate deltas.
			DELTA FRINGE	DELTA CONSTRUCTION / PROGRADATION	Fine-grained sandstone and interbedded siltstone and shale, well-bedded, transport ripples, oscillation ripples at top of beds, growth faults in lobate deltas, some contorted beds at base.
		PRODELTA	PROXIMAL	DELTA CONSTRUCTION / PROGRADATION	Silty shale and sandstone, graded beds, flow rolls, slump structures common, concentrated plant debris.
			DISTAL	DELTA CONSTRUCTION / PROGRADATION	Laminated shale and siltstone, plant debris, ferruginous nodules, generally unfossiliferous near channel mouth, grades downdip into marine shale/limestone, grades along strike into embayment mudstones.

Log pattern / lithology labels (top to bottom): Fossiliferous; Thin barrier bars and sheet sandstones; Intertidal mudstones; Point bar; Coal/underclay splays/floodbasin; Distributary channel fill; Peat/coal splays/interdistributary bay; Oscillation ripples; Flow rolls and graded beds.

ALL OR PART OF SECTION MAY BE ERODED BY FLUVIAL CHANNEL

Fig. 128. Depositional phases of delta sequence. From Brown, 1979. Permission to publish by the Tulsa Geological Society.

106   *Basic Petroleum Geology*

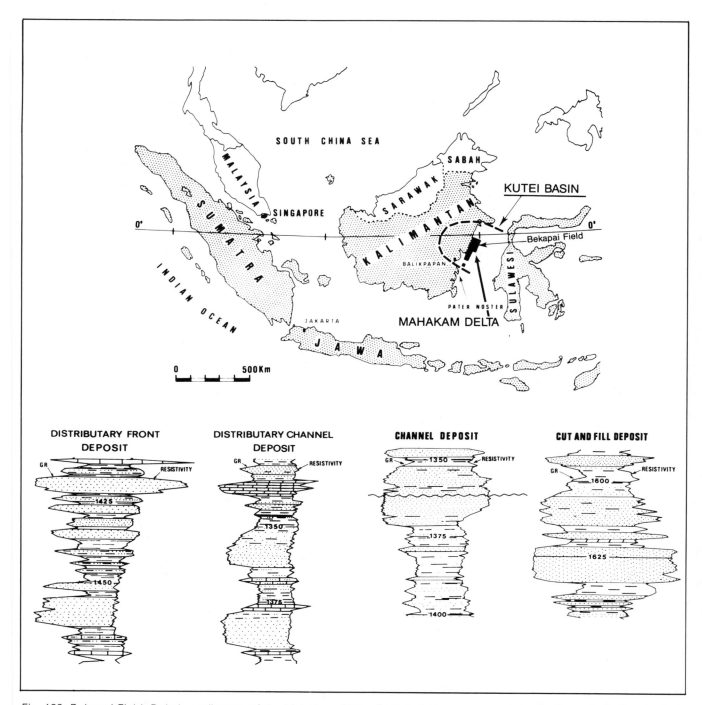

Fig. 129. Bekapai Field. Deltaic sediments of the Mahakam Delta, East Kalimantan reservoir types. From DeMatharel, et al., 1980. Permission to publish by AAPG.

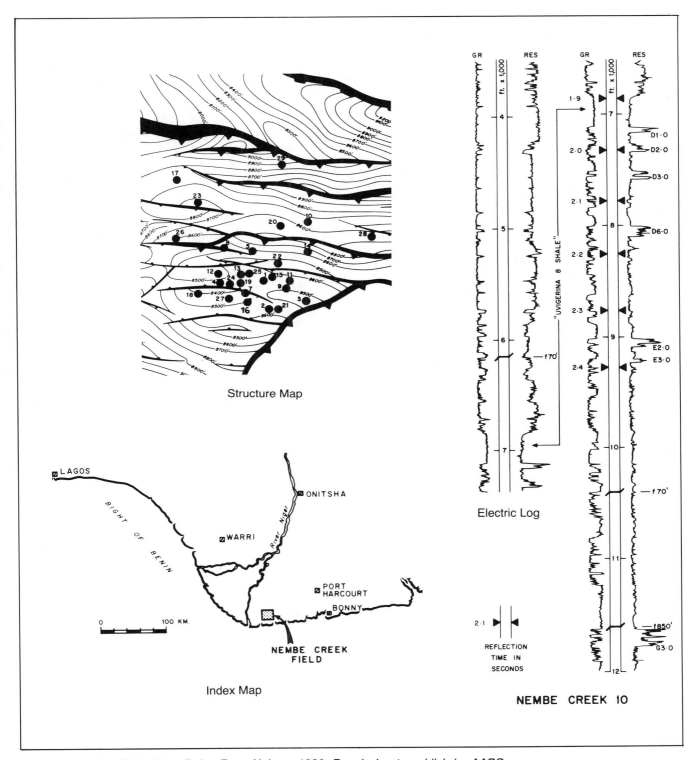

Structure Map

Index Map

Electric Log

NEMBE CREEK 10

Fig. 130. Nembe Field, Niger Delta. From Nelson, 1980. Permission to publish by AAPG.

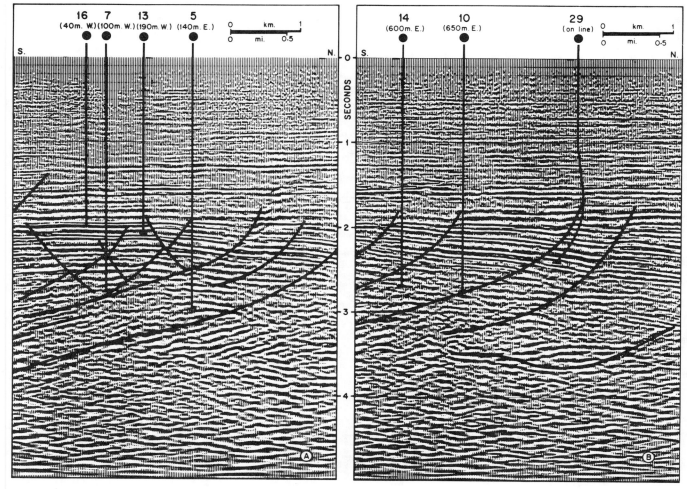

Seismic Sections

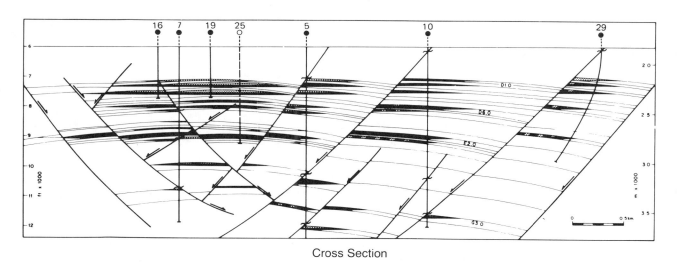

Cross Section

Fig. 131. Nembe Field, Niger Delta. From Nelson, 1980. Permission to publish by AAPG.

5   Marine Erosion and Deposition

Marine Environment

Depth zones, currents, waves, and marine life are all part of the marine condition and with sediment source bear directly upon how and what sediments are produced, how and where they are distributed and deposited, how they are preserved, and whether they will be important as petroleum source and reservoir rocks. Marine processes are intricate and complex, and represent an environment consisting of interrelated factors that combine to produce a variety of conditions favorable to generation and accumulation of oil and gas.

Marine Depth Zones

The marine environment has been subdivided into four depth zones, from the surf zone at the beach to the greatest oceanic depths (Fig. 132). Environments of deposition, erosion, and marine life vary greatly between these zones.

The littoral zone lies between the highest high tide and the lowest low tide. In this zone, maximum erosion occurs as waves break upon a beach or against a seacliff depending upon the type of coast. Life forms include animals and plants that live in both marine and terrestrial environments.

Below the littoral zone and extending to the 200 meter isobath at the approximate edge of the continental shelf, the neritic zone contains the majority of marine life and is representative of a large share of marine deposition. Wave action in the top of the neritic zone diminishes as wave base is reached with increasing depth. Light penetration into the neritic zone is greatest at the top and diminishes toward the bottom. Coral, algae, and numerous other bottom dwellers inhabit this segment of the marine environment.

At 200 meters, the top of the bathyal zone, considerable light can penetrate clear water. However, light becomes progressively dimmer with depth until darkness is reached considerably prior to the 2,000 meter isobath at the base of the bathyal zone. This interval is associated with the continental slope and varies in abundance of life and depositional environment between the top and bottom.

Fine-grained sediments accumulate on the continental slope and often are deposited under conditions of very fragile equilibrium. Storm waves or earthquakes can disturb these sediments, which break away and slide down the slope. Submarine canyons are often developed on the continental slope and can act as sediment conduits.

The abyssal zone lies below the 2,000 meter isobath and extends to the greatest depths of the ocean floor. It is devoid of light and contains animals that are designed to withstand great hydrostatic pressure. Sediments in the abyssal zone are deposited by turbidity currents from the shelf and slope, and as oozes that accumulate slowly from atmospheric dust. Organic sediments in the abyssal zone consist of shells of tiny animals that live at the ocean surface and settle to the bottom when they die.

Marine Currents

Movements of currents within the hydrosphere result from a variety of factors. Currents transport sediments and erode the sea bottom. They create environments of deposition as well as ecosystems and promote thermal exchange within the hydrosphere.

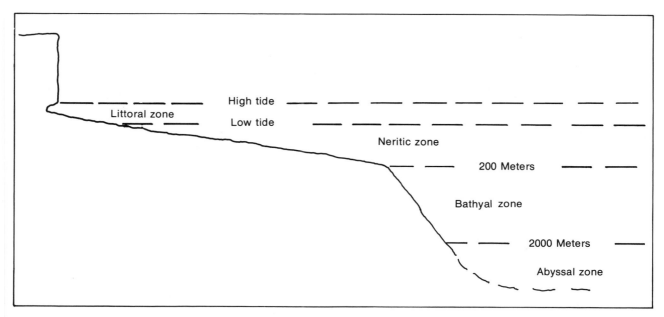

Fig. 132. Marine depth zones

1. Waves are a function of wind except for those that result from earthquakes or volcanic explosions. Wave currents develop along a beach or surf zone. Because waves rarely strike a shore at right angles, a current is developed along the shore in the direction the waves are moving (Fig. 133). This results in transport of sedimentary material in the wave zone by the longshore current. Longshore transportation causes the formation of beaches and terraces seaward from the beach as terrestrial material is carried down the coastline. Without the longshore current, sand materials would not move along the coast and there would be no beaches.

2. Tides are constantly in operation as the earth rotates under the gravitational effects of the moon and sun. This results in changes in sea level and in laterally moving currents as the sea level fluctuates. In some places, tides are very large and their related currents are strong. Coastal areas of western Korea along the Yellow Sea have tides of 33 feet that cause strong currents in inlets and bays. A tide of more than 50 feet in the Bay of Fundy of New Brunswick causes exceptionally strong currents. Tidal variations of 20 feet or more are common along the coasts of the British Isles.

Large displacement tides cause strong currents with considerable erosive and sediment transport capacities. These currents are important marine phenomena in various localities around the world.

3. Unstable sediments deposited on the edge of the continental shelf or on the continental slope are subject to relocation by storm waves or earthquakes. They slide turbulently down the slope, mix with the sea water, and flow as turbidity currents to the sea floor to spread out and settle in the quiet water. Turbidity currents deposit turbidites with graded beds consisting of coarse grains on the bottom and progressively fine grains at the top. Repeated turbidity currents result in bundles of turbidites deposited one after another.

Turbidity currents can also occur where a stream or river carries muddy water into a lake or the ocean. Muddy water that is heavier than clear water moves along the bottom until the sediment particles settle, forming turbidites.

4. Turbidity-type currents that move because of higher density can be the result of difference in temperature and salinity. Density currents may be clear and carry no sediment but will move in response to gravity because they are heavier.

5. Density currents motivated by temperature differences are properly called convection currents because heat exchange is involved. Heating in one area of the sea and cooling in another creates a temperature imbalance that causes sea water to flow. Convection currents can carry and redistribute sediments.

6. Rotation of the earth upon its axis is a strong moti-

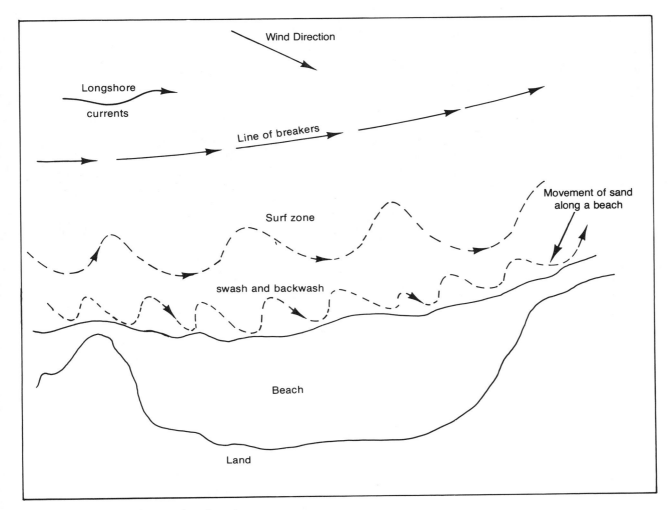

Fig. 133. Movement of coastal sediment

vator of planetary currents, which diverge north and south from the equator. Planetary currents rotate clockwise in the northern hemisphere and counterclockwise in the southern where they are responsible for major oceanic water movements (Fig. 134). Important ocean streams such as the Gulf Stream, Labrador Current, Humboldt Current, and the Japanese Current are examples of planetary currents.

The variety of marine currents developed in the oceans of the earth is responsible for shaping shorelines, developing marine profiles, distributing sediment, and creating depositional and ecological environments. Marine processes related to currents have an important role in the formation, distribution, deposition, and preservation of petroleum source and reservoir rocks.

Wave Types

Waves vary according to whether they are progressing through the open sea or breaking upon a shoal or the shore. Movement and shape of waves are affected by bottom conditions, as well as the shape of the coastline and speed and direction of the wind.

Oscillation Waves. Waves in deep water move in response to wind, moving through the water without appreciably moving the water forward (Fig. 135). As a wave passes, individual water particles move in a circular or oscillatory orbit and return to virtually the same initial position with only slight forward motion. Objects floating on the water are not carried forward by oscillation waves but are moved along by the force of the wind.

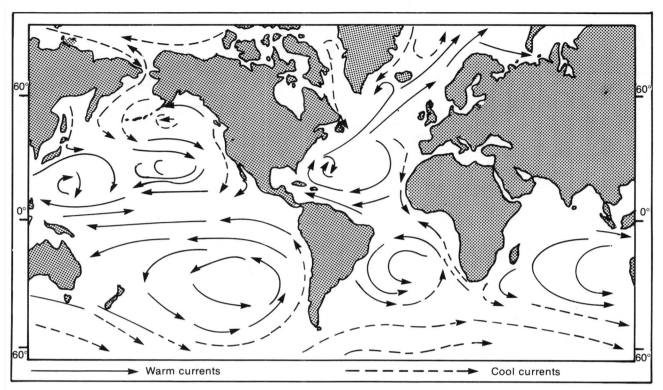

Fig. 134. Oceanic circulation

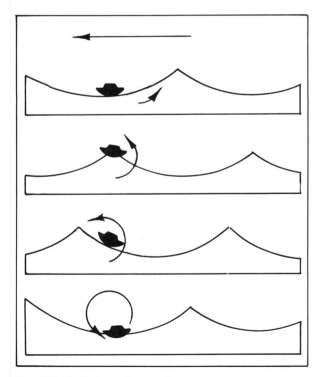

Fig. 135. Particle motion and wave passage

Oscillatory movement of water particles in deep water is greatest at the surface. Movement diminishes downward and dies out at a depth of about half the wave length of the waves (Fig. 136).

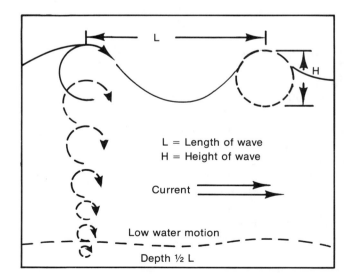

Fig. 136. Diminution of effects of oscillation waves with depth to wave base.

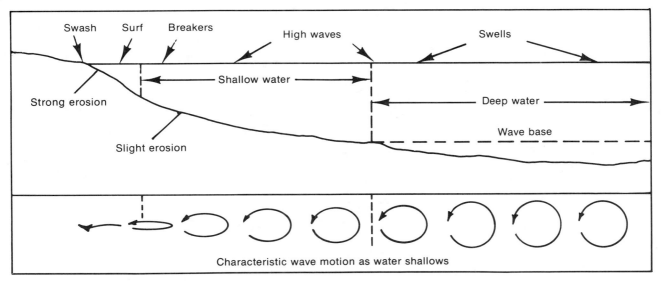

Fig. 137. Deep to shallow water wave transition

Translation Waves (Breakers). Waves of oscillation moving toward the shore encounter the sea bottom, which interferes with their rotary movement by friction, causing drag at the wave base. Friction slows the wave, allowing successive waves behind it to catch up and crowd it. The wave becomes higher and steeper as the sea bottom gets shallower and the interval between the waves shortens. Slowing of the wave and increasing its height causes the top of the wave to extend forward beyond its base, and the wave breaks or collapses upon itself in a forward motion to become a wave of translation (Fig. 137). Here the wave, a forward-moving water front, expends its energy and then moves up the slope of the beach as a sheet. Beach sediments are carried up-slope by the surge of the broken wave as erosion takes place. Wave morphology is described as follows (Fig. 138):

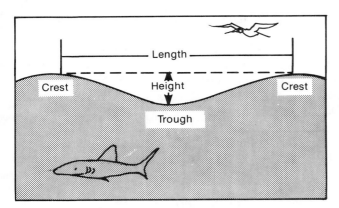

Fig. 138. Wave morphology

1. Crests are the highest points of oscillation waves.
2. Troughs are the lowest points between the crests.
3. Wave height is measured vertically from the lowest part of a trough to the highest part of the adjacent crest.
4. Wave length is measured horizontally from crest to crest or trough to trough.
5. Wave period is the time of passage between two successive crests.
6. Wave base is the depth below which no effect from oscillation waves is felt. This distance is about one half the wave length.

Wave Refraction

Inasmuch as waves usually approach the shore at an angle, and the sea bottom and shoreline are irregular, waves rarely encounter uniform conditions as they move from deep water. As parts of the wave encounter the ocean bottom, those portions begin to slow down, causing the wave to swing around to become parallel to the bottom contours. Refraction of the wave is accomplished as it bends to accommodate the irregularities of the bottom that it encounters on its way into the shore (Fig. 139). As a wave approaches a headland, it is refracted toward the headland against which erosion is concentrated. At the same time, waves are refracted away from bays and inlets, which are protected from erosion. Therefore, headlands erode faster than bays in the process of reducing an irregular coast to a smooth coast.

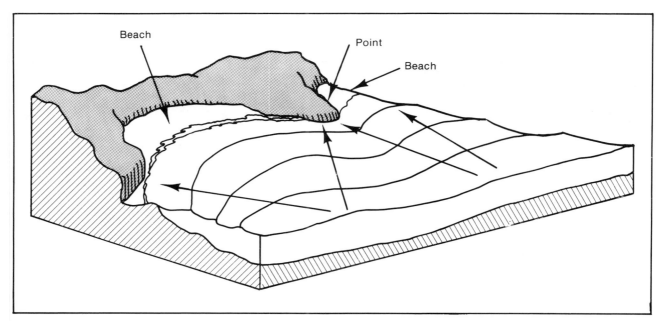

Fig. 139. Wave refraction

Marine Erosion

Most marine erosion is caused by waves and results from hydraulic action and abrasion. When waves arrive at the shore, they exert pressure upon the beach or sea cliff against which they break. Beach deposits move easily when struck by waves and are redistributed according to the sea conditions and size of the waves. Storm waves, however, are able to move large blocks of bedrock, slabs of concrete, and seaside buildings by powerful hydraulic action.

Waves that carry sedimentary material grind away bedrock exposed in the surf zone by abrasion. Cobbles, pebbles, gravel, and sand are very effective abrading agents in the surf zone, which can extend to a depth of about 30 feet. The net effect of waves is to reduce a coastline to a smooth, linear configuration with a gently sloping beach.

A marine shore zone is located between the points where the highest storm wave can reach and where the lowest low tide recedes (Fig. 140). It includes:

1. The back-shore zone, which extends from where the highest storm wave can reach to the high tide level (Fig. 140). Here, storm and high tide surf waves erode the shore.
2. The fore-shore zone, which extends from the high tide level to the lowest low tide and includes the littoral marine depth zone (Fig. 140). Wave erosion is strong in this zone as well.

The shore face lies seaward of the shore and beneath the lowest low tide level (Fig. 140). Deposition of longshore transported sediments occurs here. Beyond the shore face zone of longshore transport is the offshore zone (Fig. 140). This zone extends seaward to include the neritic zone and beyond.

Erosional Features

Irregular shorelines are subject to irregular erosion as long as the shoreline remains uneven. Wave action is directed toward transforming an uneven coastline to an even one, and in the process forms a variety of erosional features (Figs. 141, 142, and 143).

1. The *sea cliff* is an area of very active wave erosion as waves work toward wearing away the rock (Fig. 141). It is located inland from the surf zone and indicates the farthest reach of the waves. Waves eroding the sea cliff undercut it and cause blocks of rock to fall into the surf zone.

2. Wave erosion forms a *notch* at the base of the sea cliff (Fig. 141). The more the notch is developed, the more the sea cliff tumbles down to be broken down in the surf zone.

3. Waves abrading the bedrock at or below sea level

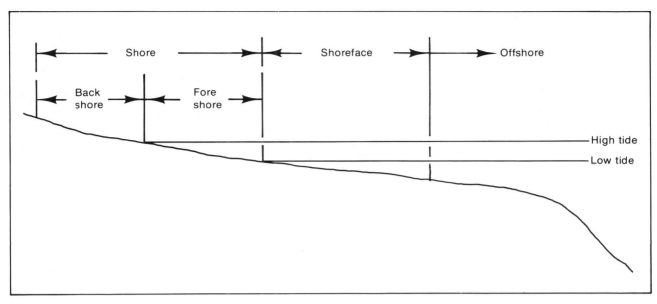

Fig. 140. Marine shore zones

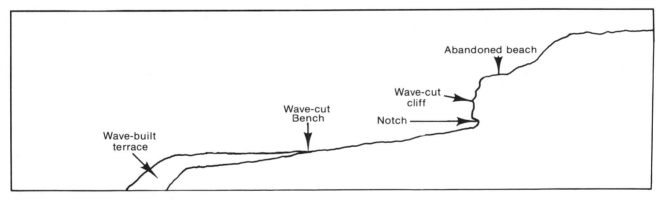

Fig. 141. Profile of wave erosion features

form an abrasion platform called a *wave-cut bench* (Fig. 141). The surface is smooth and extends inland to the notch at the base of the wave-cut cliff.

4. A promontory of land extending seaward (Fig. 142) is known as a *headland*. Because of its exposed position and wave refraction, a headland will erode more quickly than adjacent inlets and bays.

5. Bays and inlets adjacent to headlands are called *coves* (Fig. 142). They are protected and erode less rapidly than do headlands.

6. Waves attacking a sea cliff or a headland can selectively undermine the cliff and form a *sea cave* (Fig. 143).

7. Continued wave action against a sea cave can completely penetrate a narrow headland and form an *arch* through which water can pass (Fig. 143).

8. A *stack* results when an arch collapses as the result of additional wave erosion and leaves an isolated pillar that rises from the wave-cut bench (Fig. 143). Stacks are eventually eroded down to the level of the surrounding wave-cut bench.

Inasmuch as wave erosion wears away an irregular coastline and degrades whatever sea cliff might be present, the ultimate wave effort is directed toward a gently sloping concave shore surface. This represents the profile equilibrium in the marine cycles of erosion. However,

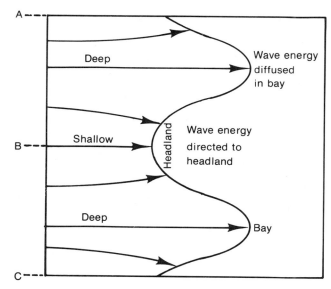

Fig. 142. Wave refraction and energy concentration

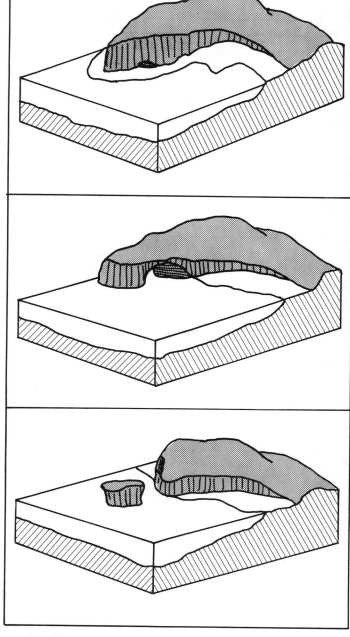

Fig. 143. Sea cave, arch, and stack development

crustal movements disturb levels of the sea, as they do base level in stream erosion, and a marine profile of equilibrium is rarely, if ever, reached.

A marine profile of equilibrium is characterized by the concave sloping surface of the shore and by a sedimentary regime that consists of enough energy to transport shore deposits but not enough to cut back the shore. If there is a sufficient sediment supply, the longshore current and wave action transport the sediment down the coast as a river of sand. Should the sediment supply cease, however, the beach is removed and disappears.

Marine Cycles of Erosion

Irregular coastlines subjected to wave erosion, with sufficient time, ultimately become smooth with gentle beaches (Fig. 144). This sequence represents the marine erosion cycles, which, if allowed by steady sea level to progress, terminate with a profile of equilibrium.

A youthful coast is irregular and subject to considerable erosion. Erosional features include headlands, sea cliffs, stacks, caves, arches, and wave-cut benches.

A long, smooth wave-cut cliff is characteristic of a mature coastline. A beach is developed on the wave-cut bench, which continues to progress inland.

A profile of equilibrium and a gently sloping, slightly concave beach comprise the old age shoreline. Sediment supply and longshore transport are in essential equilib- rium, and there is no erosion of the coast. A wide wave-cut platform disperses wave energy so that there is no additional erosion of whatever sea cliff remains. However, mass wasting on the land can further reduce the sea cliff and the shoreline.

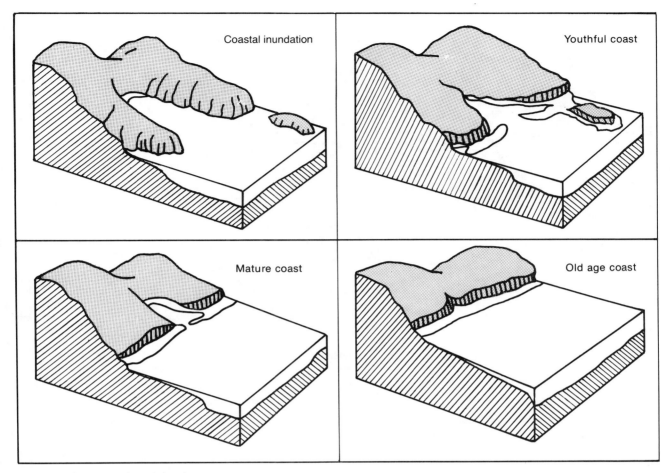

Fig. 144. Marine cycles of erosion

Marine Shorelines

Sea level moves up and down with the activity of the crust. This creates different kinds of coastlines and superimposes characteristics of several types upon each other. Submergent, emergent, and neutral shorelines are possible as sea level moves up, down, or remains stable.

As the sea level rises, it may do so more rapidly than the rate of erosion of the coastline. A submergent coast will result, causing the drowning of the mouths of rivers or glaciers discharging into the sea (Fig. 145). Headlands and inlets are common as erosion attacks the coast. Drowned river valleys are called estuaries or rias. Chesapeake Bay is a well-known estuary. A drowned glacial valley is known as a fiord. Fiords are common along the coasts of British Columbia and Alaska.

Exposure of the ocean bottom by a receding sea results in an emergent coast. An emergent coast is often very smooth, regular, and gently sloping, as it is along the Texas Gulf Coast. Wave-cut terraces high above present sea level along the California and Oregon Coasts indicates emergence of those areas where mountains come to the sea (Fig. 146). Several levels of the sea are represented by these terraces which are also common along the coasts of the British Isles.

Where lava flows and debris slides alter the coast and more or less keep pace with erosion, a neutral coastline is developed. Characteristics can be widely variable and depend upon the type of neutralizing activity occurring along the coast. Lava flows coming to the coast in the Hawaiian Islands produce a neutral coastline.

Marine Deposition

In a normal shore-to-basin sequence profile along a depositional coast, rock materials change with water depth, wave action, and amount of sediment available.

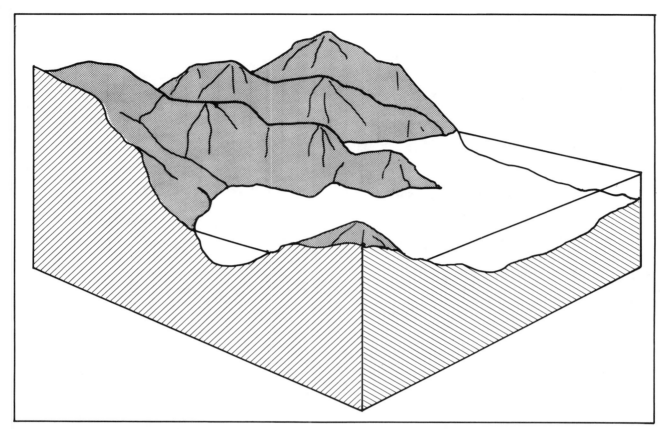

Fig. 145. Submergent coastline

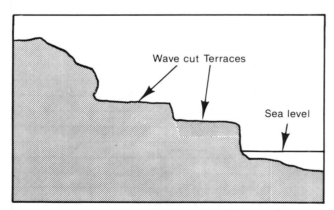

Fig. 146. Emergent coastline

Where sedimentary clastic material is available, coarse sediments are deposited in the active surf zone along a beach where they grade to progressively finer sediments offshore in a normal marine facies change or change of rock type or lithology (Fig. 147). Deeper water facies changes occur in the distribution of submarine fan sediments, for example. Coarse clastic sediments at the base of the continental slope might be graded and will become finer with seaward distance from the slope (Fig. 147).

Conglomerate along the beach zone grades quickly into sandstone in the foreshore, and to siltstone, and ultimately clastic shale, in the offshore. Excellent examples of coarse to fine facies changes occur along southern Trinidad and the west coast of the Isle of Lewis in the Outer Hebrides. Onshore boulders and cobbles, in these localities, grade into sand and finer sediments within a very short distance from the beach to offshore. Where sediment supply is limited, clastic shale in the offshore may grade laterally down the shelf into calcium carbonate mud, which might grade into a reef at the edge of the shelf. Facies changes of this type are a function of available clastic sediment. Carbonate development depends on water depth, temperature, and clarity, particularly where there are reef organisms.

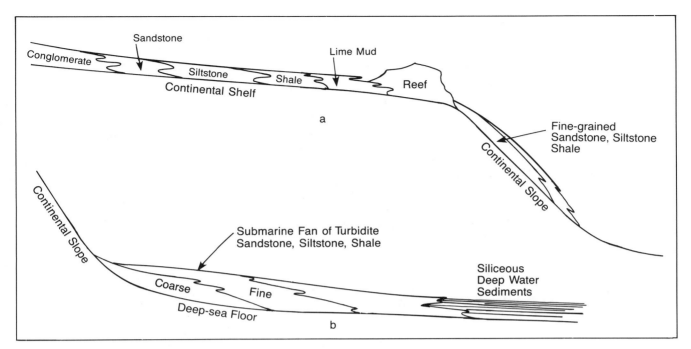

Fig. 147. Facies changes of the continental shelf and slope and near shore deep sea floor.

Because clastic shale can be carried to points far off-shore, source beds can be deposited beyond a reef system in deep water as well as in shallower water on the shelf. Shales deposited in both localities can act as source beds for clastic reservoirs near-shore and for reef developments at the shelf edge.

Clastic Depositional Features

Clastic deposition under marine conditions utilizes sediments that are largely terrestrial in origin and are provided by streams discharging into the sea. Most coarse clastic deposition occurs relatively near-shore, although sandstone can be deposited on the deep ocean floor by turbidity currents and by flow down canyons cut into the shelf and slope.

Marine deposition in the near-shore environment is often the direct result of longshore distribution of stream-introduced sediments and assumes specific shapes according to the environment. Coarse clastic sedimentation in the near-shore produces depositional bodies with good porosity and permeability and favorable reservoir potential.

1. *Beach deposits* are directly correlated with discharge of terrestrial sediments into the marine environment where they are transported and distributed by the long-shore current (Fig. 148). Gravel and sand on the beach become well-sorted and rounded, and acquire excellent porosity and permeability. They are excellent reservoir sediments.

2. *Barrier islands* lie offshore and parallel to the coast (Fig. 149). They consist of sand and gravel in the surf zone and windblown sand in the dune area behind the beach. Islands of this type are often long and narrow. Padre Island along the south Texas Coast is 110 miles long and varies from less than one mile to five miles in width. Well-developed barrier islands separate the southern coast of Nigeria from the Atlantic Ocean. Like beach deposits, barrier island sands are excellent reservoirs and produce petroleum in Texas, Louisiana, Colorado, and Wyoming in the United States, and in the North Sea.

3. Behind the barrier island, and between it and the mainland, is the *lagoon* (Fig. 149). The lagoon is a shallow body of water with access to the open sea limited by the barrier island. Lagoon deposits consist of fine-grained clastics, evaporites, and in some cases lime mud. Fine-grained lagoon clastics have some source bed potential, although in most cases they are thin.

4. Longshore distribution of terrestrial sediments from stream discharge can cause the formation of a *sand spit* (Fig. 149) on the down-current side of a headland. A spit will grow in the down-current direction if sediment sup-

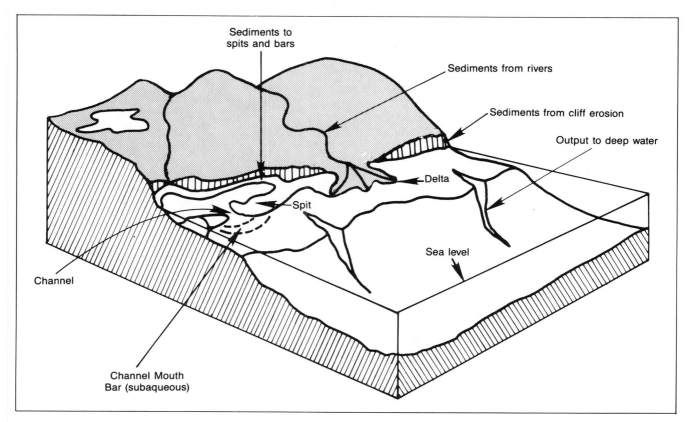

Fig. 148. Shoreline depositional system

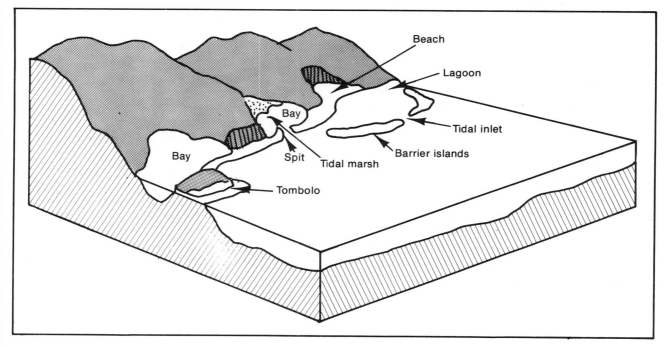

Fig. 149. Shoreline depositional features

ply and longshore drift continue. Clastic sediments that make up a spit can be sufficiently porous and permeable to have good reservoir characteristics.

5. As a spit builds down-current, it may curve in response to the drift and become a *hook*. A spit may curve at an inlet because of current into the inlet and wave refraction around the end of the spit. A hook consists of the same type of clastics as spits, beaches, and barriers and has good reservoir potential.

6. A *baymouth bar* is formed when a spit or hook grows (Fig. 150) all of the way across an inlet and seals it off from the open sea forming a lagoon in the process (Fig. 149). The barrier in this case becomes a baymouth bar with good reservoir characteristics.

7. River or tidal flow can move rapidly through a channel in a barrier island. Water velocity will increase in the channel, which is kept sediment-free by the venturi effect of forcing a large volume of water through a narrow opening. Decrease of water velocity after passage through the channel permits deposition of a channel mouth bar (Fig. 148). Channel mouth bars are commonly associated with the mouths of river or distributary channels and channels in barrier bars where they provide good reservoir potential.

8. An island or stack can be connected by a sand spit or *tombolo* to another island or the mainland (Fig. 151). Tombolos form in much the same way as spits, hooks, and baymouth bars and have similar composition and reservoir potential.

9. Sand transported by the longshore current also forms

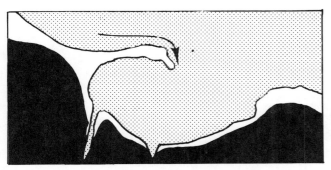

(A) Sediment moves along the shore and is deposited as a spit in the deeper water near a bay.

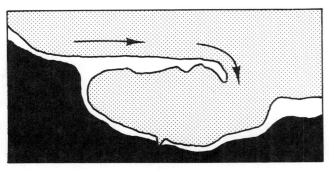

(B) The spit grows parallel to the shore by longshore drift.

Fig. 150. Spit development

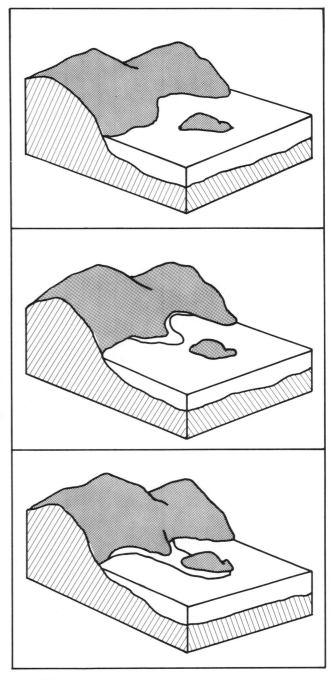

Fig. 151. Tombolo development

terraces just outside the beach zone. They are called *shoreface terraces* and are often en echelon to the shoreline in response to the longshore current and waves striking the shore at an angle. Along shallow water coasts, multiple shoreface terraces arranged in echelon can develop as they do along the barrier islands of the central and south Texas coasts. As offshore extensions of the beach, shoreface terraces have good reservoir potential.

10. A *continental terrace* can be built along the edge of the continental shelf in the offshore zone. It usually consists of fine-grained clastics that are water saturated and unstable. Storm waves or earthquakes can cause con-

tinental terrace material to slide down the slope as a submarine slide or a turbidity current. Terrace materials can be source or reservoir beds.

11. Terrestrial clastic material brought across the continental shelf, across a delta or through a submarine canyon is graded and stratified as it is distributed on the ocean bottom as a *submarine fan* at the base of the continental slope or other marine slope (Figs. 152, 153, and 154). Submarine fan sediments are deposited by slumping from the slope and turbidity currents that provide graded turbidites. Slumped material is poorly stratified to unstratified and poorly sorted. Turbidites are graded from coarse

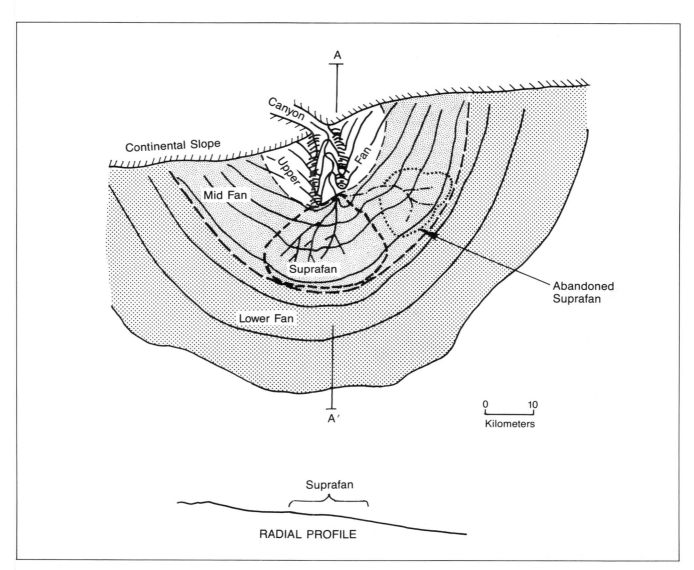

Fig. 152. *Model for submarine-fan growth with active and abandoned depositional lobes (suprafans). From Normark, 1978. Permission to publish by AAPG.*

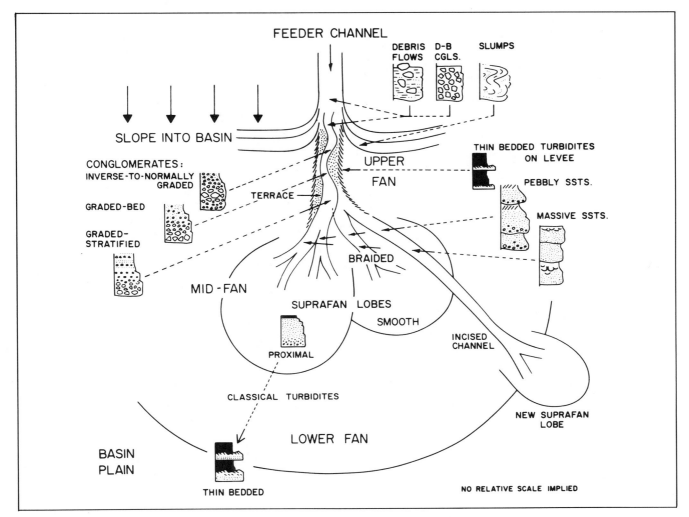

Fig. 153. Submarine-fan deposition, relating facies, fan morphology, and depositional environment. From Walker, 1978. Permission to publish by AAPG.

at the base of and near the slope to fine at the top and at distance from the slope. They can be moderately well to well sorted and provide good reservoir potential. Fan deposits and accompanying tubidites are important reservoirs in the North Sea, southern California and Texas.

Clastic deposits in the nearshore environment are often coarse-grained and have good reservoir potential. They are deposited as individual bodies that have limited lateral extent and can form good petroleum traps when sealed by impermeable beds. Exploration for nearshore clastic bodies relies heavily upon depositional geometry and established sedimentation trends that orient the individual features according to regional considerations.

Marine Carbonate Deposition

Marine deposition in the carbonate realm involves a different environment than does the clastic although carbonate rocks can be clastically derived from other carbonates. Primary carbonate deposition results from precipitation and deposits formed by plants and animals that utilize carbonates in their life processes. Clastic sedimentation of carbonates involves deposition of particles derived from organic and inorganic sources which are ultimately cemented together to form rocks.

Carbonate rocks, which consist of organic-rich algal mats, can act as source beds on occasion but for the most

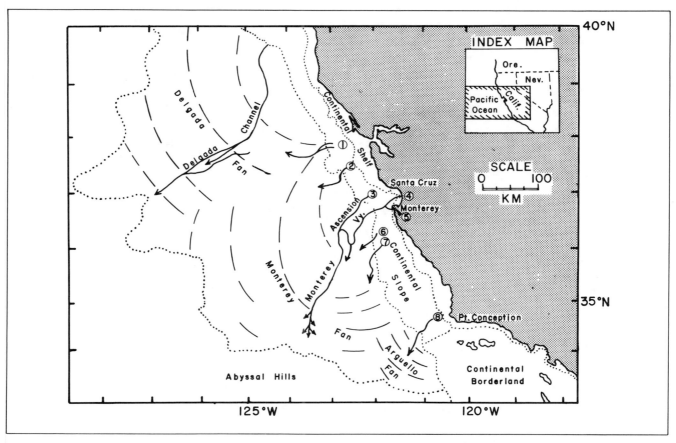

Fig. 154. *California continental rise showing Delgada, Monterey, and Arguello submarine fans. Dotted lines mark outer edge of fans and limits of continental shelf and slope areas. Circled numbers indicate positions of canyons tributary to Monterey fan: 1, Farallon; 2, Pioneer; 3, Ascension; 4, Monterey; 5, Carmel; 6, Sur; 7, Lucia-Partington; 8, Arguello. Adapted from Wilde, et al., 1978. Permission to publish by AAPG.*

part certain carbonates are better as reservoirs. Varying carbonate environments offer different types of carbonate rocks and control their suitability as reservoirs or sources.

Reef Deposits. Reef formation occurrs in several ways, most of which involve metabolic activity by plants or animals. Some marine algae remove carbon dioxide from seawater and decrease the solubility of dissolved calcium carbonate. The calcium carbonate precipitates on the sea floor as banks, which by their shape and position can be called reefs.

Corals are colonial exoskeletal animals that build large calcium carbonate structures that grow and recede in response to sea level. Reefs formed in this way assume a variety of shapes according to the substrate upon which they are built, the water depth, temperature, clarity, oxygen content, and salinity, stability of sea level, and amount of nutrients available. Reefs are normally warm water phenomena that grow in tropical waters between

30° North and 25° South. Coral organisms require clear water, abundant sunlight, and shallow depths of less than 250 feet. Marine areas in which rivers discharge large volumes of muddy sediments normally do not grow coral reefs. Fresh water from rivers is usually deterrent to reef growth as well and can kill established reefs if suddenly increased in the reef environment to the extent that salinity is appreciably reduced.

Reefs consist of many different types of corals, which vary according to locale. For the most part, the skeletal framework of a reef is very porous and permeable. Reefs produce hydrocarbons abundantly in many fields around the world. Zelten in Libya and the Golden Lane in Mexico are important examples of reef fields.

Reefs that grow directly on the shore of the mainland or an island from which they extend seaward are *fringing reefs* (Fig. 155). Fringing reefs grow where there is little or no clastic contribution from muddy streams that run off the land mass to which they are attached.

A reef separated from the mainland or an island by a lagoon is called a *barrier reef* (Fig. 155). Reefs of this kind can prevent the lagoon from having free access to the open sea. Barrier reefs often grow as linear features on tectonic lineaments or at the edge of the continental shelf. The Great Barrier Reef of northeast Australia is the best example of this feature.

An *atoll* is a barrier reef that grows around a subsiding volcanic island (Fig. 155). As the island sinks, the reef grows at the same rate that the water rises. Eventually, the island disappears beneath sea level, leaving the circular atoll consisting of a ring of coral islands. Atolls are common in the South Pacific Ocean.

The oval *patch reef* grows randomly on the shelf. It can occur in the lagoon or in areas where there are no barriers. Patch reefs grow on the shelf areas in the Bahamas where other reef types also grow.

Rapid rise in sea level causes a correspondingly rapid upward growth of *pinnacle reefs*, which can be many hundreds of feet tall (Fig. 155). Pinnacle reefs are usually relatively limited in areal extent because growth is concentrated on keeping pace with rising sea level. Pinnacle reefs are productive of oil and gas in Devonian rocks of Alberta, Canada.

Wave action bearing upon exposed barrier reefs breaks

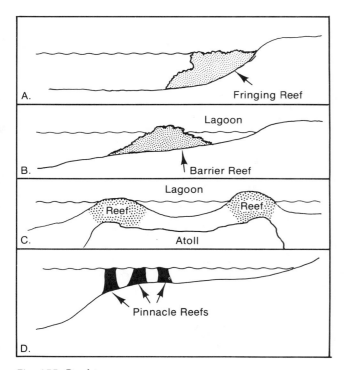

Fig. 155. Reef types

off fragments of talus, which settle at the base of the reef (Fig. 156). Most reef talus settles at the base of the reef on the seaward side in response to maximum wave energy there. Some talus is broken off and thrown over the reef to the backside and settles at the base of the reef as well. Reef talus is porous and augments the reservoir character of the reef itself.

Carbonate Shelf Deposits. Carbonate beds on the shelf (Fig. 157) can be of several types, some of which have good reservoir potential, some of which do not. Lime mud on the shelf, for instance, has limited reservoir potential unless it is dolomitized.

Oolitic limestone is formed by precipitation of calcium carbonate on minute depositional nuclei as they are washed back and forth by waves and currents. Sphericity of the ooliths is established by their mobility. Oolitic limestones are often highly porous and permeable.

Pelletal limestone comes from the excretion of burrowing organisms that ingest lime mud and pass it through their digestive systems as sand-size particles. Pelletal limestone can have good porosity and permeability.

Carbonate algal mats grow in shallow shelf waters which can be moderately to highly saline and protected from wave action to preserve their often rich organic content. Petroleum source potential of algal mats can be high with low oxygen content and high organic preservation.

Lagoon Deposits. Deposition in lagoons located behind barrier reefs is often reflective of limited access to the open sea (Fig. 157). Evaporation of sea water and precipitation of dissolved constituents can be important depositional factors. Terrestrial clastics are not abundant in carbonate lagoons but can combine with evaporites and layered carbonates in which case they may provide petroleum source potential.

Lime mud precipitates out of seawater as calcium carbonate. As a rock it is usually hard and dense and has limited porosity and permeability. Diagenetic re-crystallization (dolomitization) and addition of the magnesium ion results in a volume decrease and often improves porosity in hard, dense limestones.

Evaporites such as gypsum and rock salt can precipitate out of seawater as it evaporates. They are often interbedded with dense lime mud and thin, fine-grained terrestrial clastics. Primary dolomite precipitated directly from seawater occurs in few places in the world.

Marine Deposits and Petroleum

1. Sand and gravel deposits associated with beach sedimentation are normally deposited upon fine-grained sediments over which they transgress or regress with changes

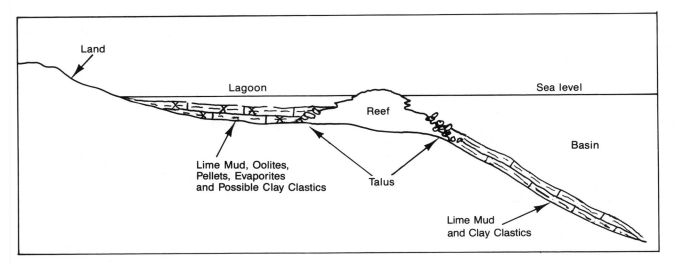

Fig. 156. Barrier reef cross-section

in sea level. They can be very coarse and consist of pebbles or cobbles on English, French Mediterranean and South Trinidadian beaches. Typical beach sands are well-sorted and well-rounded and can be excellent reservoirs.

Electric log response of beach deposits is the inverse of that for stream channel sands (Fig. 158). Beach sands are deposited upon fine-grained sediments that have little porosity and reduced spontaneous potential and resistivity response. Some Cretaceous sandstones comprise regressive and transgressive beach and nearshore deposits in the San Juan basin of New Mexico (Fig. 159). This area is well known for tight gas sands, which can be stimulated to enhance production.

Coarse beach sands with high resistivity and self potential characteristics and which overlie fine-grained shelf deposits produce the pattern shown in Figure 158.

2. Sands deposited on barrier islands are very similar

to those deposited on beaches. They are often difficult to distinguish from beach deposits and produce a similar electric log response (Figs. 160 and 161). Typical electric log responses illustrate a variety of terrestrial marine and nonmarine depositional environments (Fig. 162).

Barrier island sands are porous and have excellent reservoir potential (Figs. 163, and 164). They are often en-

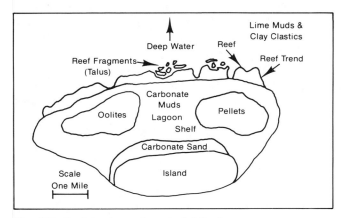

Fig. 157. Carbonate sediment distribution

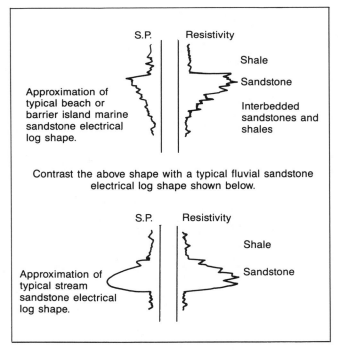

Fig. 158. Electric log responses for beach/barrier island and stream channel sandstones.

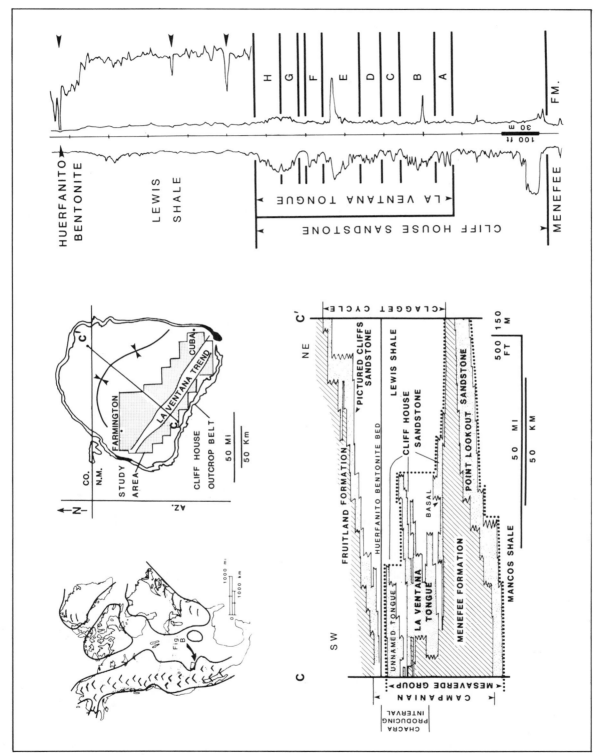

Fig. 159. Transgressive-regressive sandstones of the San Juan Basin. From Palmer and Scott, 1984. Permission to publish by AAPG.

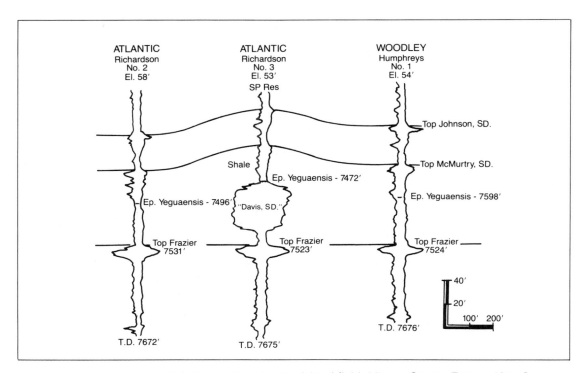

Fig. 160. North-South cross-section of Hardin beach or barrier island field, Liberty County Texas. After Casey and Cantrell, 1941. Permission to publish by AAPG.

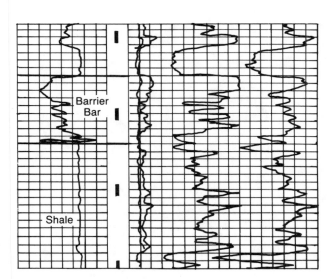

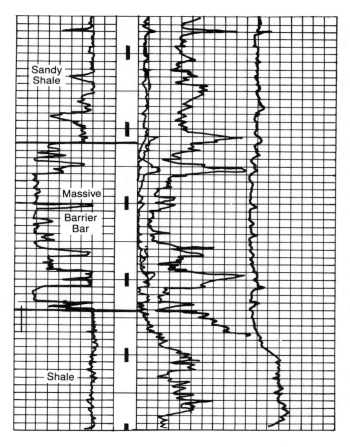

Fig. 161. Electric logs of barrier bar sandstones. Above: Thin barrier bar underlain by shale. Right: Thick barrier bar underlain by shale.

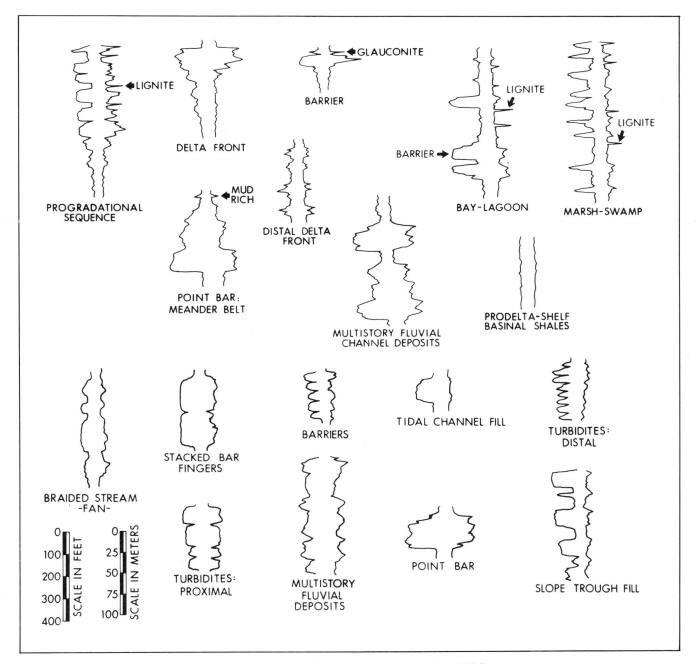

Fig. 162. Facies and log patterns. From Garcia, 1981. Permission to publish by AAPG.

closed by shale source beds, which seal the sandstone body into which their hydrocarbons have been expelled.

3. Spit, hook, and baymouth bar deposits are similar to beaches and barrier bars. They consist of the same porous sands and have the same good reservoir properties. Their electric log responses are similar as well.

4. Submarine fan sediments consist of slump, channel and turbidite components. There are extensively productive reservoirs in the North Sea Frigg Field (Figs. 165, 166, and 167), where reserves amount to over seven trillion cubic feet of gas, and the North Sea Forties Field with reserves of over four billion barrels of oil.

5. Organic-rich, deep-water basin deposits have very good source bed potential (Fig. 156). They are deposited

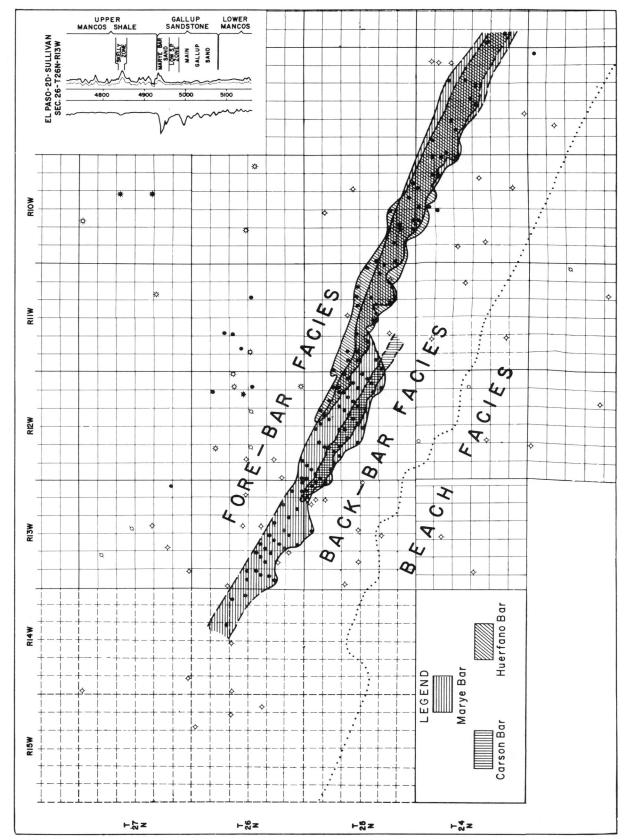

Fig. 163. The Bisti Field, New Mexico, a barrier bar. From Sabins, 1963. Permission to publish by AAPG.

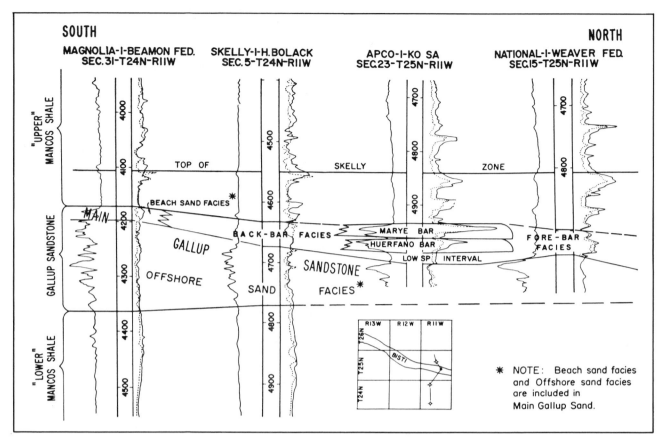

Fig. 164. South-north electric log section, Bisti Field, New Mexico. From Sabins, 1963. Permission to publish by AAPG.

well below the wave base where their carbonaceous content is preserved. These beds provide hydrocarbons to up-dip reservoir rocks.

6. Fine-grained sediments deposited in a lagoon (Fig. 156) between the mainland and a barrier bar or baymouth bar can have source bed potential if they contain organic material and are deposited under quiet water conditions. Lagoon sediments are not necessarily very thick or laterally extensive.

7. A marine swamp between distributary channels on a delta may provide good source beds for channel sand deposits. These beds can be extensive when considered in comparison to the Mississippi and Nile deltas, which are quite large.

Back island or back beach marine swamps can provide hydrocarbon source beds for beach and barrier island sandstones. They can be extensive along the depositional strike of the coast.

8. Porous reef carbonates can be highly productive reservoir rocks (Figs. 168, 169, 170 and 171). In many areas they are easily identified on seismic sections (Fig. 172). Reefs in Mexico (Fig. 173), Canada, West Texas (Fig.

174), and North Africa are well known for their high rates of oil and gas production. Reef carbonates can be good source beds in some producing areas. However, since reef organisms require oxygen and reef development is associated with oxygen, oxidation of organic material can significantly reduce petroleum generation potential in reefs.

9. Porous algal and dolomite banks can act as reservoir rocks. Some fine-grained dolomites can act as source beds in sedimentary environments consisting entirely of carbonate rocks that may be associated with some evaporites as well. A number of south Florida oil fields produce from carbonates that presumably derived their hydrocarbons from adjacent organic-rich dolomite banks. Rates of production from these fields are marginally economic onshore, but would be uneconomic offshore. Some mideastern carbonates are considered to be excellent source beds in a number of very large fields. Carbonate algal mats in the salty, shallow, marginal waters of the Red Sea are often very rich in preservable organic material and appear to have significant petroleum source potential.

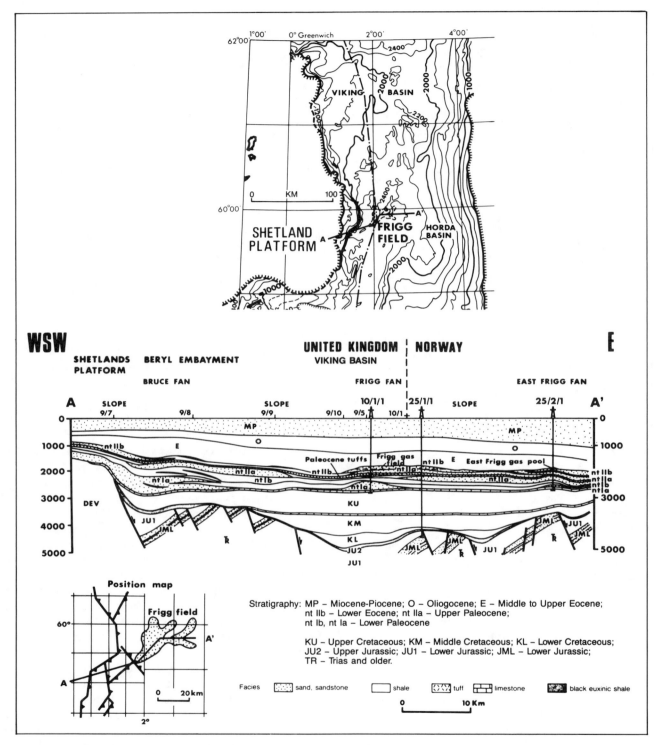

Fig. 165. Viking Basin, Frigg Field. Above: Submarine fan deposits. Below: West to East section across Viking Basin. From Heritier, et al., 1980. Permission to publish by AAPG.

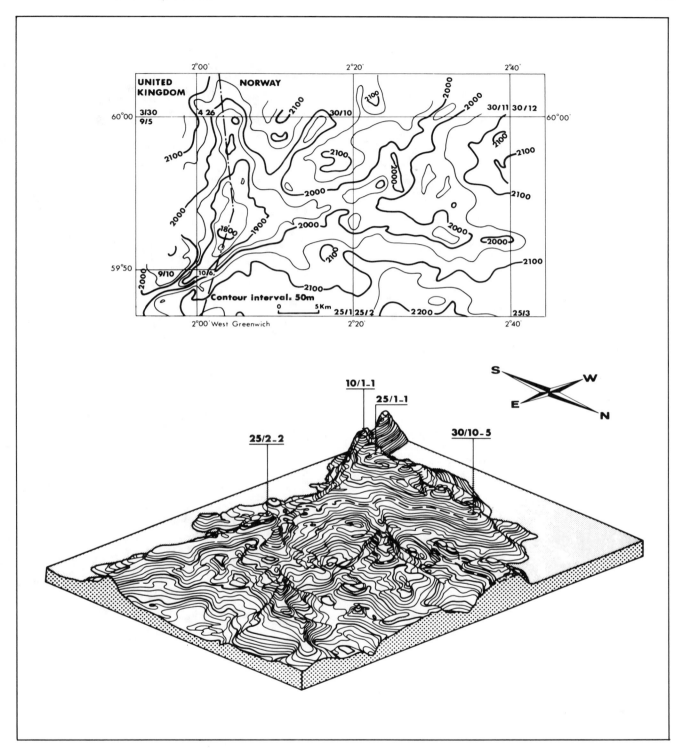

Fig. 166. Fan Shaped Frigg Field. Above: Seismic structure of Frigg field at top of Frigg sand. Note fan shape and southwest sediment source. Below: Computer generated perspective view from northeast of Frigg field with several wells. From Heritier, et al., 1980. Permission to publish by AAPG.

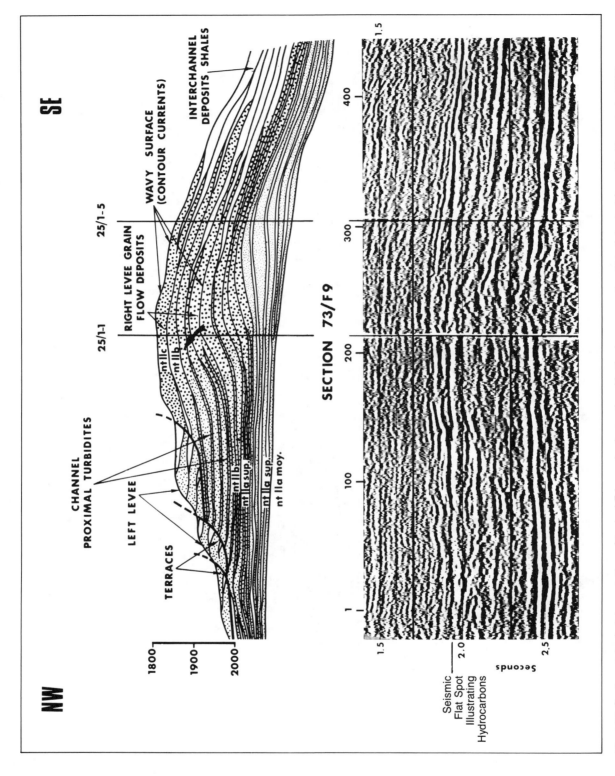

Fig. 167. Sedimentologic interpretation of typical Frigg seismic section. Vertical exaggeration 2×. Adapted from Heritier, et al., 1980. Permission to publish by AAPG.

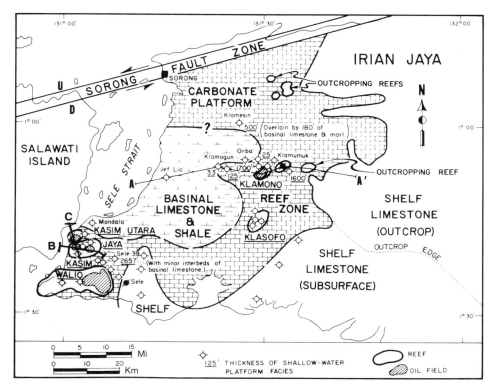

Fig. 168. Carbonate facies in Salawati basin, Irian Jaya, New Guinea. From Vincelette and Soeparjadi, 1976. Permission to publish by AAPG.

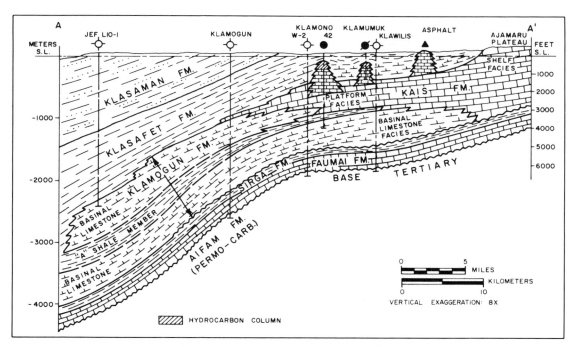

Fig. 169. Cross-section of reef production, Indonesia. From Vincelette and Soeparjadi, 1976. Permission to publish by AAPG. See Figure 165.

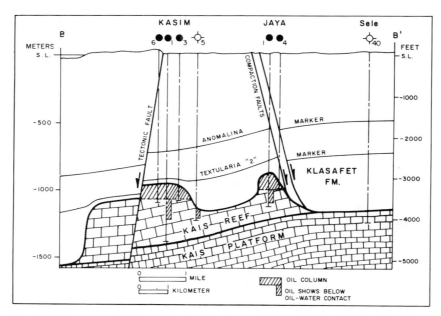

Fig. 170. Cross-section through the Kasim and Jaya Fields, Indonesia. From Vincelette and Soeparjadi, 1976. Permission to publish by AAPG. See Figure 168.

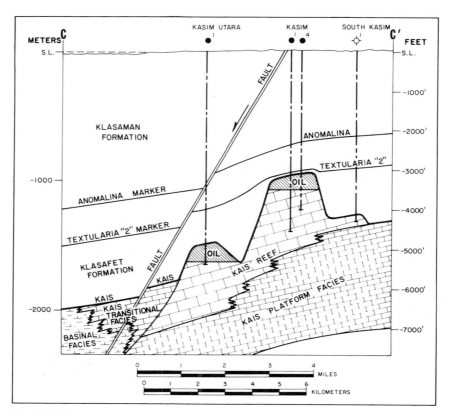

Fig. 171. Cross-section through the Kasim Utara and Kasim Reefs, Indonesia. From Vincelette and Soeparjadi, 1976. Permission to publish by AAPG. See Figure 168.

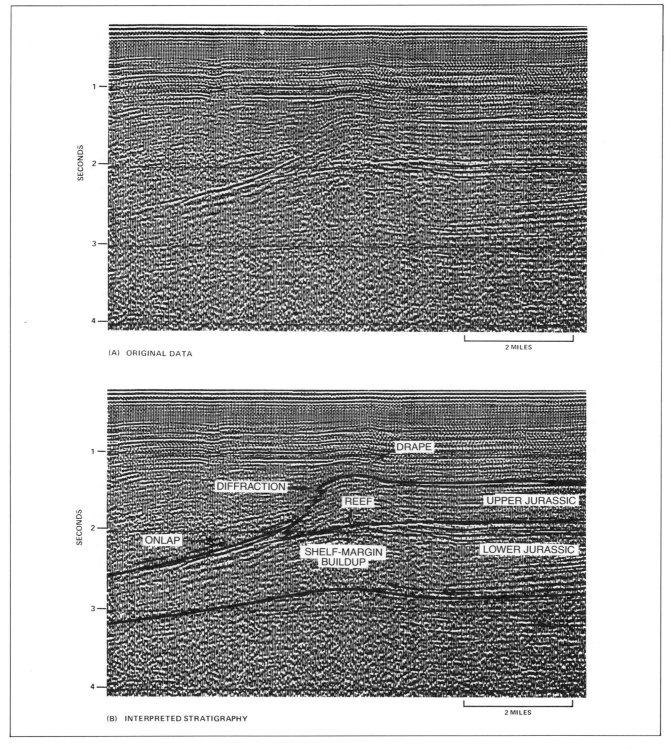

(A) ORIGINAL DATA

2 MILES

(B) INTERPRETED STRATIGRAPHY

2 MILES

Fig. 172. Original and interpreted reef seismic line. From Bubb and Hatlelid, 1977. Permission to publish by AAPG.

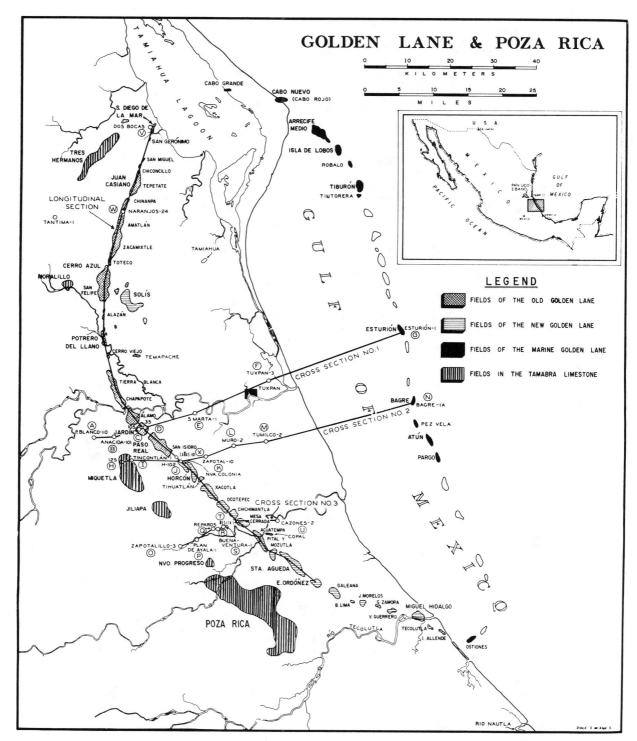

Fig. 173. Golden Lane and Poza Rica carbonate fields, Mexico. From Viniegra and Castillo-Tejero, 1970. Permission to publish by AAPG.

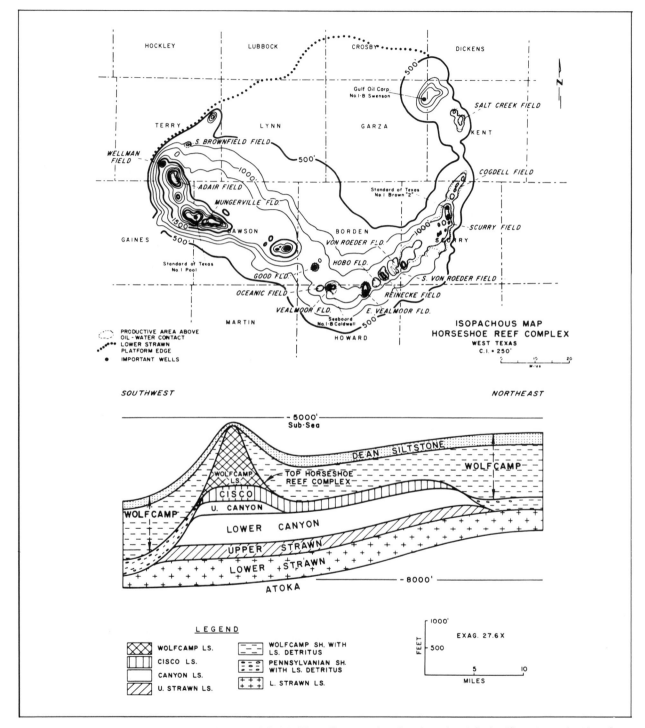

Fig. 174. Above: Isopach map of Horseshoe reef field, West Texas, showing location of significant production from reef limestone along crest of atoll. Below: Southwest-northeast schematic cross-section through thickest known part of Horseshoe atoll. From Vest, 1970. Permission to publish by AAPG.

6   Depositional Basins

Depositional Basins

There can be no absolute criteria for the specific definition of a depositional basin because of the multitudes of parameters involved. Treatment of the concept of depositional basins, as a generality, is substantially more realistic since it connotes a place where sedimentary deposition takes place. Beyond that, however, depositional basins do not lend themselves to detailed classification because of wide-ranging variations of depositional environments and tectonics related to both basin formation and the types of sediments involved.

Depositional basins occur on continental margins, and conform generally to certain conceptual styles, within continental limits, and in oceanic settings diverge in the details of their sediments and structure. Some are basins and some are embayments. Some contain clastics, some contain carbonates, some contain evaporites, and some contain all of these.

The importance of depositional basins is that they are places in which sediments that generate and accumulate petroleum are deposited. Source and reservoir beds are products of deposition under many types of environments, which themselves are functions of the basins in which they occur.

Basin Types

A general basin classification depends upon the locations in which they occur rather than the type of sedimentation. This pertains to locations relative to interior continental masses, continental margins and oceanic locations.

Geosyncline. Extensive linear troughs in which thick sediments accumulate are called geosynclines (Fig. 175). They subside slowly as they receive sediments and remain active for long periods of geologic time. Marine sediments are characteristic of geosynclines, which are usually marginal to continents or to island arc complexes.

Modern geosynclines occur along zones of subduction where the sea floor moves under continental crust. Deformation in the geosyncline can be appreciable because of the interaction between oceanic and continental crust.

Subduction zones or trenches are places where rapid and chaotic sedimentation occurs (Figs. 176 and 177). This part of the geosyncline is called the *eugeosyncline* and is located on the seaward side of the subduction system (Fig. 175). Eugeosynclinal sediments are deposited in deep water, are thick and heterogeneous, and usually are not considered to contain good source beds. Therefore, eugeosynclinal beds are not particularly prospective of oil and gas.

On the landward side of the geosyncline, marine sedimentation occurs under more tranquil conditions than in the eugeosyncline. The *miogeosyncline* (Fig. 175) represents the shelf of the continent where deposition is in shallow water, is slower, and sediments are not as thick as in the eugeosyncline. Miogeosynclinal rocks consist of sandstone, shale, and carbonates, which are considered prospective of oil and gas and produce in various parts of the world.

Geosynclinal development and history are complex and are not always the same. Examples range from the active type of system in the Aleutian Islands or Japanese Islands to the passive tectonic environment of the Atlantic Coast of the United States.

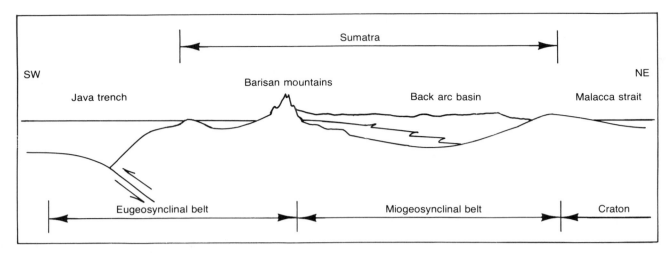

Fig. 175. Cross-section of Indonesian geosyncline

The Aleutian and Japanese Islands separate their respective geosynclines into the eugeosynclinal subduction zone (Fig. 178) and the miogeosynclinal basin or shelf on the continental side of the system. The eugeosyncline and miogeosyncline are separated by a volcanic island arc in each case.

Petroleum production in this system comes from the shelf edge of the miogeosyncline behind the island arc. An example of this is the petroleum production from the Tertiary sediments in the back arc basin in Sumatra.

Geosynclinal development along the Atlantic Coast of the United States is not as tectonically active as in the island arc system. Here, the separation between eugeosyncline and miogeosyncline is a result of water depth, with the former located in the deep water and the latter in shallow water (Fig. 179). In fact, different terminology is used to emphasize the dichotomy, and substitutes miogeocline and eugeocline for miogeosyncline and eugeosyncline, respectively. No volcanic rocks are deposited in a geosyncline of this type, which contains red beds, evaporites, marine clastics, and carbonates (Fig. 180).

Petroleum production in this type of geosyncline usually comes from sediments on the shelf where marine clastics and reef carbonates provide good reservoirs. The production from parts of offshore Brazil, the Grand Banks, and the North Sea are related to this type of geosyncline. Most highly prospective parts of the passive geosyncline are on the shelf and the coastal plain, which lie within the miogeocline and represent the type of shallow marine deposition normally associated with petroleum production.

Intracratonic Basin. Most continental masses have a stable interior, or craton. Cratonic areas are not tectonically very active and beyond the formation of areas of local subsidence develop little or no deformation. Basins that form in response to local cratonic subsidence are known as intracratonic basins. They are repositories for marine sedimentation when the craton is undergoing oceanic transgression. In some cases, intracratonic basins contain sedimentary sections which might exceed 14,000 feet in thickness.

Marine deposition in intracratonic basins relates to shallow water environments associated with shelf conditions. Marine clastics, carbonates, and evaporites are typical of some basins from which good amounts of pe-

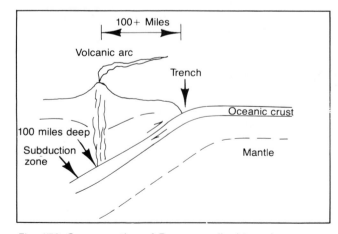

Fig. 176. Cross-section of Eugeosynclinal trench

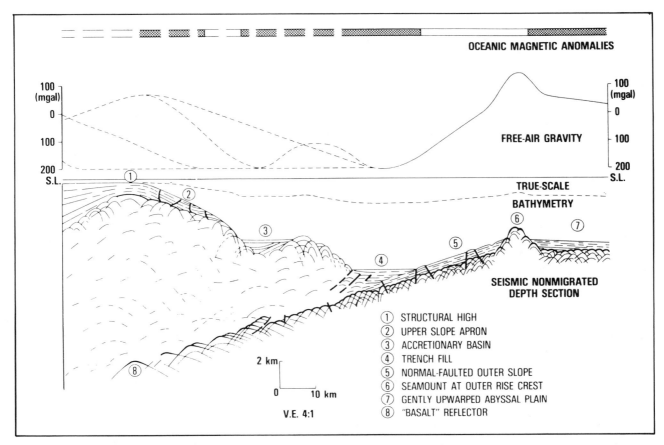

Fig. 177. Geophysical expressions of a trench. From Dickinson and Seely, 1979. Permission to publish by AAPG.

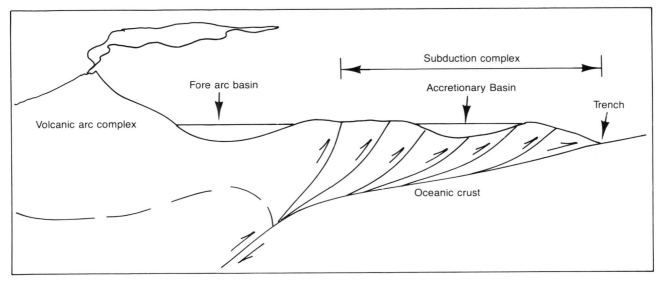

Fig. 178. Structures associated with trenches

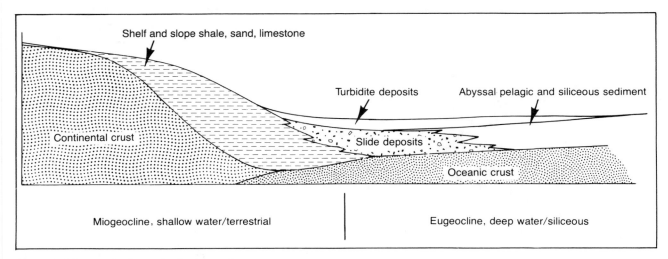

Fig. 179. Miogeoclinal-eugeoclinal couple

troleum have been produced. The Michigan Basin is a good example of an intracratonic basin (Fig. 193), which has produced from barrier bar and beach deposits, reefs, and salt structures.

Continental Margins. Basins of many types develop along the different types of continental margins examined in the section on plate tectonics. These are tectonic basins, which are produced according to the type of continental margin involved. Most continental margins combine marine sedimentation with stages of basin development concomitant with the progress of the development of the margin. This means that sediments along continental margins range from continental, through evaporite, to shallow clastics and carbonates, to deep-water turbidites and pelagic deposits, depending upon the stage of margin development.

Structural development of basins during margin evolution creates traps for petroleum accumulation as deposition occurs. Such a continuing history of deposition and associated deformation creates numerous source and reservoir bed possibilities throughout the development of the margin.

Basin Classification. Basin-forming tectonics, depositional sequences, and basin-modifying tectonics provide criteria to establish a general classification system (Figs. 181 and 182). Continental, continental margin and oceanic basins can be represented by eight tectonic/sedimentary cycles that illustrate the deformational and de-

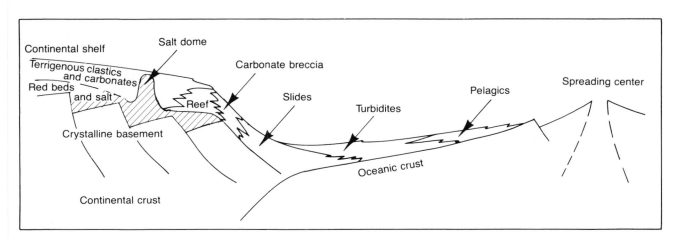

Fig. 180. Atlantic continental margin of North America

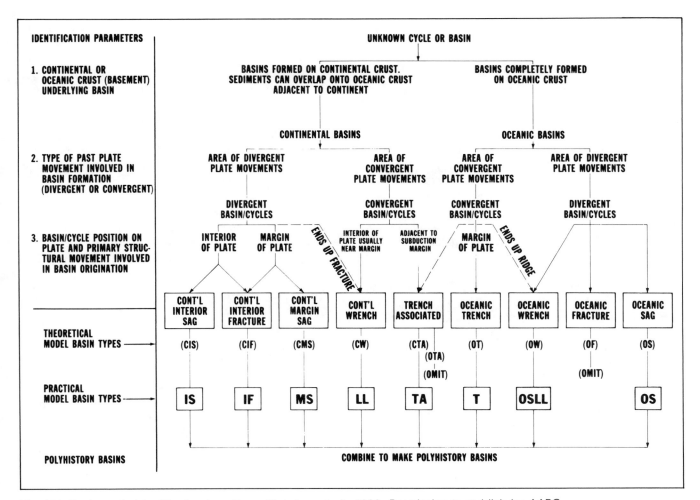

Fig. 181. Basin-cycle identification key. From Kingston, et al., 1983. Permission to publish by AAPG.

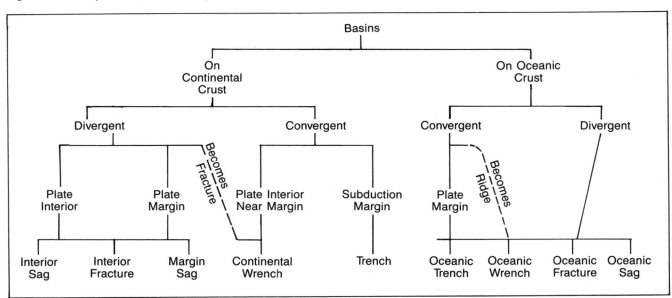

Fig. 182. Basin type. Adapted from Kingston, et al., 1983. Permission to publish by AAPG.

145

positional histories of each according to their tectonic positions and the types of sediments deposited. Significant major simularities between analogous basins in different locations permit comparison and classification. Basin classification establishes tectonic and depositional patterns that permit identification of exploration parameters according to a fundamental framework. What is more difficult or impossible to address in such a classification is the cluster of factors that separate a producing basin of one style from a non-producing basin of the same style. Certainly, divergent continental margin basins are recognized in many places around the world. Not all of them are productive, however. The factors that control productivity in one basin may be quite different from those that control it in another similar basin elsewhere.

Evolution of basins involves cycles of deformation and deposition (Figs. 183, 184, 185, 186, 187 and 188). Most basins are modified by subsequent geologic history following initial development (Fig. 189). Some are favorably affected and retain or enhance their petroleum potential. Others are adversely affected or eventually destroyed and lose their petroleum potential. All basins, however, ultimately lose their petroleum potential if affected by sufficiently prolonged geologic processes. Multiple history basins are productive in many places (Fig.

190) because of a variety of depositional and deformational processes following initial development (Fig. 189). As a basin begins to form, sedimentation accompanies the initial stages that become deformational as development continues. Deformation exercises control over subsequent sediments and folds and faults them. With each successive basin-forming cycle the deposition and deformation are superimposed on previous cycles enhancing the overall petroleum potential of the basin (Fig. 191).

Basin classification establishes parameters that permit tectonic-sedimentary recognition and lead to appropriate analysis and exploration (Fig. 192). Multiple detailed factors hold the key to success or failure of the exploration program, which correctly or incorrectly interprets them (see following).

The primary uses of the global basin classification are summarized as follows: (1) to locate and identify all basins of the world in one framework (the system can expand the explorationist's viewpoint to include all possible basin types, and not just those with which he or she has had personal experience—it is an aid to exploration thinking); (2) to permit the separation of complex basins into their simple component parts, for analysis as simple units; (3) to compare plays within one or two

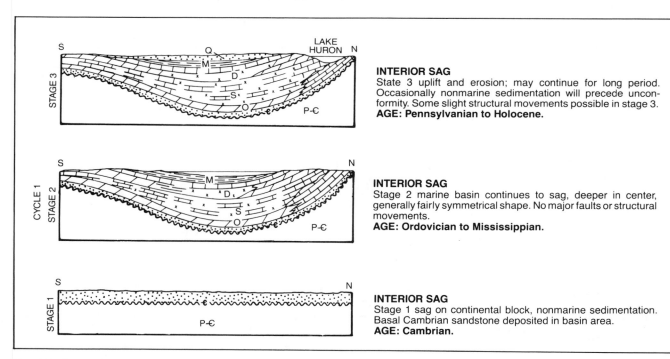

Fig. 183. Evolution of the Michigan Basin an interior sag basin. From Kingston, et al., 1983. Permission to publish by AAPG.

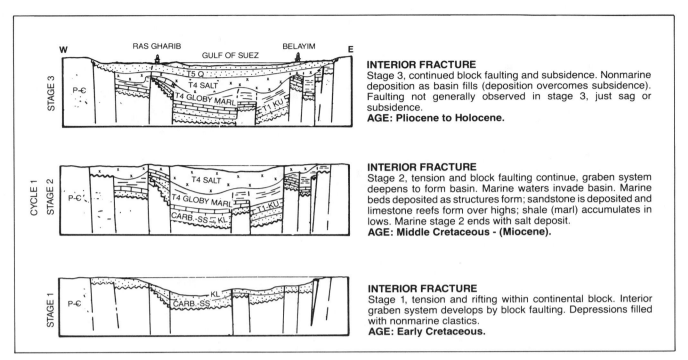

Fig. 184. Evolution of the Gulf of Suez interior fracture basin. From Kingston, et al., 1983. Permission to publish by AAPG.

The content of the figure includes the following labels and text:

INTERIOR FRACTURE
Stage 3, continued block faulting and subsidence. Nonmarine deposition as basin fills (deposition overcomes subsidence). Faulting not generally observed in stage 3, just sag or subsidence.
AGE: Pliocene to Holocene.

INTERIOR FRACTURE
Stage 2, tension and block faulting continue, graben system deepens to form basin. Marine waters invade basin. Marine beds deposited as structures form; sandstone is deposited and limestone reefs form over highs; shale (marl) accumulates in lows. Marine stage 2 ends with salt deposit.
AGE: Middle Cretaceous - (Miocene).

INTERIOR FRACTURE
Stage 1, tension and rifting within continental block. Interior graben system develops by block faulting. Depressions filled with nonmarine clastics.
AGE: Early Cretaceous.

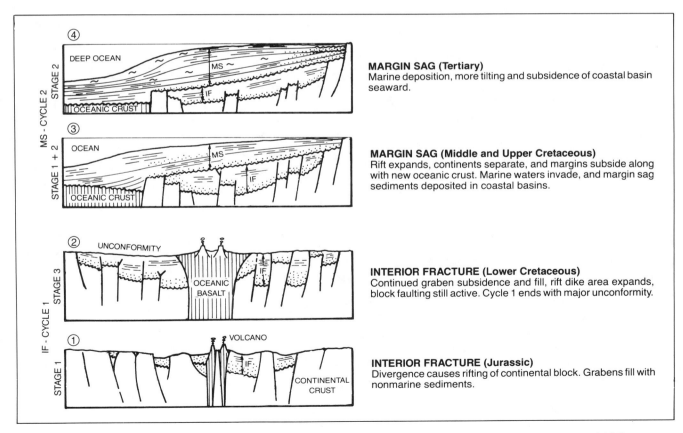

Fig. 185. Evolution of continental margin sag basin. From Kingston, et al., 1983. Permission to publish by AAPG.

The content of the figure includes the following labels and text:

MARGIN SAG (Tertiary)
Marine deposition, more tilting and subsidence of coastal basin seaward.

MARGIN SAG (Middle and Upper Cretaceous)
Rift expands, continents separate, and margins subside along with new oceanic crust. Marine waters invade, and margin sag sediments deposited in coastal basins.

INTERIOR FRACTURE (Lower Cretaceous)
Continued graben subsidence and fill, rift dike area expands, block faulting still active. Cycle 1 ends with major unconformity.

INTERIOR FRACTURE (Jurassic)
Divergence causes rifting of continental block. Grabens fill with nonmarine sediments.

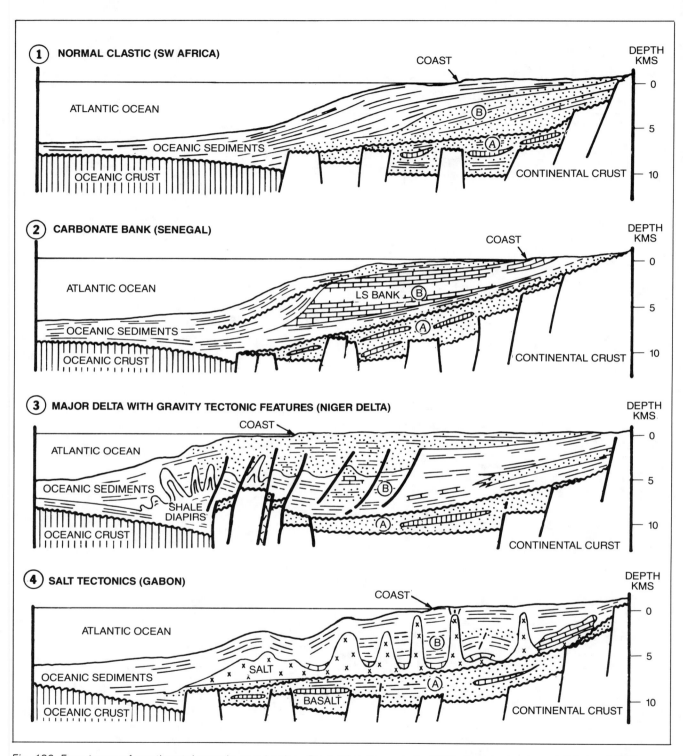

Fig. 186. Four types of continental margin sag basins. From Kingston, et al., 1983. Permission to publish by AAPG.

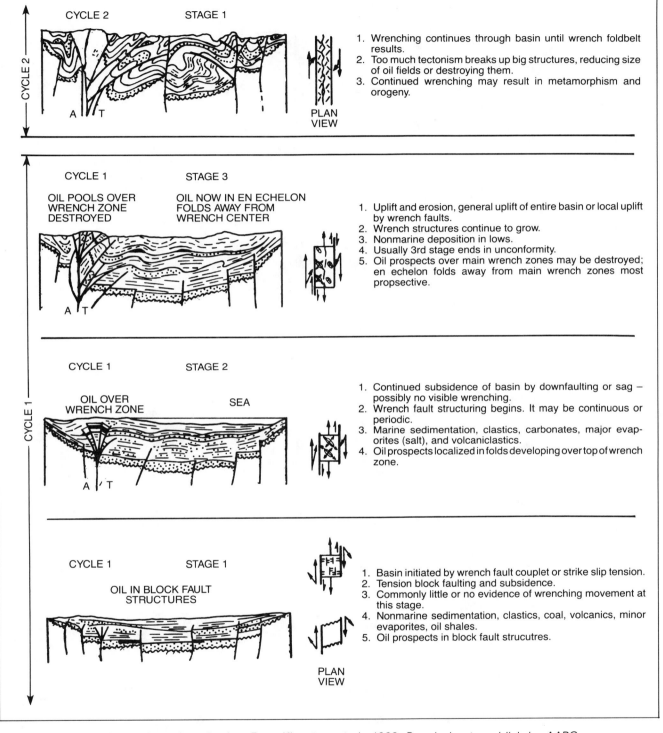

CYCLE 2   STAGE 1

1. Wrenching continues through basin until wrench foldbelt results.
2. Too much tectonism breaks up big structures, reducing size of oil fields or destroying them.
3. Continued wrenching may result in metamorphism and orogeny.

PLAN VIEW

CYCLE 1   STAGE 3

OIL POOLS OVER WRENCH ZONE DESTROYED

OIL NOW IN EN ECHELON FOLDS AWAY FROM WRENCH CENTER

1. Uplift and erosion, general uplift of entire basin or local uplift by wrench faults.
2. Wrench structures continue to grow.
3. Nonmarine deposition in lows.
4. Usually 3rd stage ends in unconformity.
5. Oil prospects over main wrench zones may be destroyed; en echelon folds away from main wrench zones most propsective.

CYCLE 1   STAGE 2

OIL OVER WRENCH ZONE   SEA

1. Continued subsidence of basin by downfaulting or sag — possibly no visible wrenching.
2. Wrench fault structuring begins. It may be continuous or periodic.
3. Marine sedimentation, clastics, carbonates, major evaporites (salt), and volcaniclastics.
4. Oil prospects localized in folds developing over top of wrench zone.

CYCLE 1   STAGE 1

OIL IN BLOCK FAULT STRUCTURES

1. Basin initiated by wrench fault couplet or strike slip tension.
2. Tension block faulting and subsidence.
3. Commonly little or no evidence of wrenching movement at this stage.
4. Nonmarine sedimentation, clastics, coal, volcanics, minor evaporites, oil shales.
5. Oil prospects in block fault strucutres.

PLAN VIEW

Fig. 187. Evolution of wrench or shear basins. From Kingston, et al., 1983. Permission to publish by AAPG.

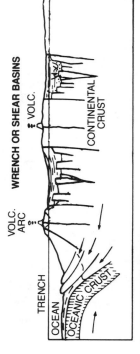

TRENCH ASSOCIATED BASINS

WRENCH OR SHEAR BASINS

MALAY BASIN, SUMATRA BASIN, VOLC. ARC, MENTAWI BASIN, JAVA TRENCH, OCEAN, OCEANIC CRUST (COLD), CONTINENTAL CRUST

4. PRESENT DAY
A. Trench sediments deformed. Thrusting produces new non-volcanic arc and associated basin.
B. Volcanoes inside arc.
C. Strike-slip basins continue to fill and deform by wrench-fault couplet.

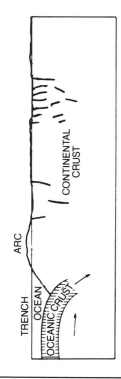

WRENCH OR SHEAR BASINS

VOLC. ARC, TRENCH, OCEAN, OCEANIC CRUST, CONTINENTAL CRUST

3. PLIOCENE TIME
A. Trough continues to collect sediments which are continuously folded and thrust faulted. This is only area of compression.
B. Volcanoes inside arc.
C. Strike-slip tension basins fill first with nonmarine clastics, later with marine sediments. Basins sporadically wrenched (structured) as they fill.

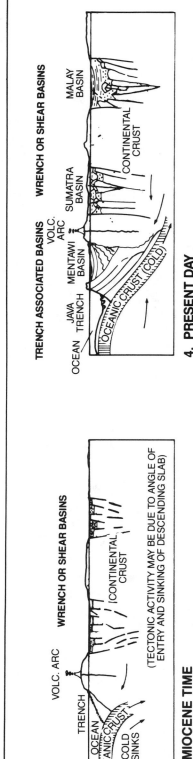

VOLC. ARC **WRENCH OR SHEAR BASINS**

TRENCH, OCEAN, OCEANIC CRUST, COLD SINKS, CONTINENTAL CRUST

(TECTONIC ACTIVITY MAY BE DUE TO ANGLE OF ENTRY AND SINKING OF DESCENDING SLAB)

2. MIOCENE TIME
A. Arching of upper slab. Basin formation.
B. Trough collects deep-water sediments (some volcanic).
C. Volcanoes on inside arc.
D. Wrench or shear basins initiated by block faulting and differential plate movement.

ARC, TRENCH, OCEAN, OCEANIC CRUST, CONTINENTAL CRUST

1. MIDDLE TERTIARY
A. Covergence of 2 plates (oceanic + continental) subduction begins.
B. Trench and arc are formed (downbending of oceanic plate causes tension in overriding plate; or cold oceanic plate simply sinks under upper plate, and no compression results).

Fig. 188. Evolution of Java-Sumatra convergent basins (shear and trench). From Kingston, et al., 1983. Permission to publish by AAPG.

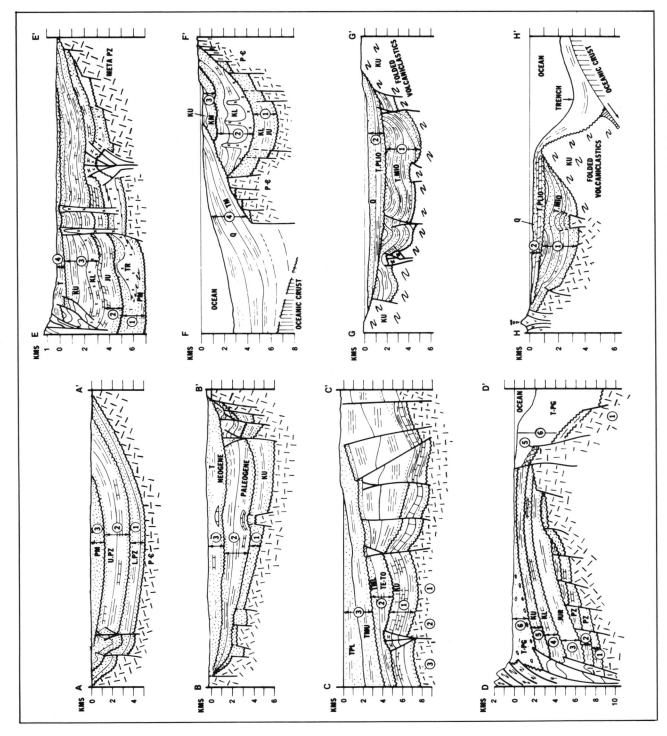

Fig. 189. Multiple history basins. Sequences are numbered. From Kingston, et al., 1983. Permission to publish by AAPG.

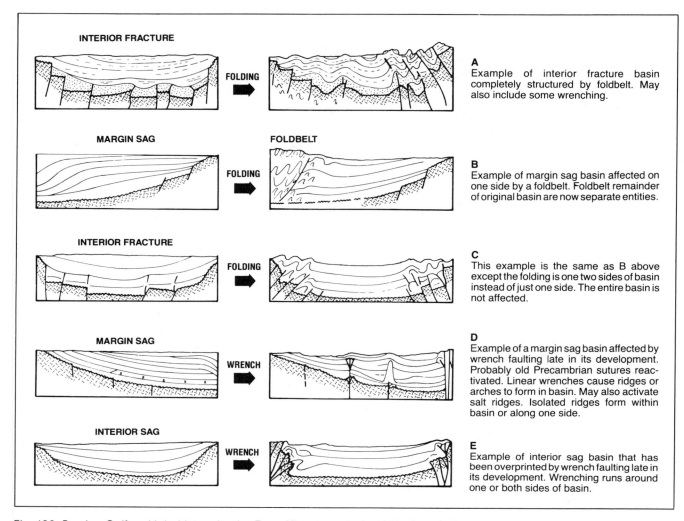

Fig. 190. Persian Gulf multiple history basin. From Kingston, et al., 1983. Permission to publish by AAPG.

basins of the same type, or two or more basins classified as different types; (4) to provide a system for evaluating favorable plays and risks for each basin type, because risks should be understood before venture decision; (5) to predict what geologic events must be found in a basin to improve oil prospectiveness; (6) to enhance the prediction of oil potential in unknown or little known basins by referring to known basins of the same classified type; (7) to provide a system where the paleontology, seismic stratigraphy, geochemistry, and sedimentary history of like (and different) basins can be compared and evaluated; (8) to permit location and assessment of the best specific-play areas in a basin, not just total sediment volume; and (9) to act as a vehicle for comparative assessment of hydrocarbon basins, worldwide. (From Kingston, et al., 1983)

Generalized Depositional Basin. Inasmuch as depositional basins are different in their depositional histories, deformational patterns, and their geographic locations, it is somewhat difficult to generalize by presenting a single basinal motif. However, concepts of depositional basins that come from marine deposition factors are applicable to any basin, provided the appropriate conditions are extant.

Broad categories of differentiation derive from the types of deposition within the basin. These result directly from the amount of sediment contribution from terrestrial sources. This means that there is either appreciable clastic sedimentation that defines a certain marine environment or little or no clastic sedimentation which defines a different marine environment.

The schematic diagram (Fig. 194) shows a hypothetical basin bounded by clastic-producing highlands on one

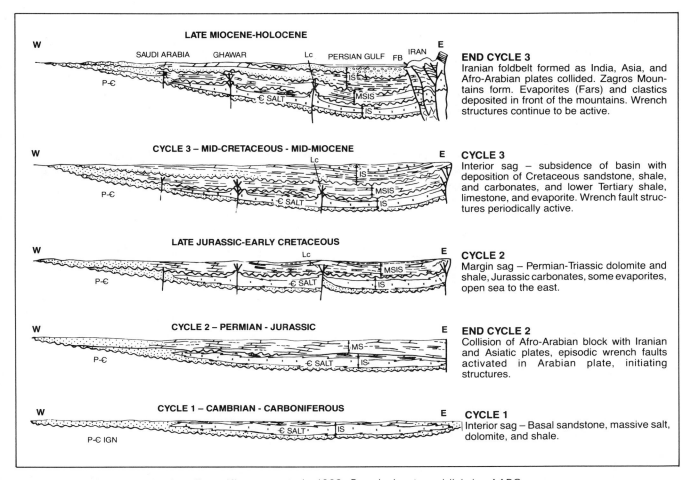

Fig. 191. Multiple history basins. From Kingston, et al., 1983. Permission to publish by AAPG.

side and lowlands on the other, which provide little or no clastic contribution. Water conditions are expected to be clear and lacking in turbidity adjacent to the lowlands. Muddy, turbid water is to be expected along the mountainous coast.

Sediment contribution from the land is expected to be high along the mountainous coast. Conglomerate, sandstone, siltstone, and shale are deposited as beaches, barrier bars, spits, hooks, and terraces and in lagoons and the deep part of the basin. Source beds consisting of thick organic shale are deposited in the basin and in lesser amounts in the lagoon between the mainland and the barrier bar. If deltas develop along this coastline, source beds develop in the deltaic swamps as well.

Reservoir beds are deposited as barrier bars, beaches, shoreface terraces and delta distributary channels, as well as hooks and spits along irregular shorelines. Where barrier bars and baymouth bars restrict access to the open sea, evaporation of seawater can occur, and deposition of marine evaporites with interbedded clastics might take place.

An important adjunct note to this scenario cautions against the assumption that a highland coast is the only place where clastic sedimentation can occur. The contrary is true since large rivers traverse flat, extensive coastal plains and introduce large volumes of terrestrial sediments to the marine environment. The Mississippi River is not associated with mountainous terrain yet provides abundant clastics to the Gulf of Mexico.

Terrestrial clastic contribution along a lowland coast can be extremely low, particularly if the land mass comprises carbonate terrain. In any event, low clastic marine environments have clear water and are conducive to carbonate deposition where water depth and temperature are appropriate.

Carbonate deposition consists of reefs on the shelf and at the shelf edge and lime mud on the shelf and for some distance down the continental slope. Evaporites, if present, will be deposited in the lagoon behind the barrier reefs along with some lime mud and some thick clastic shales, which may contain organic matter. The shale in the lagoon is possible if there is some minor clastic contribution to the predominantly carbonate environment. An excess of clastic materials precludes reef development because reef organisms require clear marine water.

Lime mud on the slope seaward of the reef grades to marine shale in the basin. The marine shale in the basin and the shale in the lagoon are potential source beds if there is sufficient included organic material.

Reefs on the shelf and shelf edge are potential reservoirs. If the marine shales transgress the reefs they can provide the seal necessary to retain the oil that they generate as source rocks.

Additional caution is needed to prevent generalization about lowland coasts and carbonate deposition. Carbonate deposition is well-documented along the mountainous coasts of Central America and around mountainous volcanic islands in the Pacific Ocean. The schematic diagram (Fig. 194) is intended to illustrate erosion and deposition in the terrestrial and associated marine environments.

Basins and Petroleum Occurrences

A general commentary on marine depositional environments in depositional basins comprises a review of source bed and reservoir rock considerations. These involve a development of source beds high in organic material and porous and permeable reservoir rocks.

Source beds consist of black shale deposited in the basin and in the lagoon behind a barrier bar or reef. Black shale in the basin is thicker and better developed than in the lagoon. Deltaic black shales are excellent source beds.

Reservoir rocks comprise reefs on the shelf and shelf edge, barrier bars, beaches, hooks and spits, and delta distributary channels. They can be interbedded with, overlain by, or underlain by source beds.

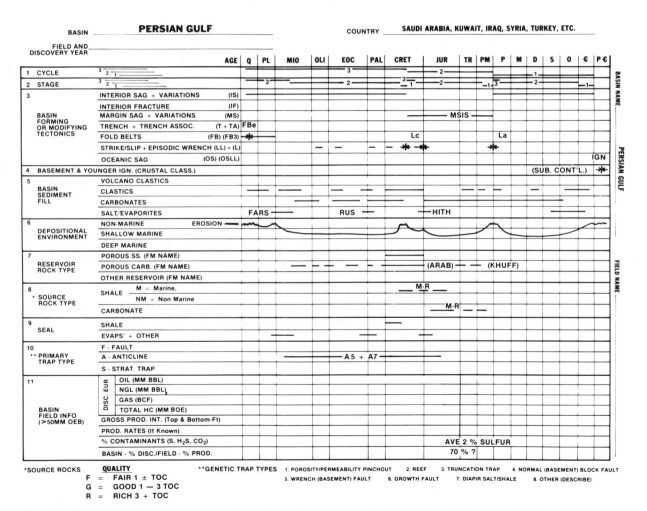

Fig. 192. Basin classification summary form. From Kingston, et al., 1983. Permission to publish by AAPG.

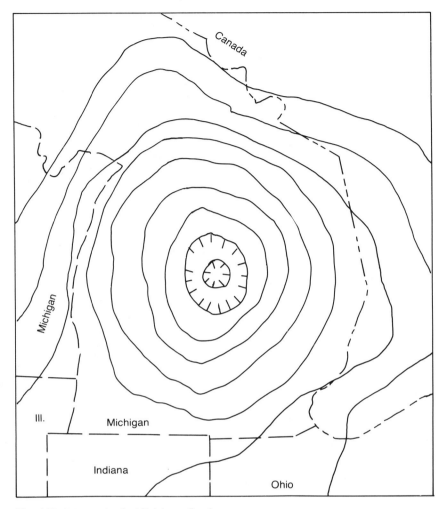

Fig. 193. Intracratonic Michigan Basin

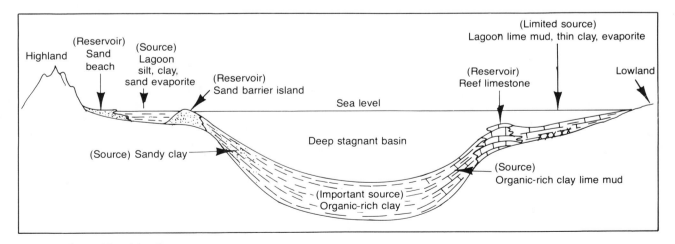

Fig. 194. Depositional basin

Lacustrine, Desert, and Glacial Environments

Lacustrine Environment

Lake beds are deposited in many different kinds of lakes around the world. Some lake deposits can act as source beds and others as reservoirs. They consist of the same lithologic types as stream and marine rocks and include clastics, carbonates, and evaporites.

Lake Basins

Lake basins form wherever natural or man-made dams occur and impound bodies of water (Figs. 195 and 196). They can be large or small, deep or shallow, seasonal or permanent.

1. Closed irregularities on the sea floor can act as lake basins following marine withdrawal. Basins formed this way are subject to modification by subsequent drainage and erosion patterns. Some Mesozoic deposits of the western United States were deposited in inherited depressions.

2. Crustal movements often form lake basins and are related to the type of tectonics affecting the area under consideration. Faulting and folding occur in tectonically active areas in response to the prevailing stress conditions that control the shapes of the lake basins. Lakes in the Great Basin of the western United States and in the East Africa rift system were formed by faulting.

3. Glacial action is the most powerfully effective, currently operational eroding agent. Lake basins are scoured out of bedrock by alpine and continental glaciers. The Great Lakes were formed by glacial erosion.

4. Sand, gravel, and silt effectively disrupt drainage patterns when dumped by glaciers on the landscape.

Lakes formed in this way are abundant in Minnesota, Wisconsin, and Michigan.

5. Earthquake-induced landslides in mountainous areas can interfere with stream drainage and impound water. Recent landslides in Montana created Hebgen Lake.

6. Where volcanic activity occurs, lava flows can interfere with drainage patterns in a manner similar to the action of landslides. Volcanism in the Pacific Northwest and Hawaii has potential for forming lake basins by damming streams.

7. Violent volcanic explosions can remove much of the upper portion of a volcano and form a caldera, or large crater, in which water may accumulate. Crater Lake in Oregon is a beautifully photogenic example of a caldera.

8. The meandering of a mature stream often results in cutting off a meander and isolating it as an oxbow lake. Numerous oxbow lakes dot the floodplain of the southern part of the Mississippi River.

9. Solution of limestone terrain results in collapse structures called sink holes, which act as lake basins in many places. Sink hole lakes are abundant in central Florida.

10. Wind erosion and landscape deflation often create low areas that can fill with rainwater. Blow-out lakes are usually temporary during the rainy season and occur in desert regions.

11. Where permafrost thaws, differentially low areas can result and subsequently fill with melt water or rainfall. Thawed permafrost lakes occur in arctic areas.

12. Playa lakes occur in basins of central drainage. They are seasonal with rainfall and are usually very shallow. Playa lakes are abundant in the intermontane valleys of the Great Basin.

157

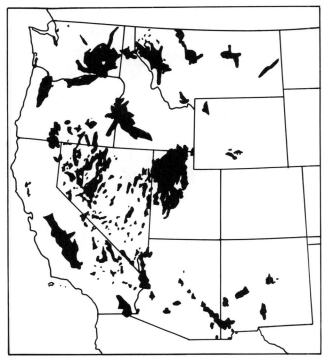

Fig. 195. Glacial lakes of the western United States

Fig. 196. Large glacial lakes of North America

Extinction of Lakes

Lake extinction occurs when erosion eliminates the impoundment or the lake basin fills and the water is displaced.

Lakes fed by sediment-laden streams fill with sediment deposited where the stream enters the lake and slows down. This occurs at the head of the lake and results in a delta that encroaches upon and eventually fills the lake. Lake Mead, a deep artificial lake east of Las Vegas, Nevada, is filling at the head with sediment from the Colorado River.

A stream-fed lake responds to erosion at its outlet, which lowers its level. Ultimately the lake is eliminated as the lake basin becomes part of the drainage system of the stream. Lakes in glacially disrupted drainage systems in Minnesota and Wisconsin may be eliminated by erosion of their outlets. Erosion of the outlet of Lake Erie at Niagara Falls will eventually eliminate the lake.

Lakes in much of the central United States were filled in by fine-grained, glacially derived windblown loess during glacial times. Lakes in desert areas may also be eliminated by windblown deposits.

Climatic Change. When the water source of a lake is reduced or eliminated by climatic changes, the lake disappears or is diminished in size. Early Tertiary lakes in Nevada and Utah disappeared because of climatic changes. Many Pleistocene lakes in the Great Basin have also disappeared. The Great Salt Lake (Fig. 197) is the highly saline remnant of freshwater Pleistocene Lake Bonneville. Lake Lahontan in northwestern Nevada shrank to become Pyramid Lake as the Pleistocene ended.

Seasonal lakes in dry climates disappear because of evaporation. Playa lakes evaporate and leave a residue of salt and other evaporite minerals.

Vegetative Growth. Lakes in Florida, Minnesota, Wisconsin, and Michigan often contain fast-growing vegetation which can become so dense that the lake will eventually fill completely. Vegetative fill in lakes in far northern and southern latitudes often changes to peat. Lakes filled with vegetation also become victims of sediment entrapment by the network of plants. This results in the lake filling with a combination of plant material and sediment, which eliminates the lake.

Subsurface Drainage. Lakes that drain underground can disappear when the efficiency of the drainage system increases. Drainage exceeds the water supply and the

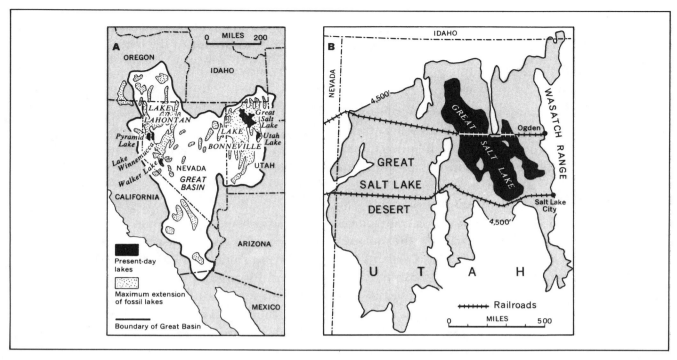

Fig. 197. Late Pleistocene lakes in the Basin and Range region of Nevada and Utah. From Kay and Colbert, 1965. Permission to publish by John Wiley and Sons.

lake becomes extinct. Areas underlain by soluble rocks, such as limestone, are subject to this type of lake extinction.

Lacustrine Deposition

Lacustrine depositional conditions can be very similar to those in the marine environment. In fact, large lakes like the Great Lakes have sediment systems of provenance transport, distribution, and deposition that are essentially marine in physical character and differ only in variability of water salinity and identity of seasonal effects. Each lake, however large or small, has its own set of environmental parameters, which result in the lake representing a microcosm of depositional and erosional completeness (Figs. 198 and 199). For these reasons, lakes, which can vary with climate and sediment source, can offer a variety of environments for hydrocarbon generation and accumulation.

Most lacustrine environments consist of fresh water conditions that do not support marine fauna. However, some lakes have very high salinities.

Facies variations under lacustrine conditions can be of considerable magnitude within a substantially limited lateral extent. Lacustrine facies can change abruptly be-

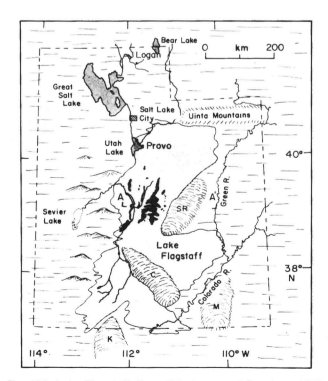

Fig. 198. Lake Flagstaff. From Friedman and Sanders, 1978. Permission to publish by John Wiley and Sons.

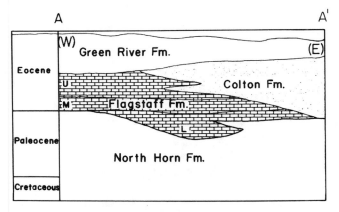

Fig. 199. Lake beds of Lake Flagstaff, Utah. From Friedman and Sanders, 1978. Permission to publish by John Wiley and Sons. Courtesy of the Geological Society of America.

cause of limitations on the lateral extent of individual environments.

Temperate latitude lakes are strongly affected by seasonal changes from winter to summer. Summer deposits contain more sediment and organic matter than organic-deficient and sediment-poor winter deposits. Summer deposits are thicker than winter deposits. Seasonal sediment laminations in lake deposits are known as varves.

Lacustrine deposits include clastics, carbonates, and evaporites. They have characteristics similar to marine rocks. Lacustrine sedimentary structures are also similar to marine types.

Lacustrine Deposits and Petroleum

Lacustrine environments include all sediment types found under marine conditions. Water salinities vary under lacustrine conditions from virtually no salinity to very high salinity. Lakes are usually considered to contain fresh water because most of them do. Some lakes, however, contain waters that are much more saline than seawater.

Lacustrine conglomerate and sandstone are universal and can be excellent reservoir rocks. Tertiary lake beds of Utah (Fig. 200) and Wyoming contain clastic reservoirs. Sediments that include clastic reservoirs can be up to 45,000 ft thick in the Tertiary lake basins of mainland China (Figs. 201, 202, and 203) where they are often tilted and faulted in varieties of traps.

Shales and silts of Tertiary lake deposits in Utah and Nevada act as good source beds. Some lacustrine shales have been very effective as source beds.

Lacustrine limestone and dolomite are similar to their marine counterparts and have the same potential to act as reservoirs. Some are finely crystalline, hard, and minimally porous. Others are coarsely crystalline, contain fossils, have an algal origin, and good porosity. Lacustrine carbonates do not commonly occur as coral reefs, however. Some lacustrine carbonates can be good reservoirs; others have no reservoir potential unless fractured.

Certain units of the Tertiary Green River Formation of Utah and Wyoming contain kerogen-rich dolomitic marl. These beds are siliceous carbonates that contain calcite, dolomite, fine-grained quartz, and clay minerals with small amounts of evaporite minerals. Kerogen in the Green River beds is extractable as petroleum under certain retorting processes. Reserves of the Green River hydrocarbons are extensive.

Lacustrine evaporites include salt, gypsum, anhydrite, and trona. Borax is mined from lake beds in Death Valley, California.

Lacustrine sediments are productive of oil and gas in many areas of the western United States (Figs. 204, 205). These are source beds for other rocks, including fractured volcanics. Tertiary lake beds contain kerogen in many western United States localities and are potentially a rich hydrocarbon source. Oil from lacustrine sources is common in mainland China where it comprises the bulk of Chinese production. Lacustrine oil is produced in Brazil and Indonesia where it is less important. Coarse lacustrine clastics are good reservoir rocks, as are some porous carbonates. Highly saline lacustrine conditions may produce important organic concentrations in carbonate precipitates, which appear to have considerable source potential.

Desert Environment

Desert conditions can produce excellent reservoir rocks, which include sand dunes, alluvial fans, and talus. Source bed generation under desert conditions seldom occurs since oxidation potential is high. Highly saline lake deposits that develop locally in desert environments may produce high organic concentrations and may occasionally provide laterally limited petroleum source beds.

Desert erosion and deposition are primarily sub-aerial processes. Playa lake and alluvial deposition are ephemerally subaqueous but may be subject to considerable sub-aerial modification and undergo significant oxidation.

Desert Erosion

Erosional processes operational under desert conditions involve substantial wind abrasion and periodic

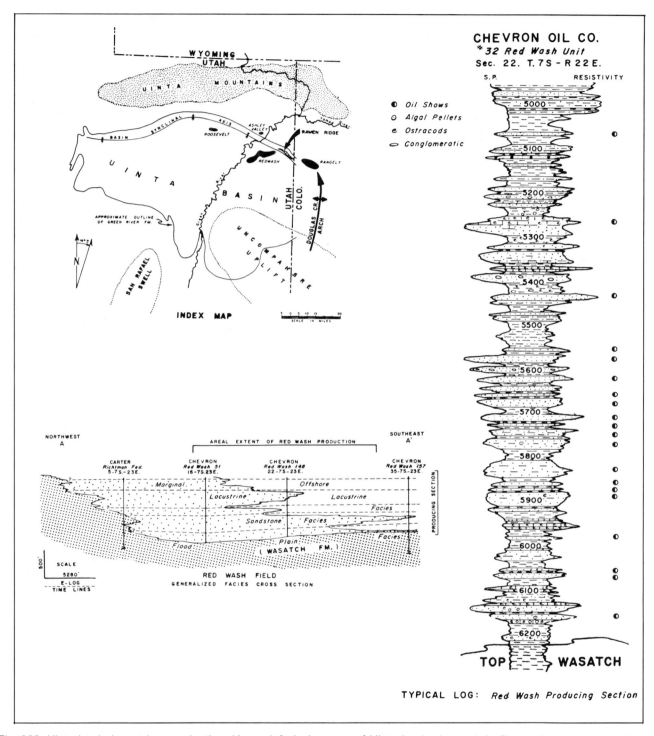

Fig. 200. Uinta basin lacustrine production. Upper left: Index map of Uinta basin. Lower left: Generalized stratigraphic section across Red Wash area showing distribution of facies changes. Right: Typical electric log and lithology of producing section at Red Wash field. From Chatfield, 1972. Permission to publish by AAPG.

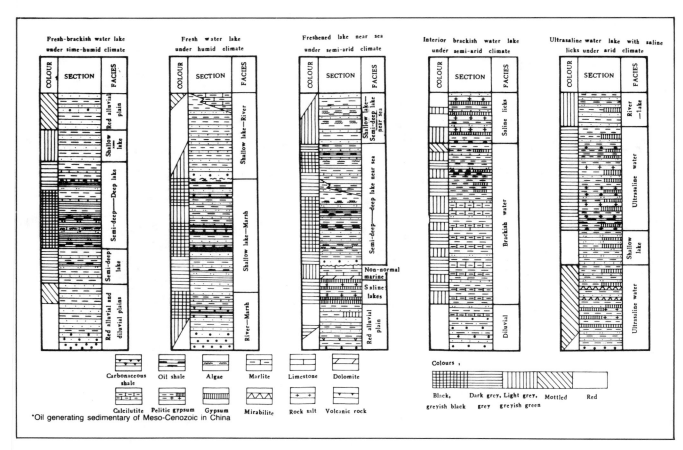

Fig. 201. Chinese lake sediment environments. From Tian, et al., 1983. Permission to publish by Oil and Gas Journal.

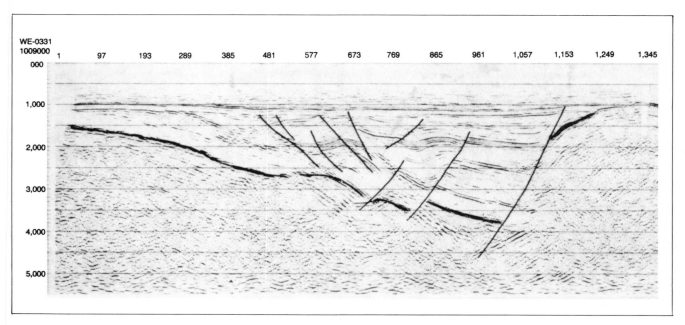

Fig. 202. Seismic section illustrating faulted Chinese lake basin sediments. From Lu, et al., 1982. Permission to publish by Oil and Gas Journal.

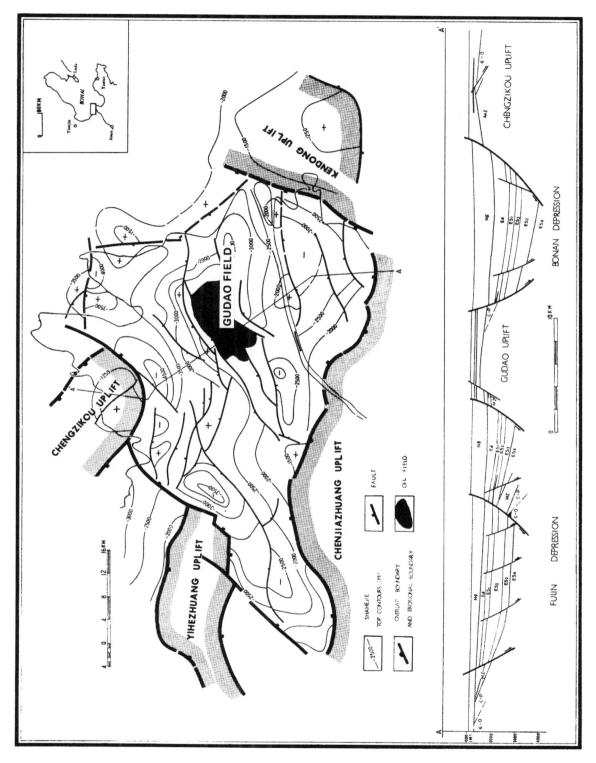

Fig. 203. Gudao Field, Zhanhau Basin, China. From Chen and Wang, 1980. Permission to publish by AAPG.

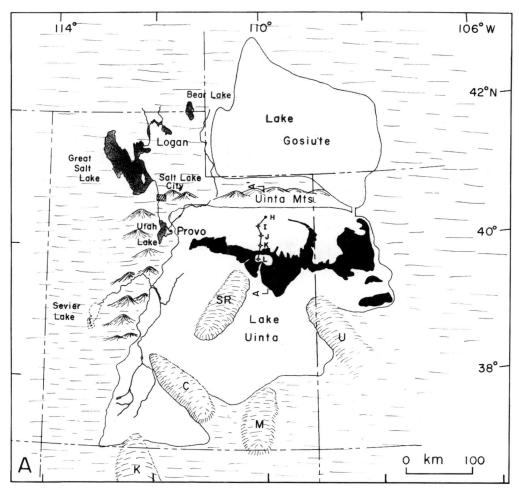

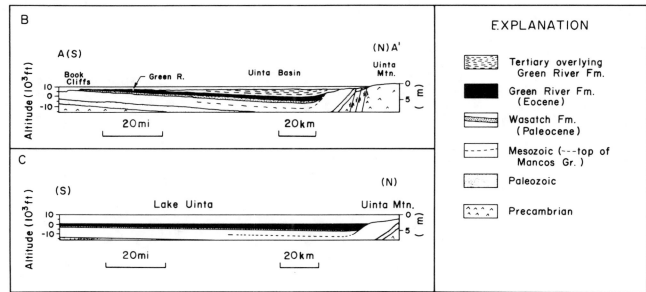

Fig. 204. Map and sections of Lake Uinta, Utah. From Friedman and Sanders, 1978. Permission to publish by John Wiley and Sons and Harper and Row. Original Material compiled by Orlo Childs and P. T. Walton.

164

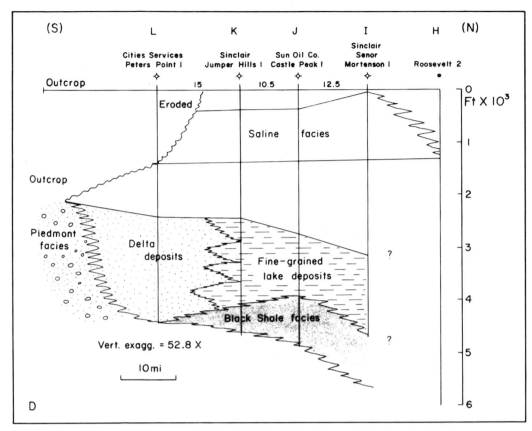

Fig. 205. Green River lake deposits, Uinta Basin, Utah. From Friedman and Sanders, 1978. Permission to publish by John Wiley and Sons.

seasonal water erosion. Deserts in temperate regions are affected by winter freezing and thawing, which can be an important weathering agent.

Wind Erosion. Increased wind velocities result in the increased ability of the wind to carry abrasive materials as well as an increased ability to remove particles from the desert floor. Wind removal of particles on the desert floor causes reduction of the amount of soil material and results in lowering the level of the landscape. As fine-grained materials are removed, coarser, heavier particles remain as lag gravels and become concentrated as desert pavement. Establishment of desert pavement imposes a degree of stability to the soil surface and reduces the amount of erosion. Selective deflation in small areas can result in circular depressions or blowouts. Deflation can be forestalled by desert pavement, removal of all soil to bedrock, and establishment of vegetative cover.

Wind-transported particles are effective abrading agents under desert conditions and act in a manner similar to sandblasting. Where winds prevail from a preferred direction, pebbles and cobbles on the desert floor become faceted in response to abrasion. Facets face the direction from which the wind originates and can represent several wind directions on a single cobble or pebble.

Rocks exposed in cliffs or slopes are shaped by abrasive winds that erode shale and siltstone beds more rapidly than sandstone, limestone, or igneous outcrops. As wind abrasion erodes desert topography, it produces more particles that contribute to additional abrasion.

Water Erosion. Periodic seasonal desert rains contribute to erosion. However, water erosion in true desert areas proceeds very slowly because of the absence of rain. Annual rainfall averages less than 0.10 inch in some parts of the Sahara Desert where many years no rain falls at all. Occasionally, however, a rainfall of several inches may occur, causing considerable erosion for a very short time. Many years may elapse before another rainfall occurs, with no water erosion taking place in the interim.

Desert Deposition

Much desert deposition is the result of wind action. Water deposition can be important in many desert areas as well. Wind and water modify their deposits and often combine to work in the depositional process.

Wind Deposition. Wind deposits are usually fine-grained but can range up to coarse sand. Most wind deposits are finer and are well-sorted and comprise well-rounded particles. Dunes are the usual mode of wind deposition (Fig. 206). They assume different shapes in response to wind direction and velocity, geologic structure, and topography. Dune sand is among the best sorted of clastic sediments and offers excellent reservoir potential.

Barchans are crescent-shaped features with horns

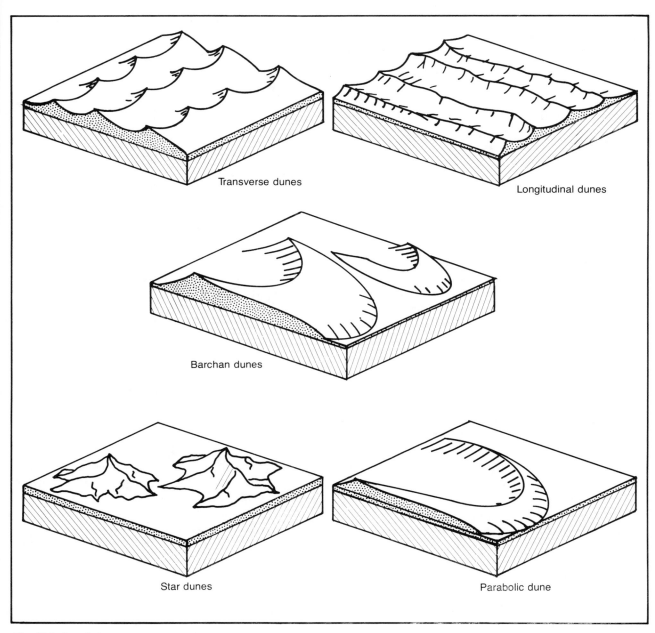

Fig. 206. Sand dune types

pointing downwind (Fig. 206). Limited sand supply and persistent wind from one direction are necessary to form barchans.

Transverse dunes form as ridges normal to the wind direction (Fig. 206). A plentiful sand supply and sparse or no vegetation are important factors.

Long sand ridges, parallel to the wind direction, are *longitudinal dunes* (Fig. 206). They may be over 300 feet high and 50 miles long. Limited sand supply and winds varying within one particular direction form these dunes.

Horseshoe-shaped dunes with horns pointed into the wind are *parabolic dunes* (Fig. 206). They form where winds are moderate and there is some vegetation, and usually occur along sea coasts.

Dune fields comprise areas of extensive sand or sand seas. They can develop in response to outcrop patterns, according to prevailing winds. Some dune fields in the Sahara Desert are over 100 miles wide and several hundred miles long.

Loess consists of wind-deposited silt containing some clay and fine sand. It is derived from mechanical weathering in deserts and is deposited in broadly extensive blankets that can cover hills and valleys. A glacial form of loess is windblown material that comes from glacial melt-water streams.

Water Deposition. Arid areas often have a poorly developed drainage system because stream erosion is limited. Basins characterized by interior drainage and no external outlet can form in these areas. Interior drainage basins are common in the Great Basin where fault block activity contributes strongly to disrupting any erosional tendencies to establish regional drainage patterns (Fig. 207). *Playa lakes* form in these basins in response to periodic seasonal rains and accumulate clastic sediments and interbedded evaporites. Some longer-lived playa lakes in less arid areas can contain sediments with relatively limited petroleum source potential.

Desert-stream deposits are affected by periodic seasonal rainfall and can be poorly sorted with subrounded grains. Some desert-stream sandstone grains can be moderately well-rounded, however.

Alluvial fans form where mountain drainages meet the valley floor and ephemeral desert streams abruptly slow down and deposit their suspended bed loads (Figs. 207, 208). They represent rapid deposition derived from steep mountain stream gradients and sudden, intense, periodic rains of high volume and short duration. Alluvial fans are normally cut by deep distributory channels that break up the primary channel in a manner similar to what occurs on a delta. An alluvial fan can be described as a delta-shaped detrital accumulation that forms under subaerial

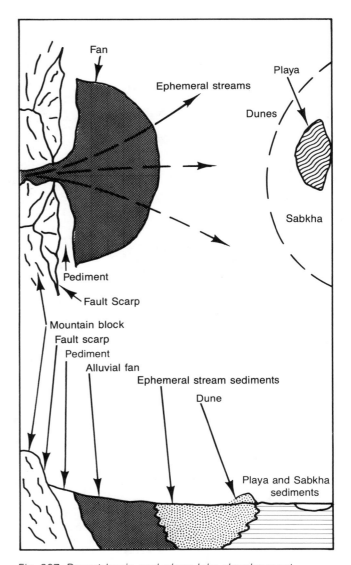

Fig. 207. Desert basin and playa lake development

arid or semi-arid conditions in mountainous or piedmont areas (Fig. 209). Large alluvial fans are characteristic of the desert southwest. Several well-developed examples occur along the flanks of the fault-block mountains bounding Death Valley in California. The well-known red rock monoliths of The Garden of the Gods at Colorado Springs, Colorado, are arkosic, quartz-rich, Permo-Pennsylvanian alluvial fan deposits eroded from the rising ancestral Rocky Mountains.

Alluvial fans in the geologic record are not often productive because of poor grain sorting, locally low permeability and lack of lateral distribution. However, the Cutler Formation (Fig. 210), a Permo-Pennsylvanian al-

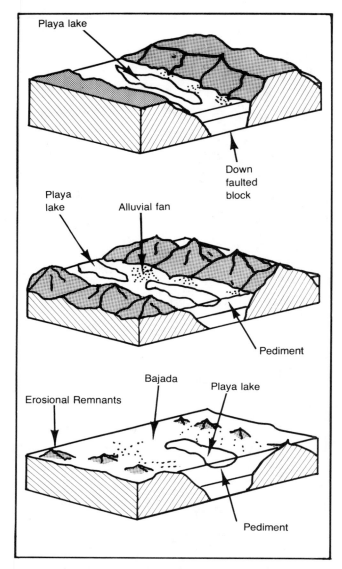

Fig. 208. Desert basin deposition

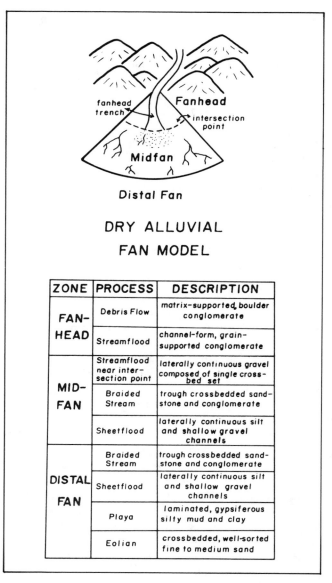

DRY ALLUVIAL
FAN MODEL

ZONE	PROCESS	DESCRIPTION
FAN-HEAD	Debris Flow	matrix-supported, boulder conglomerate
	Streamflood	channel-form, grain-supported conglomerate
MID-FAN	Streamflood near inter-section point	laterally continuous gravel composed of single cross-bed set
	Braided Stream	trough crossbedded sand-stone and conglomerate
	Sheetflood	laterally continuous silt and shallow gravel channels
DISTAL FAN	Braided Stream	trough crossbedded sand-stone and conglomerate
	Sheetflood	laterally continuous silt and shallow gravel channels
	Playa	laminated, gypsiferous silty mud and clay
	Eolian	crossbedded, well-sorted fine to medium sand

Fig. 209. Depositional model for alluvial fan. From Mack and Rasmussen, 1984. Reprinted by permission.

luvial fan, in western Colorado and eastern Utah contains several relatively clean, locally productive sandstone units. Alluvial fan deposits produce in the Quiriquire Field, Venezuela (Fig. 211).

Fan-deltas form where alluvial fans are deposited in standing water (Fig. 212). They often form along mountainous stream-eroded or glacially-incised shorelines where erosion of the terrain produces abundant alluvium that is transported to the fjords and inlets along the coast. Deposition, which is often rapid, is controlled by topo-

graphic relief and the rate of erosion. Deposits are coarse near the sediment source and become progressively fine away from it.

Fan-delta sediments produce petroleum in the Mobeetie oil and gas field in the Texas Panhandle (Figs. 213 and 214). The Brae oil field produces from a block-faulted fan-delta in the North Sea (Fig. 215).

Valley Fill. Erosion of mountain ranges in arid or semi-arid environments transports clastic material to valleys

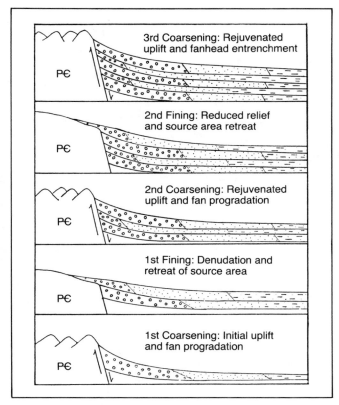

Fig. 210. Alluvial-fan deposition of the Cutler Formation near Gateway, Colorado. From Mack and Rasmussen, 1984. Reprinted by permission.

adjacent to mountain ranges (Fig. 208). Linear mountain ranges in the Great Basin provide continental alluvial fan and fluvial deposits to the down-dropped fault block valleys between the ranges. Many of the valleys are very deep and accumulate up to 20,000 feet of gravel, sand, silt, clay, and evaporites.

Rock Fall and Talus Deposition. Rock fall and talus deposits accumulate under arid conditions that permit only reduced modification by occasional running water. These deposits are not extensive because they accumulate in mountainous areas and are locally distributed.

Desert Deposits and Petroleum

Inasmuch as desert sediments are deposited under virtually exclusive subaerial conditions, they include no source beds. Thick deposits of black, organically rich shale are not found in desert deposits. However, porous

reservoir clastics are common in desert sediments and produce in various localities around the world.

Sand dunes consisting of well-sorted, well-rounded, clean, medium to fine-grained sand are probably the best reservoir rocks deposited under any environment. Porosity and permeability are excellent and are reduced only by the amount of intergranular cement. Sand dune, or windblown sand deposits, produce hydrocarbons from the Permian Rotliegendes Formation of the North Sea (Figs. 216 and 217). Its onshore equivalents of Yorkshire, England, are well sorted, rounded and loosely cemented, and offer important reservoir potential.

Considerable production is realized from the Jurassic Nugget-Navajo sandstones of Wyoming and Utah in the western United States. Nugget-Navajo beds are considered to be sand dunes that were deposited over much of the central Rocky Mountain area (Figs. 218 and 219). The upper Jurassic Norphlet Formation consists of subarksoic coastal dune sediments and produces oil in the Hatters Pond Field in southern Alabama. Jurassic Entrada dune sandstones produce petroleum in the Colorado and Wasatch plateaus of Colorado and Utah.

Alluvial fan and valley fill deposits similarly have no source potential. However, although they are poorly sorted, they can have sufficient porosity and permeability to be reservoirs. Rock fall and talus deposits have limited distribution, and, although variably permeable, are of limited reservoir potential.

Glacial Environment

Glacial environments derive from several different types of glaciers according to their locations and lateral extent. Some glaciers occur at high altitudes, others come to the sea, and others remain exclusively on continental masses. Glaciers develop where snowfall exceeds melting, and movement of ice occurs.

Glacier Types

Mountain or valley glaciers occur in mountainous areas and flow downhill as linear ribbons of ice (Figs. 220 and 221). They gouge out U-shaped valleys as they proceed downhill and leave a considerably incised topography when they melt. Mountain glaciers occur in high latitudes or at high altitudes. Several valley or mountain glaciers that join at the foot of a mountain range coalesce to form a *piedmont glacier* (Fig. 222). Piedmont glaciers occur at high latitudes.

Continental glaciers cover large portions of continental

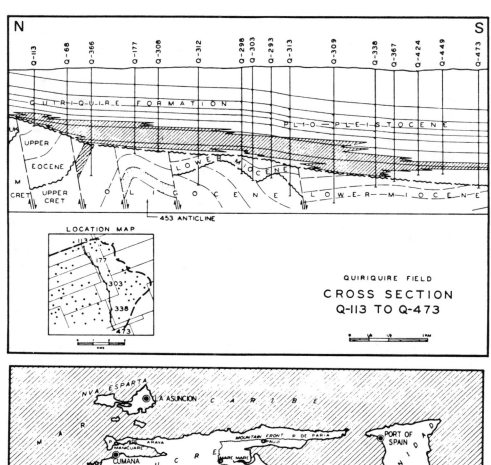

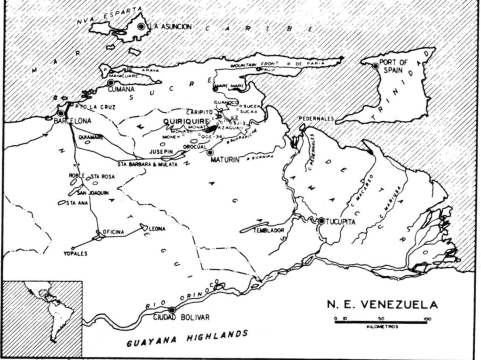

Fig. 211. Quiriquire field, Venezuela. Above: Index map of Northeast Venezuela showing Quiriquire field, Venezuela. Below: North-south cross-section through eastern portion of Quiriquire field. Quiriquire alluvial fan is shaded area above unconformity. From Borger, 1952.

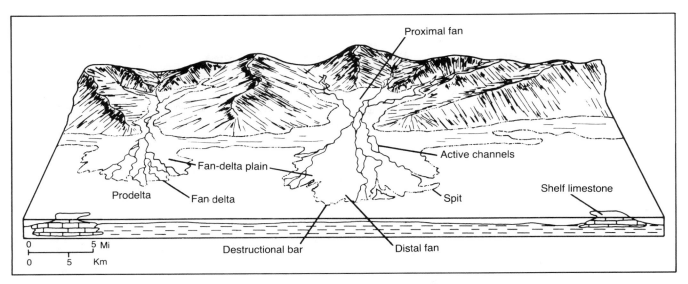

Fig. 212. Fan-delta system. From Dutton, 1982. Permission to publish by AAPG

land masses and most often occur during periods of glaciation (Fig. 223). They usually emanate from high latitude areas and move toward lower latitudes as climatic conditions permit. Continental glaciers covered most of the northern half of North America during the Pleistocene. They retreated and disappeared when world climatic conditions became warmer.

An ice mass, thickest in the middle and flowing radially over the underlying land surface, is called an *ice cap* (Fig. 224). The Greenland ice cap is over 10,000 feet thick in the central part of Greenland.

Glacial Erosion

Erosion by glacial action is the most effective natural landscape degradation process known. Glacial erosional processes proceed more rapidly and more efficiently than running water or marine wave action.

Glacial ice picks up cobbles and boulders, which become frozen into its base. As the glacier moves, its rock-studded base grinds down and smoothes the bedrock (Fig. 225). Striations caused by rocks in the basal glacial ice are common on glacially eroded bedrock surfaces.

Glacially polished bedrock is typical in central Wisconsin, where glacial ice moved toward the south-southwest. Extraordinarily well-developed glacial striae and grooves illustrate the westward movement of mountain glaciers from the Teton Mountains into eastern Idaho.

Meltwater penetrating bedrock fractures freezes and breaks out rock fragments (Fig. 226). They freeze into the base of the glacial ice and are carried away. Plucking in this manner continuously lowers the surface of the bedrock, which is also undergoing perpetual abrasion.

Landforms associated with mountain glaciers differ appreciably from those developed by continental glaciers. Mountain glaciation is limited to individual valleys and locally developed features, which include the U-shaped valley, amphitheater, cirque, horn, serrated divide, hanging valley, and the fiord.

The normal form of the cross-section of a mountain or valley glacier is in the shape of a U. Glacial *U-shaped valleys* are very distinctive in appearance and differ greatly from stream-cut valleys.

A U-shaped glacial valley often terminates upstream in a topographic bowl or *amphitheater*. This area is the originating point of the glacier and represents the locus of maximum snow and ice accumulation.

The bowl-shaped depression, or cirque (Fig. 220), in the floor of an amphitheater is often occupied by a *cirque lake*. Plucking and ice wedging are responsible for cirque and amphitheater development.

Headward erosion of glaciers from several directions toward a common high area in a mountain or mountain range often isolates a *horn*, which comprises a pyramidal erosional remnant (Fig. 220). Continuing erosion of a horn ultimately will eliminate it.

Headward glacial erosion of a linear mountain range by glaciers on opposite sides of the range produces a

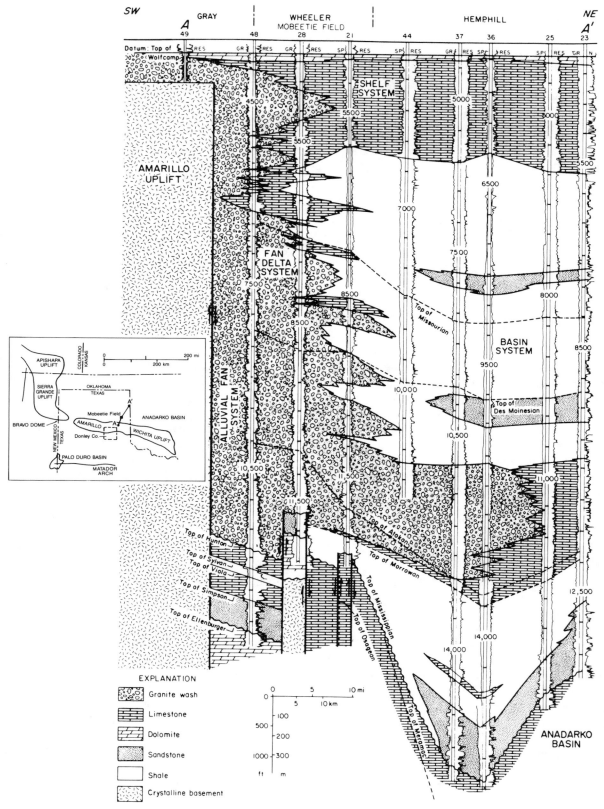

Fig. 213. Mobeetie field fan-delta system, Panhandle, Texas. From Dutton, 1982. Permission to publish by AAPG.

172

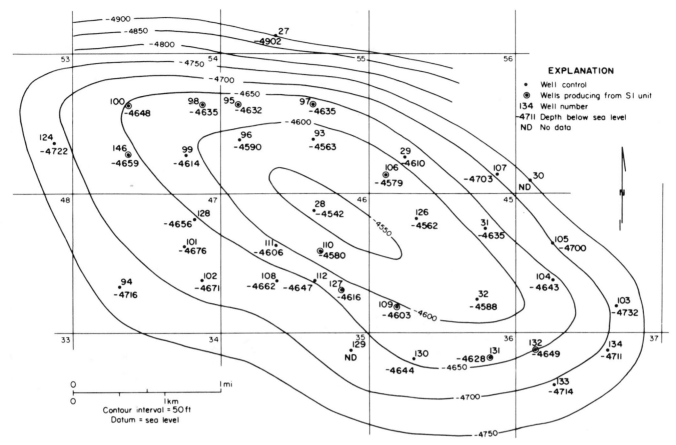

Fig. 214. Missourian sandstone structure map, Mobeetie field. From Dutton, 1982. Permission to publish by AAPG.

sawtooth ridge or *serrated divide*. Further erosion of a serrated divide can form horns at various locations along the ridge.

Since valley glaciers cut downward rapidly, they leave tributaries, occupied by smaller glaciers or streams that are unable to maintain the same erosional pace, as *hanging valleys* (Fig. 220). Waterfalls are formed where the tributary streams of the hanging valleys enter the main glaciated valley.

Fiords are formed by mountain glaciers that enter the sea and cut deep U-shaped valleys below sea level, which are invaded by seawater after glacial melting. Fiords are typical of sea coasts in Greenland, the Arctic Islands, and much of Scandinavia.

Continental glaciers, as opposed to mountain glaciers, cover large areas and form regional terranes. Glaciated bedrock topography is rounded, subdued, and often flat, and the bedrock surface is often grooved, striated, and polished. During abrasion, soil is removed and the bed-

rock surface is left bare. Continental glacial gouging can produce lakes that are elongated in the direction of ice flow. Some elongate lakes are due to differences in bedrock geology, however.

Glacial Deposition

Glacial deposits range from those carried directly by glaciers to those transported by glacially related streams and to others that are carried by the wind. Materials deposited by glacial action are collectively referred to as *drift*. Drift occurs in a variety of forms and is described according to its origin. *Till* is unsorted material deposited directly by the ice. Glacial deposits have limited petroleum potential worldwide, but are locally productive.

Moraines. Direct deposition of glacial material is accomplished during the formation of several kinds of moraines. They occur as blanket deposits or in shapes peculiar to certain types of deposition.

Blanket deposits of unsorted boulders, gravel, sand,

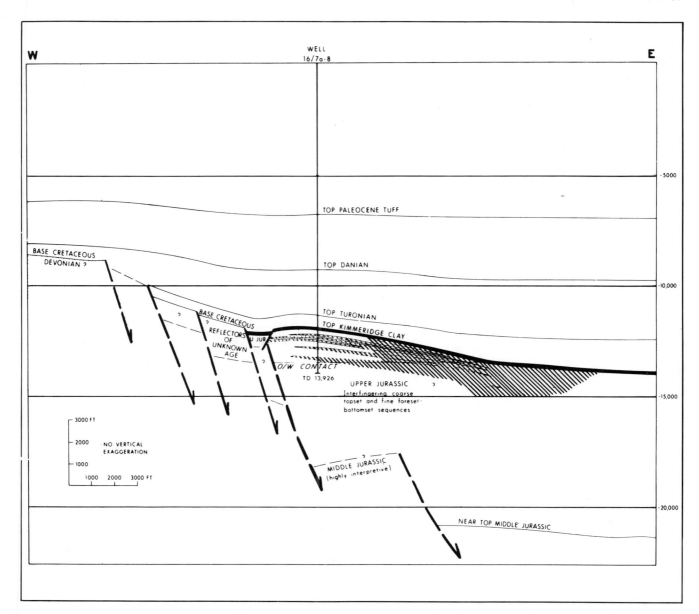

Fig. 215. Geologic section of the Brae fan-delta oil field, North Sea. From Harms, et al., 1981. Permission to publish by Institute of Petroleum, London.

and clay are carried on top of or within a glacier and are deposited as a *ground moraine* during retreat (Fig. 227).

Unsorted glacial debris deposited as a ridge at the farthest advance of a glacier comprises a *terminal moraine* (Fig. 227). The size of the terminal moraine is related to how much debris was being carried and how long melting and glacial flow were in equilibrium prior to retreat.

Glacial retreat is often interrupted by periods of equilibrium between melting and glacial flow. When this occurs, a moraine, analogous to and behind the terminal moraine, is formed (Fig. 227). Several periods of equilibrium can result in several *recessional moraines*.

Mountain glaciers collect ridges of material along their margins, which fall from steep bounding valley walls.

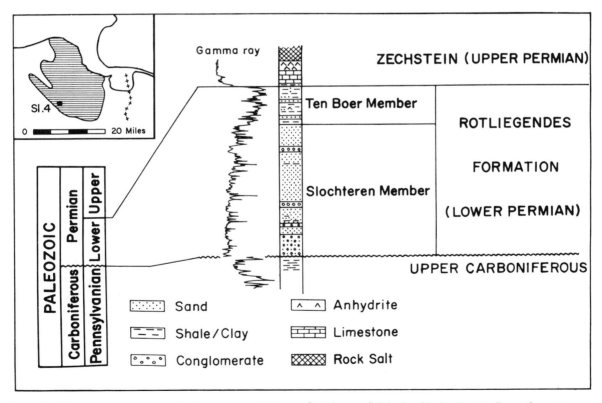

TIME-STRATIGRAPHY			THICK NESS	LITHOLOGY	ENVIRONMENT OF DEPOSITION	GAS OCC.	ROCK STRATIGRAPHY			MAIN TECTONIC EVENTS
CENOZOIC	Quaternary		900'		Continental					
	Tertiary		1000'-3000'		Deltaic to Marine					Laramide
MESOZOIC	Cretaceous	U	2200'-3500'		Marine					Late Early Kimmerian
		L								
	Triassic		0-1200'		Arid Continental to Restricted Marine		Muschelkalk	M		
							Buntsandstein	L Triassic		
PALEOZOIC	Permian	U	2000'-4800'		Restricted Marine		IV Leine III Aller II Stassfurth I Werra		Zechstein	
		L	400'-900'		Arid Continental to Coastal	☼	Ten Boer Member		Rotliegendes	Saalian
							Slochteren Member			Asturian
	Pennsylvanian		>3000'		Humid Deltaic to Marine					Variscan

Early Alpine

Variscan

▤ Shale/Clay	⊙⊙⊙ Conglomerate	⊠ Salt	⊟ Limestone	Ⅲ Chalk	
░ Sand	⋀⋀ Anhydrite	▭ Coal	⊞ Dolomite	⫽ Removed by diff. truncation	

Gamma ray

ZECHSTEIN (UPPER PERMIAN)

Ten Boer Member

Slochteren Member

ROTLIEGENDES

FORMATION

(LOWER PERMIAN)

UPPER CARBONIFEROUS

SI.4

0 ▬▬▬ 20 Miles

PALEOZOIC

Permian | Upper

Carboniferous | Pennsylvanian | Lower

░ Sand	⋀⋀ Anhydrite	
▤ Shale/Clay	⊞ Limestone	
⊙⊙⊙ Conglomerate	⊠ Rock Salt	

Fig. 216. Rotliegendes gas producing desert deposit, Groningen field, the Netherlands. From Stauble and Milius, 1970. Permission to publish by AAPG.

175

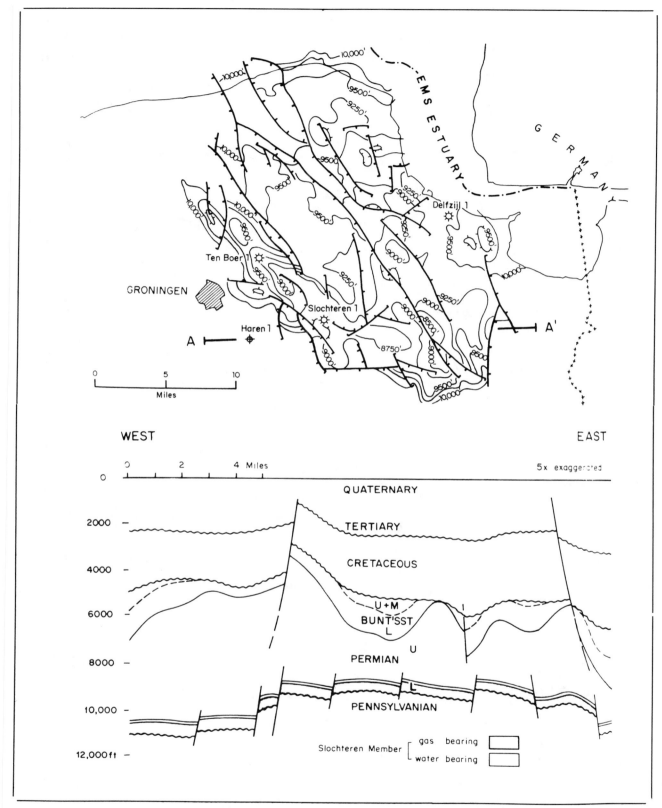

Fig. 217. Cross-section and structure map, Groningen field. From Stauble and Milius, 1970. Permission to publish by AAPG.

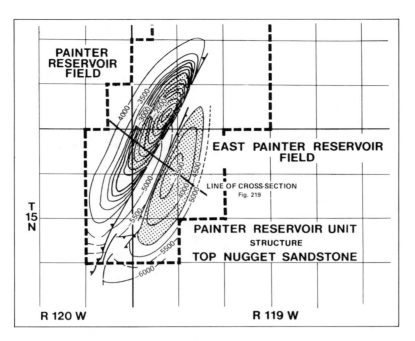

Fig. 218. *Nugget structure, Painter Reservoir, Wyoming. Permission to publish by Chevron.*

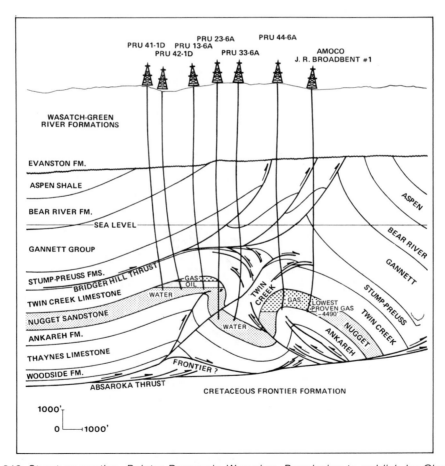

Fig. 219. *Structure section, Painter Reservoir, Wyoming. Permission to publish by Chevron.*

177

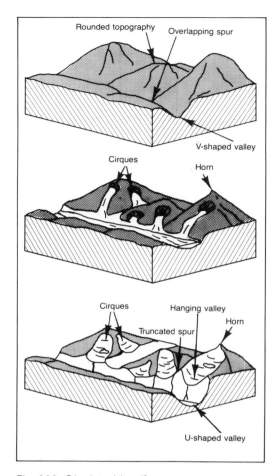

Fig. 220. Glaciated landforms

When mountain glaciers melt, they leave these ridges of unsorted materials along the boundaries of their valleys as *lateral moraines* (Fig. 228).

Where lateral moraines of tributary glaciers meet, a *medial moraine* is formed in the central portion of the main valley they feed (Fig. 228). Multiple tributaries form multiple medial moraines.

A *drumlin* is a glacially-deposited hill streamlined in the direction of ice movement (Fig. 229). Drumlins are common in portions of North America subjected to glacial deposition.

Fluvio-glacial. Fluvio-glacial deposits are some of the best-sorted, glacially originated sediments. Glacial streams account for several different types of deposits, including kame, esker, and outwash deposits.

Stratified sediment deposited by a subglacial stream as a fan or delta is called a *kame* (Fig. 229). These features occur where a rapidly moving sediment-laden stream loses its velocity and in spreading out deposits its load where it passes through a gap in a terminal or recessional moraine or from under an ice sheet.

Eskers are sinuous ridges of sub-glacial stream deposits laid down in the channel of the stream (Fig. 229).

Outwash is stratified drift deposited by subglacial streams beyond the terminal moraine (Fig. 229). Broadly distributed outwash is deposited on an outwash plain.

Glacio-eolian. Glacio-eolian deposits, or loess, consist of wind-deposited, glacially-derived silt with little or no stratification. There are thick loess deposits in China that probably originated during glaciation in Asia. Loess de-

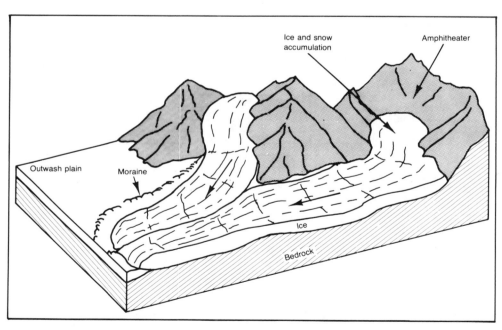

Fig. 221. Mountain glacier system

178

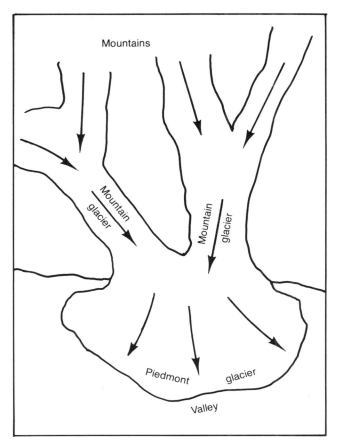

Fig. 222. Piedmont glacier system

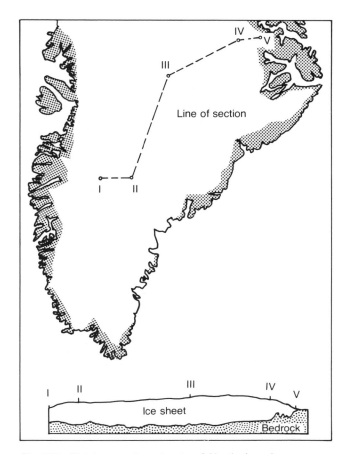

Fig. 223. Pleistocene ice sheets of North America

Fig. 224. Greenland ice cap

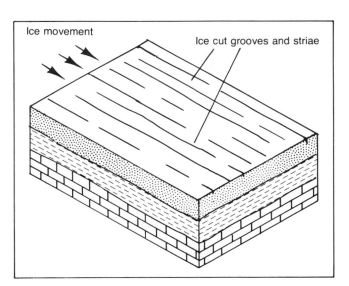

Fig. 225. Glacially cut grooves and striations

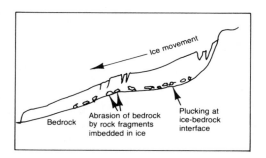

Fig. 226. Plucking of bedrock surface

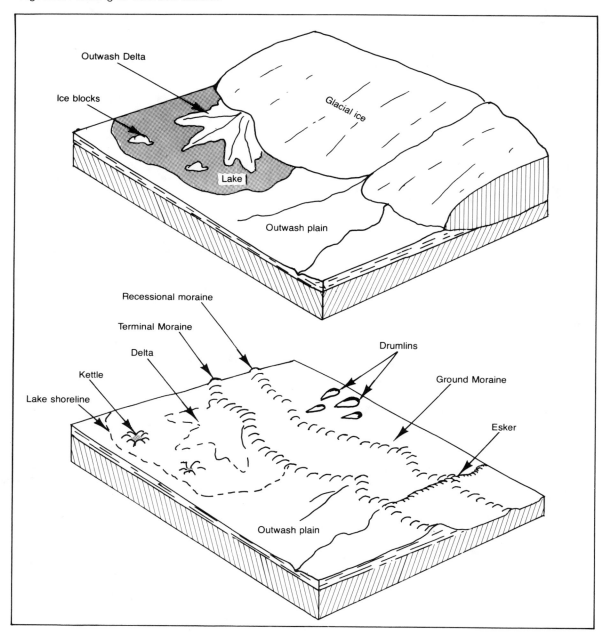

Fig. 227. Continental glacier features

180

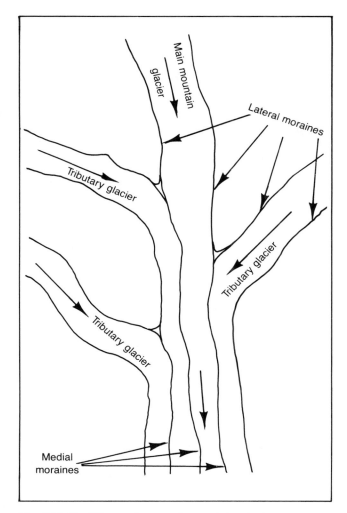

Fig. 228. Medial moraines and mountain glaciers.

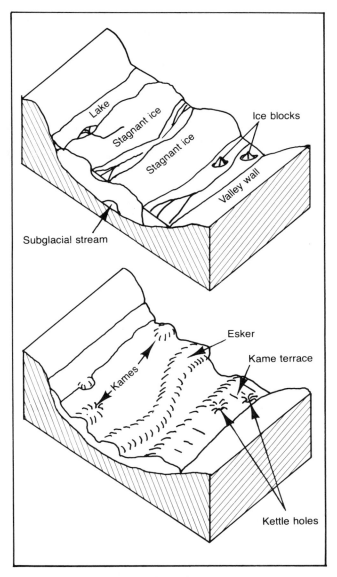

Fig. 229. Stratified glacial deposits

posited in the central plains of the United States came from glaciation in northern North America.

Glacio-marine. Glacial sediments deposited and reworked under marine conditions can come from any of the glacial sources, such as moraine, fluvial, or ice-rafted deposition, provided they reach the sea. The glacio-marine environment is common to arctic and sub-arctic regions where glacial processes meet the sea. Some of the prospective sediments in the central Gulf of Alaska have a glacio-marine origin.

Moraine deposition occurs under marine conditions where a glacier terminates at the sea. Piedmont and valley glaciers come to the sea in the Gulf of Alaska, and moraine deposits form barriers to the entrances of several fiords.

Glacially-fed streams flow to the sea and deposit their

sediments in the marine environment. Sediments from the Copper River in the northwestern Gulf of Alaska are distributed as deltas, beaches, and barrier bars.

Debris-carrying blocks of ice break off glaciers that meet the sea and deposit sediments at distances away from their sources. Ice-rafted sediments can be abundant near the snouts of glaciers but become less of a factor elsewhere.

Glacial Deposits and Petroleum

Rocks deposited under glacial environments are not normally associated with petroleum production. Glacial sediments rarely produce environments in which organic deposits can be preserved in sufficient amounts to generate appreciable volumes of petroleum. Accordingly, glacial sediments are not considered suitable to be source beds.

A variety of glacial sediments have enough porosity and permeability to act as reservoir beds. Certain fluvio-glacial or glacio-marine sediments have enough sorting to have reservoir characteristics. However, the lateral extent of most individual glacial deposits is limited and reservoirs can be restricted to very local distributions.

Sediments with a glacial origin that are redeposited under marine conditions have the potential for wider distribution than those in the terrestrial environment. Their opportunities for association with marine source beds is evident.

Glacial sediments have produced little petroleum. Glacio-marine sediments are prospective in the Gulf of Alaska, but so far drilling results are not encouraging. Upper Paleozoic glacial till produces oil in northwest Argentina and southeast Bolivia. Heavy oil is produced from Carboniferous-Permian glacial clastics in the Marmul Field of Southern Oman (Fig. 230). Late Cenozoic glacial clastics produce considerable amounts of oil in the Cook Inlet of Alaska.

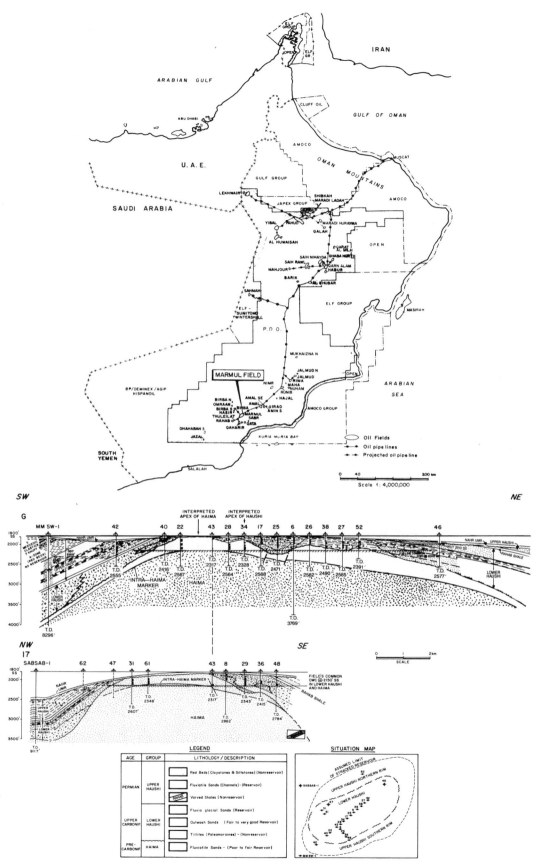

Fig. 230. Marmul Field, Oman. From de la Grandville, 1982. Permission to publish by AAPG.

183

8

Subsurface Water and Diagenesis

Subsurface Water

Water encountered in rocks beneath the earth's surface is important to the ultimate characteristics of reservoirs. Groundwater, or subsurface water, is involved in a variety of factors that relate to petroleum in its effect on porosity and permeability, distribution and migration of hydrocarbons, reservoir pressure, and the characters of fluid contacts.

Groundwater Classification

Water content and composition vary substantially but are indicative of reservoir history and therefore relate directly to petroleum potential. Subsurface water is classified initially according to its origin. Subsequently, it can be further classified by its chemical composition.

Genetic Classification. The origin and history of subsurface waters are considered in this classification. There are occasions when the origins of subsurface waters may be obscured or are combined by geologic events. In these cases, classification must consider the modifying factors if it is possible to determine them. Three classes of subsurface water representing very different origins, connate, meteoric, and juvenile, are evident in reservoir and related rocks.

Connate waters are those included within the formation at the time of deposition of the enclosing sediments. They are removed from the hydrologic cycle at the time of their inclusion and do not return until released from the formation. Connate waters occur in hydraulically closed systems that permit migration within the system but not from it or into another system. High salinity is typical of connate waters that contain abundant chloride. Salinity of connate waters usually increases with depth. Chemical activity involving recrystallization and mineral deposition in porous rock can be related to connate water.

Subsurface waters also contain *meteoric water*, which is introduced to buried rocks from the atmosphere. Meteoric water is part of the hydrologic cycle and originally comes from rain or snowfall. It finds its way into the subsurface by penetrating permeable soils and rocks and establishes geologically compatible circulation systems.

Once meteoric water is included within the subsurface circulation system, there is no assurance that it will return to the surface. However, it is more likely that meteoric water will return to the surface as springs or seepages than will connate water, which has been trapped since deposition. This does not mean, however, that connate water cannot be released at the surface by geologic events that destroy a closed system. Nor does it mean that meteoric water, once in a subsurface system, is necessarily more susceptible to release from the system.

Meteoric waters are usually low in chloride and dissolved solids. Bicarbonate is a common constituent in meteoric waters, which strongly affect areas underlain by carbonate rocks.

The specific identification of *juvenile waters* involves their derivation from materials within the earth. Water from volcanoes or deep-seated igneous masses is juvenile water that has never been part of the hydrologic cycle.

A less rigorous definition of juvenile waters might include water from hot springs and geysers. However, it is likely that most springs and geysers emit heated meteoric water that has percolated into the heating system at depth. It is difficult to differentiate between true juvenile water and heated meteoric water in many cases. As a result, they are often combined and identified as having a volcanic origin.

Chemical Classification. Classifications of subsurface waters are made according to salinity, pH, and various types of dissolved solids. In general, meteoric waters are low in chloride and sulfate but can contain considerable amounts of the latter. They also contain abundant amounts of bicarbonate. Connate waters are high in chloride and low in sulfate and bicarbonate. Connate waters are substantially more saline than seawater and have appreciably different total compositions.

Meteoric and connate waters are the most important to the petroleum industry because they relate directly to reservoir characteristics and production factors. Juvenile waters rarely influence oil field operations.

Aquifer

Water-bearing rocks, or aquifers, have many of the same characteristics as hydrocarbon reservoirs. They must have some degree of porosity and permeability and a water source. A sealing layer, such as a shale zone, is necessary above and below the porous zone to maintain the water in the aquifer. All of these factors combine to form a permeable, water-bearing aquifer.

Water Table

Surface water that falls upon or runs across the landscape penetrates the ground, percolates through the soil and bedrock, and establishes a water table or water level compatible with the local drainage system (Fig. 231). Development of the water table involves the establishment of several zones from the surface of the ground into the subsurface.

Rainfall, runoff water, or running water percolate into the ground from the surface (Fig. 232). Topography

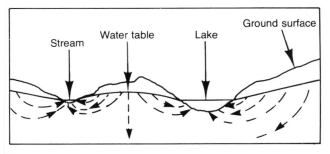

Fig. 231. Groundwater table

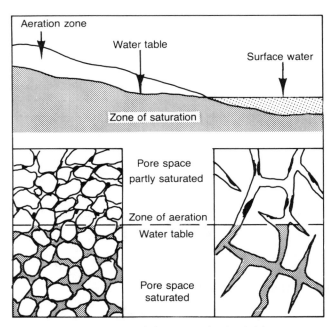

Fig. 232. Development of the groundwater table

channels surface runoff to conform to the drainage system. When surface water seeps into the ground, it passes through a *zone of aeration* in the soil, regolith, or bedrock (Fig. 232). Pores in this zone contain air except when water is passing through it. Minerals in the zone of aeration may be dissolved and carried downward by water passing through the zone.

Where percolating water accumulates below the zone of aeration, the *zone of saturation* is formed (Fig. 232). Here the openings are filled with water all of the time. The upper surface of the zone of saturation represents the water table, which can vary in elevation with seasonal changes in the water supply. The water table assumes a configuration equivalent to the topography under which it lies. However, because of capillary pressures, the groundwater table has a topography of considerably less amplitude than the ground surface.

The water table can come to the surface along the bank of a stream or the shore of a lake (Fig. 232). It usually lies within a few meters of the ground surface but can be much deeper.

A locally developed water table supported above the regional water table by an impermeable zone creates a perched water table (Fig. 233). Perched water tables are often manifest as springs in hillsides where the water seeps out over the impermeable zone supporting it.

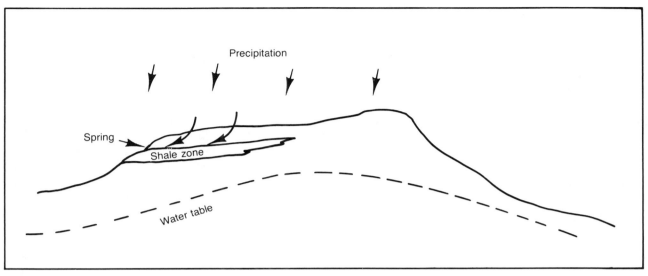

Fig. 233. Spring and perched water table

Water Wells

A water well can be dug or drilled and represents a location where water can be reached and brought to the ground surface (Fig. 234). In areas where rainfall is abundant, well water can come from the water table, which is likely to be shallow. In arid climates where there might be no appreciable water table, well water is likely to come from deep aquifers. Deep aquifers can also provide water in areas where there is a shallow water table.

A water well in which the water does not rise in the pipe and requires pumping for production is a *pumping well*. Water that moves up the wellbore and flows above the level of the wellhead comes from a flowing *artesian well* (Fig. 234). A hydrostatic head is developed when the charging area is higher than the rest of the aquifer. This causes water to flow from wells drilled in the aquifer at locations lower than the charging area.

Any time water rises in the wellbore because of hydrostatic conditions, an artesian well results. If the water

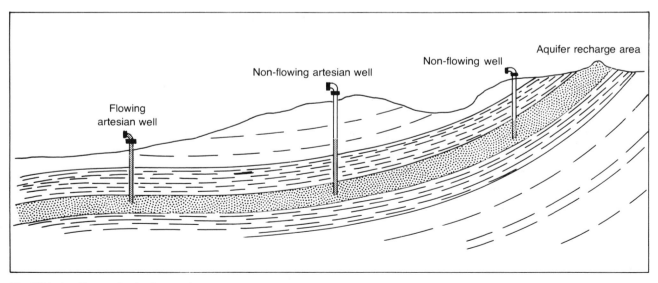

Fig. 234. Aquifer and artesian system

rises in the wellbore above the aquifer, but does not flow, a *nonflowing artesian well* is the result (Fig. 234). Non-flowing artesian wells must be produced by pumping.

How well water wells produce is contingent upon aquifer characteristics, hydrostatic head, availability of water at the charging area, and how well the well is completed (Fig. 235). An aquifer does not produce equally in all locations. Proper selection of a well site is important.

Groundwater Processes

Groundwater is involved in a number of specific processes that affect porosity, permeability, rock composition, and rock competence. Many of the processes are similar, but because they involve organic vs. inorganic materials and elements vs. minerals and rock materials vs. bone, they are considered individually.

1. *Cementation* of sediment grains into rock occurs from mineral cement precipitating out of groundwater solutions. The grains are bound together by the mineral material, which forms an intergranular matrix. Cementation reduces porosity and permeability to zero if all intergranular space is filled. Sediments are hardened and converted to rock by cementation as one of the processes.

2. *Replacement* occurs when groundwater action interchanges one mineral in a rock mass for another. In this way, entire minerals are substituted for each other. This process often results in the formation of pseudomorphs where the replacing mineral assumes the external physical shape of the crystals of the mineral being replaced.

3. The replacement of one element in a rock mineral by another mineral introduced by groundwater is called *substitution*. Many elements in the periodic table are easily interchanged to form new minerals when substituted.

4. *Petrification* takes place when inorganic mineral material replaces organic material. Replacement of wood fibers with silica produces petrified wood, which retains much of the original internal wood structure.

5. Mineral material is *disseminated* through a rock mass when it occupies pores and cracks. Groundwater percolating through a rock mass deposits precipitated disseminated mineral material in a manner similar to that involved in cementation. The primary distinction lies in the rock grains being cemented before dissemination occurs. Pore space and permeability are reduced by dissemination.

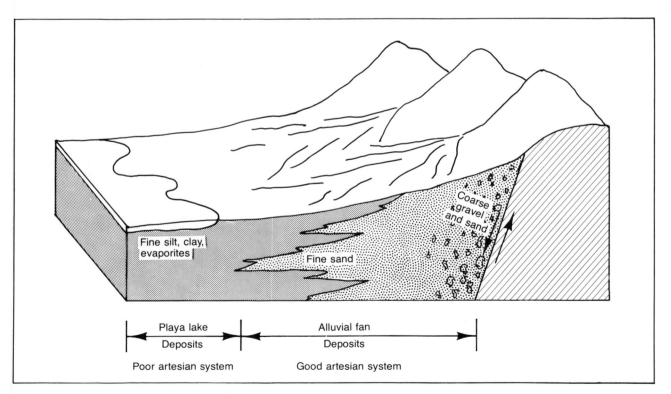

Fig. 235. Desert intermontane valley artesian system

6. The filling of pore spaces in bone by mineral material results in *permineralization*. Porosity is reduced as the bone is rendered significantly heavier by the infilling mineral material.

7. When rock and mineral materials are affected by groundwater *solution*, they are dissolved, transported, and perhaps reprecipitated elsewhere. Most mineral materials have some degree of solubility, depending upon the solutions, and they can be removed or concentrated. Limestone terrain in a humid climate is readily dissolved by reactions involving combinations of rainwater and carbon dioxide. Solution of limestone terrain results in cavern formation and the deposition of dripstone (Fig. 236). Porosity and permeability are often enhanced by solution.

Groundwater Products

Inasmuch as groundwater solutions promote chemical and physical changes in surface and subsurface rocks, these solutions and subsequential changes in the rock have economic significance. Some groundwater processes and products are economically beneficial and others are not. However, groundwater activity and its results can be important in the distribution of not only water-bearing rocks but petroleum-bearing rocks as well.

1. Surface waters affect exposed rock surfaces and combine with weathering processes to form and concentrate soil over the bedrock (Fig. 237). Water that enters the subsurface by percolating through the soil creates an interchange of mineral constituents within the soil. Distribution and concentration of the soil by groundwater enriches certain organic and inorganic constituents that are beneficial to agricultural interests.

2. Groundwater solutions concentrate clay minerals by chemically altering soil and bedrock to stable, resistant compounds. Pure residual clays with certain favorable characteristics are sought for the manufacture of ceramics.

3. Chemical groundwater concentrations of low mineral concentrations often produce enriched ore bodies, whether by oxidation or reduction. Ore enrichment results in accumulation of higher quality mineral deposits from low grade, subeconomic deposits.

4. Subsurface water strongly affects areas underlain by carbonate rocks. Most carbonate areas consist of limestone and dolomite, which are subject to alteration by carbonic acid. Dilute carbonic acid is formed by carbon dioxide combining with rainwater, runoff water, and subsurface water. Its continuing presence causes chemical breakdown of carbonate rocks, causing the formation of caves and cavern systems (Fig. 238).

Caves and caverns develop by solution along openings caused by fractures and bedding planes that slowly increase in size. Carbonate material is removed in solution to be redeposited where solubility, temperature, and evaporation permit. Deposition of carbonate materials in cavern systems produces dripstone features such as stalactites hanging from the cavern ceiling and stalagmites rising from the floor (Fig. 239).

5. Solution of limestone terrains can also form uneven karst topography. Large sinks, or solution cavities, form where cavern ceilings collapse, leaving the cavity open to the sky. Where sinks are numerous and strongly affect the physiography, a distinctive and peculiar land-surface pattern called karst topography is formed (Fig. 240). Karst topography is common in Kentucky, Florida (Fig. 241), Cuba and the Karst area of Yugoslavia, which provided its name.

Groundwater activity is very important in the alteration of reservoir rocks. Mineralization of subsurface rocks

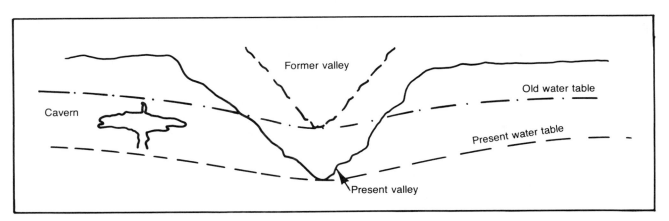

Fig. 236. Cavern development and stream erosion

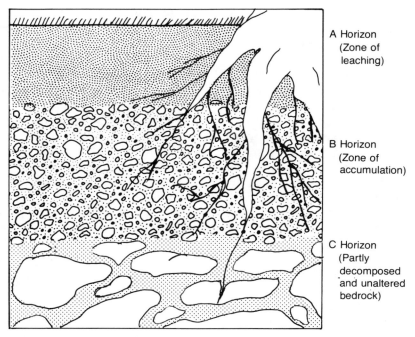

Fig. 237. Soil development

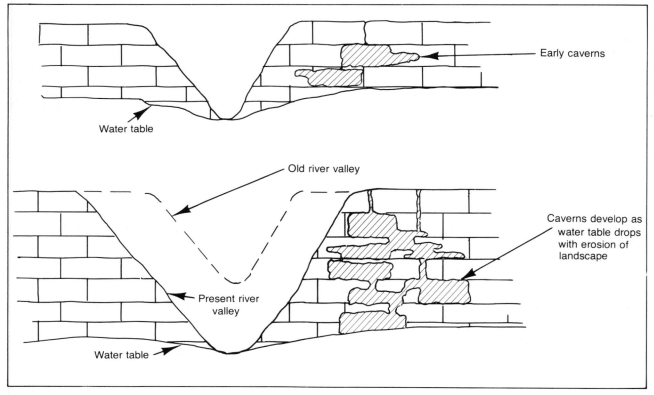

Fig. 238. Cavern development and groundwater table

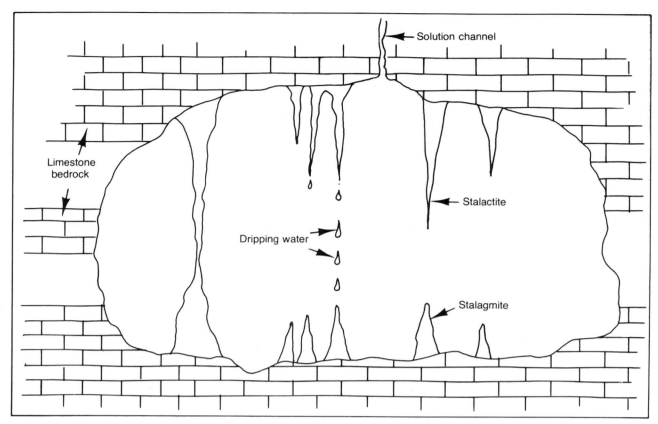

Fig. 239. Cavern features

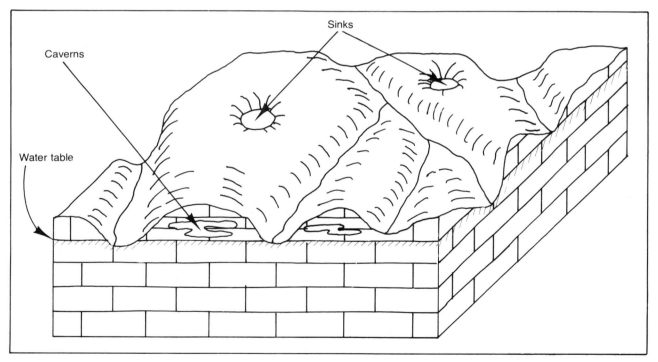

Fig. 240. Karst topography development

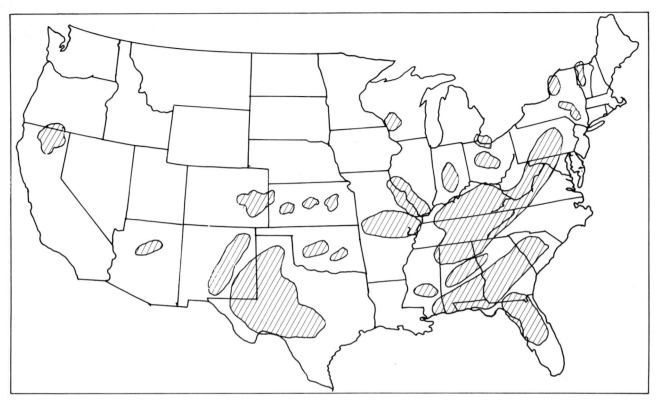

Fig. 241. Karst topography areas of the United States

can increase or decrease porosity and permeability. Cementation and dissemination can importantly reduce or eliminate porosity and permeability. Certain rocks, known as good reservoirs in some areas, can be eliminated as reservoirs in adjacent areas by the reduction of porosity and permeability due to mineral filling of pore space.

Carbonate reservoirs in parts of west Texas and New Mexico are rendered unprospective by infilling of porosity by anhydrite in adjacent areas. Porous sandstone beds in some subsurface areas of western Utah and eastern Nevada belie their appearance as hard impermeable quartzite in surface exposures in nearby areas. Alteration of intergranular clay materials in sediments in the Gulf of Alaska has substantially reduced porosity and permeability in some areas. Some previously cemented midcontinent sandstones have experienced greatly enhanced fluid transmission potential by groundwater removal of intergranular cement.

Rocks originally deposited with limited reservoir capability can be greatly improved by porosity enhancement by groundwater. On the other hand, rocks with excellent primary reservoir characteristics can be rendered unprospective by groundwater action.

Without groundwater, subsurface fluid migration would not occur, and the transport and accumulation of hydrocarbons would not take place. Fluid drives, capillary activity, filling of pore spaces, and hydraulic instability related to groundwater are fundamental to movement of hydrocarbons from source beds to reservoirs.

Groundwater and Petroleum

It has been shown that subsurface water is involved in a multitude of processes, which produce a variety of results. From the specific standpoint of petroleum, the action of groundwater has an important effect upon reservoir characteristics directly related to porosity and permeability. Beyond the cementation, solution, dissemination, and other effects upon reservoir rocks, groundwater is involved in the expulsion and migration of fluids from source beds to reservoirs, the composition of reservoir fluids, formation fluid drives, the establishment of fluid contacts, and the shape and attitude of fluid contacts. Solubility of hydrocarbons in groundwater is an important consideration as well.

1. Oil and gas occupy part of the pore space in a

subsurface reservoir. The remaining space is occupied by formation water. The formation, migration, and accumulation of the oil in the reservoir are substantially affected by the composition and behavior of the associated water. Identification of individual reservoirs is accomplished by analysis of formation waters, which have their own compositional characteristics. Reservoir pressures are often maintained by fluid content, which affects ultimate hydrocarbon recovery.

2. Maintenance of reservoir pressures by formation water has direct application to production methods, recovery programs, and ultimately to total recoverable hydrocarbons. These considerations are a function of porosity, permeability, oil field water content, and distribution of pressures related to formation fluids, as well as structural deformation.

3. Inasmuch as formation fluids, which include hydrocarbons and water, accumulate under hydrostatic and deformational pressure, they become zoned along horizontal contacts (Fig. 242). Fluid contacts develop because of the differences in specific gravity between gas, oil, and water and are horizontal unless there is fluid movement or other cause within the reservoir. Pressure partially controls the hydrocarbon phases within any reservoir and determines whether gas, dissolved gas, condensate, etc. are the predominate petroleum states.

4. Moving oil field waters can cause tilted fluid contacts (Figs. 243 and 244). However, another possible cause is the presence of an asphalt layer at an oil-water contact that creates a permeability barrier and maintains the contact after tectonic tilting. Tilted contacts are often oriented in the direction of the latest tilting of individual blocks.

A summary of the favorable aspects of the action of groundwater upon reservoirs deals primarily with reservoir enhancement by increasing porosity and permeability. This is accomplished by solution that attacks the rock mass itself or removes soluble intergranular cement. Other favorable groundwater actions include porosity enhancement by volume reduction intrinsic to certain recrystallization or mineral substitution processes.

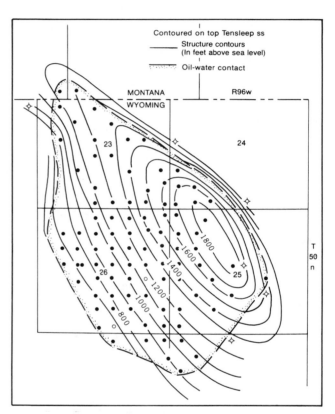

Fig. 243. *Tilted oil-water contact, Frannie Field, Wyoming. From Hubbert, 1953. Permission to publish by AAPG.*

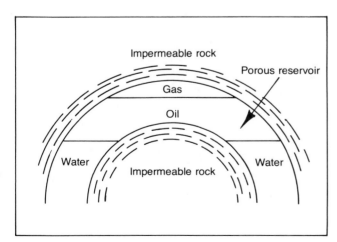

Fig. 242. *Anticlinal oil and gas accumulation*

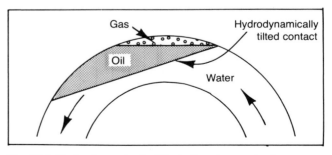

Fig. 244. *Hydrodynamically tilted oil-water contact*

Post-Depositional Processes–Diagenesis

Deposition of sediments is followed by their transformation into rocks, which is called diagenesis and involves processes that change a sediment mass into coherent rock, while physically and chemically altering the rock to its final form. Physical and chemical processes operational during diagenesis can increase or decrease volume, enhance or reduce porosity and permeability, introduce or delete mineral constituents, or reorganize mineral constituents already present in the sediments.

Diagenesis is important in petroleum-related considerations because reservoir quality is almost always affected. Expulsion of fluids from source beds is accomplished during diagenesis when oil migrates from its place of origin to where it accumulates.

Lithification (Induration)

Part of the diagenetic process hardens or lithifies sediment to rock. Several events are possible during lithification, most of which reduce porosity and permeability as consolidation occurs.

Sediment compaction is the result of burial by overburden of younger sediments. Compaction reduces porosity and permeability by reducing separation between grains. Effects of compaction increase with depth of burial (Fig. 245).

Compaction is also responsible for variations in the packing arrangement of rock grains. Open packing results in increased porosity, and closed packing results in a decrease (Fig. 246). Cubic porosity, associated with open packing of spheres, is nearly 48%.

Sufficient pressure exerted upon sediments will impart cohesiveness to the compacted material commensurate with the amount of confining pressure. However, compaction alone is not sufficient to convert sediment entirely to rock. Other factors are required.

Deposition of cementing and disseminating material

Table 21
Effects of Diagenetic Processes on Rock Forming Materials

Process	Increase	Decrease
Compaction		X
Precipitation		
Cementation		X
Dissemination		X
Solution	X	
Recrystallization	X	
Fracturing	X	

Diagenesis generally reduces porosity as rock material is consolidated.

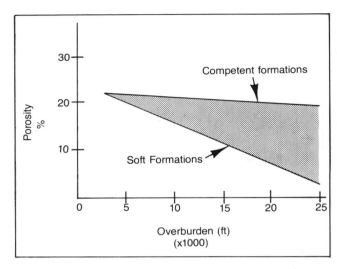

Fig. 245. Sediment compaction and porosity reduction

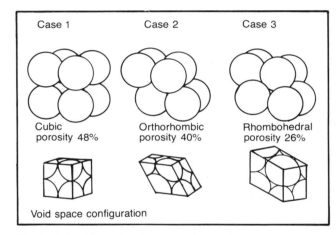

Fig. 246. Packing of spheres and porosity

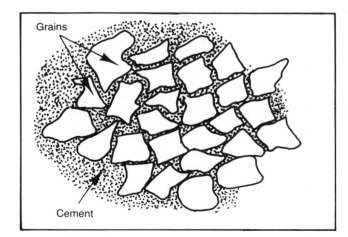

Fig. 247. Cementation of sand grains

between grains of sediment, whether compacted or uncompacted, will bind the grains together (Fig. 247). Abundant cement produces more solidity but reduces porosity and permeability. Compacted sediments that are tightly cemented produce hard, coherent rocks. Cementation furthers the process of lithification begun by compaction of sediments.

Rock materials subjected to very high pressures may be affected by *pressure solution* when grain contacts flow or blend together over extended periods of time. Intergranular spaces fill with mineral material derived from individual grains, and porosity is decreased or eliminated.

Carbonate terrains can be subjected to solution-motivated ion exchanges involving calcium carbonate and magnesium chloride and be converted to dolomite.

$$2CaCO_3 + MgCl_2 \rightarrow CaMg(CO_3)_2 + CaCl_2$$

Dolomitization is accompanied by a volume reduction that increases the porosity of the previously unporous limestone when converted to porous dolomite. Deposition of dolomite in pre-existing pores and fractures, however, reduces initial porosity by infilling the void space.

Conversion of soft lime mud to limestone is accompanied by recrystallization of the calcite to a very fine interlocking crystalline texture. Fossils in a limestone environment are often recrystallized to clear calcite during diagenesis.

Solution

Solution related to diagenesis can eliminate parts of the mineral constituents in the sediments and substantially increase porosity. Diagenetic solution can improve reservoir quality by reducing or eliminating intergranular cement or by removing some of the primary rock-forming mineral or minerals.

Fracturing

Diagenetically related fracturing, caused perhaps by sediment loading, can increase the porosity and permeability of a particular rock. However, late-stage mineral-bearing groundwaters can reclose diagenetic fractures by mineral infilling and recrystallization.

Diagenesis and Petroleum

Diagenetic effects are of considerable importance to petroleum migration and accumulation. Without diagenesis, hydrocarbon fluids are not likely to move, nor are some types of reservoirs formed. So, expulsion of fluids by compaction of sediments promotes migration, and solution and recrystallization processes improve reservoir quality. Conversely, cementation and compaction have direct bearing upon reduction of reservoir quality indicating that a single process such as compaction can have both favorable and unfavorable results.

Compaction is an important fluid migration mechanism because it expels hydrocarbons from source beds and promotes their movement into reservoirs. An advancing groundwater front of mineral cementation can also cause fluid migration by displacement through the pore system.

Solution by groundwater recrystallization and fracturing improves porosity and permeability in potential reservoir rocks. Compaction, cementation, and mineral deposition and recrystallization in fractures reduce porosity and permeability.

Diagenetic processes that involve groundwater can often be reversed by the introduction of subsequent solution or recrystallization processes. Reservoir characteristics of most rocks, regardless of diagenetic history, can be improved by fracturing.

9

Structural Geology

Structural Geology

Attitudes of sedimentary rocks are important criteria in petroleum exploration, because folded rocks form structures that are fundamental to hydrocarbon accumulations. Comprehensive exploration programs utilize careful structural studies that assist in establishing deformational history and lead to understanding timing of structural events and petroleum migration.

Recognition of structures and the primary stress-related causes of structure are important in establishing the tectonic geometry of any area undergoing exploration. Structural geometry is significant because it delineates structural trends and identifies how an exploration program should be organized to maximize available data. Prediction of structural trend and occurrences of structures is based upon knowledge of stress and patterns that derive from stress. This results in an orderly and systematic approach to an otherwise potentially disorderly exploration effort.

It is evident that without uplift and subsidence related to tectonic and structural events there would be no sediments. Sediments are derived from positive crustal elements and deposited in negative elements. Distribution and erosion of sediments is controlled by structural activity.

Petroleum traps occur as a result of structural activity. Different types of structures produce different types of traps. In fact, some structures are controlled by stratigraphic variations that are caused by structural events. In some cases, structural and stratigraphic events combine to form petroleum traps that can be localized by previous structural activity.

Tectonic and deformational patterns must be studied to unravel the structural history so that petroleum migration and accumulation can be followed and established. It is not always easy and often involves indirect evidence that bears upon the geometry and the related three-dimensional thought process that ultimately decides the location of the next prospect.

Stress and Strain

The fundamentals establish a basis upon which structural geology can be understood. Because structural geology is three-dimensional, fundamentals of descriptive geometry and stereographic projection are important for visualization of structural concepts in space. Much of detailed structural analysis involves statistical treatment of three-dimensional structural plots, which, when combined, can produce important spatially-related results. The concepts appropriate to a fundamental understanding of spatial relations are important.

Stress is force applied to cause deformation. For structural geology purposes, stress is considered to be directed and consists of three mutually perpendicular axes.

When stress is applied, deformation, or strain, is the result. If bending force, or stress, is applied to a wooden stick it will bend until it breaks. Strain in the response to the stress is manifest as the bending of the stick and is temporary unless the stick remains bent. However, when the stick breaks it is permanently deformed, and the strain is represented by the break.

Triaxial Stress. Three mutually perpendicular stress axes comprise a compressive nonrotational pure shear stress field (Fig. 248). It consists of σ_1, σ_2, σ_3, which are maximum, intermediate, and least principle compressive stresses, respectively. For all practical purposes σ_1 is compressive, σ_2 is neutral, and σ_3 is dilational or extensional. Deformation in the triaxial system occurs along two conjugate shear zones, the angle between which is bisected by σ_1. The maximum value of the angle of internal friction between the conjugate shear zones approaches but does not exceed 90° and is normally much less. Rock competence determines the angle of internal friction, which averages about 60° and increases with ductility and decreases with brittleness. The line of intersection between the conjugate shear zones is represented by σ_2.

The triaxial stress system is applicable to rotational or simple shear stress couple as well as a compressive, non-rotational pure shear stress field (Fig. 249). A stress couple causes rotation of various elements and orients its features in a predictable pattern relative to the triaxial system.

1. *Synthetic faults* develop when second order rotation is opposite first order rotation. Counterclockwise first order rotation is opposite clockwise second order rotation, and synthetic left-lateral faults result in the example shown (Fig. 249).

2. *Antithetic faults* result when second order and first order rotation move in the same direction. Second order and first order rotation are counterclockwise in the example shown and form antithetic left-lateral faults (Fig. 249).

3. Compression along σ_1 causes the formation of *fold axis* normal to σ_1. This occurs in a manner similar to the development of wrinkles in a rug when it is pushed from one end.

4. *En echelon normal faults* are the result of extension along σ_3 and perpendicular to σ_1. They manifest themselves as normal or extensional faults that segment the fold axis.

5. *First order rotation* is the direction of rotation of the rotational couple, which produces simple shear. In Figure

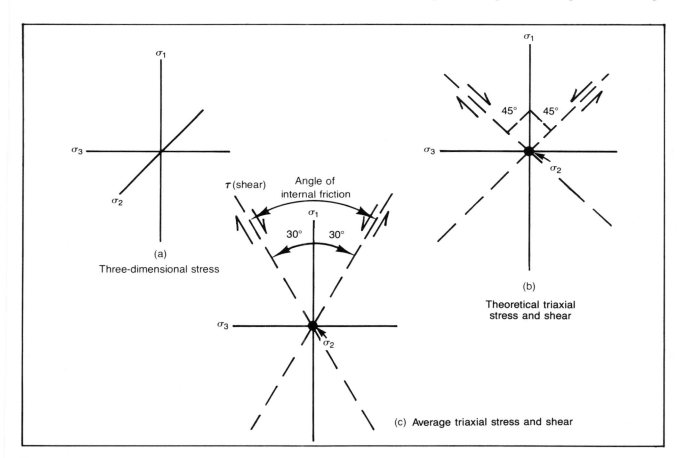

Fig. 248. Triaxial stress

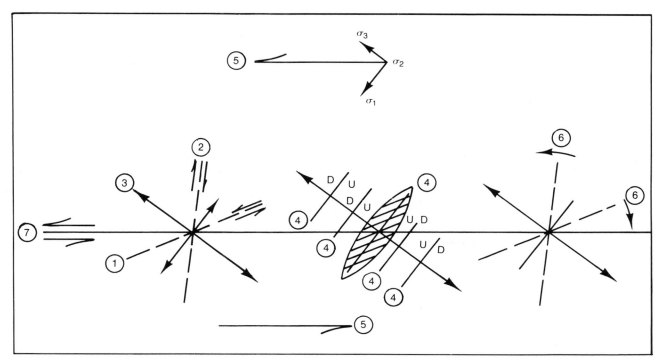

Fig. 249. Rotational shear (Circled numbers refer to reference numbers on pages 205, 206.)

249, first order rotation is counterclockwise.

6. *Second order rotation*, or internal rotation, responds to σ_1 and is clockwise or counterclockwise, depending upon the type of rotational couple and whether the affected faults are synthetic or antithetic.

7. As deformation continues in the rotational pure shear couple, the direction of the deep-seated fault responsible for the couple manifests itself at the surface as *"P" direction*. It does so along a trace that is parallel to the fault itself and shows offset in the direction of the fault offset.

Faults

A fault is a break in the earth's crust along which there is or has been discernible movement. There are several varieties of faults contingent upon the orientation of the triaxial stress field. Displacement along faults can vary from a few millimeters to several hundred miles. Some faults move blocks up or down, others move blocks laterally over one another, and others move blocks laterally to slide by each other.

Faulting is an important adjunct to structural deformation because it establishes the orientations of stress directions that develop the geometry of tectonic provinces and facilitates their exploration. Structural petroleum traps are usually accompanied by some type of faulting that creates structural relief by displacement along a fault zone. Structural features are often modified by faulting so that individual traps, comprising separate fault blocks and containing hydrocarbons, have their own intrinsic reservoir characteristics.

Fault Terminology. The accompanying diagram (Fig. 250) schematically illustrates the primary components observed in fault-related deformation.

The angle that a fault plane makes with the horizontal is the *dip* of the fault plane. Fault plane dips vary from vertical to horizontal and are oriented in response to the stress field in which they form.

The *strike* of a fault plane is represented by the line of intersection between the fault and horizontal surface. Strike line is always horizontal, and since it has direction, is measured either by azimuth or bearing, for example, N 30° E (Fig. 250).

Horizontal movement component of a fault is known as the *heave* of the fault. The *throw* of a fault is the vertical movement component of its displacement. Both heave and throw are variable with movement along the fault and the dip of the fault plane.

Actual linear movement along a fault plane represents

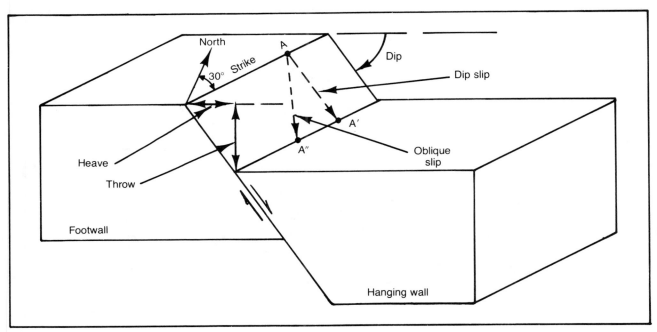

Fig. 250. Fault terminology

the *slip* of the fault. Movement down the fault plane perpendicular to the strike is dip slip (A to A'). Oblique slip occurs at an angle to the strike (A to A").

The *hanging wall* of a fault is located above the fault surface and bears upon it. The *footwall* of a fault is beneath the fault. It occupies the position beneath the fault regardless of whether the hanging wall has moved up or down.

Gravity is the principal deformational stress involved in *normal faulting*, in which the hanging wall is displaced downward and relative to the footwall (Fig. 251). The fault plane of a normal fault usually dips at a high angle (60°) or more. In the Gulf Coast, however, normal faults manifest shallow dips in many cases. Gravity acts as the maximum principal compressive stress in normal faulting and is directed vertically. Intermediate and least principal compressive stresses σ_2 and σ_3 are oriented horizontally.

Compression is usually responsible for *reverse faulting* where the hanging wall is moved up relative to the footwall (Fig. 252). A reverse fault that dips 30° or less becomes a thrust fault, which results from a horizontal maximum principal compressive stress.

Intermediate principal stress remains horizontal in reverse and thrust faults. However, σ_3 is nearly vertical in reverse faults and is vertical in thrust faults.

Strike-slip faults occur where crustal blocks slide laterally by each other along nearly vertical zones of move-

ment (Fig. 253). The triaxial stress system is oriented with σ_1 and σ_3 horizontal and σ_2 vertical. Strike-slip faults are referred to in many ways, including wrench, tear, transcurrent, and translation faults. They are identified by the sense of movement of the block on the far side of the fault from the viewer (Fig. 254). If the viewer is observing the fault, and the block on the far side of the fault has moved to the right, the fault is known as a right-lateral strike-slip fault. Right-lateral faults are also known as right-slip, right-handed, or dextral strike-slip faults.

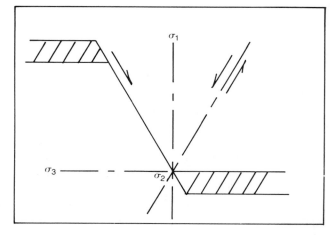

Fig. 251. Normal fault, cross-section

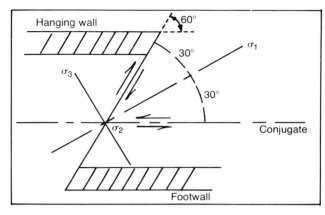

Fig. 252. Reverse fault, cross-section

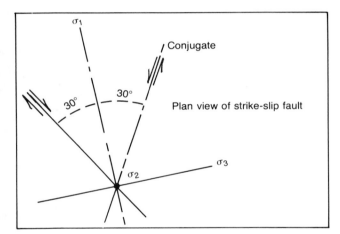

Fig. 253. Strike-slip fault, plan view

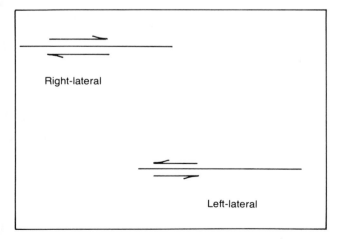

Fig. 254. Strike-slip fault movement definition

Faults with left-handed displacement are referred to with similar nomenclature. However, left-handed faults are sinistral, as opposed to dextral or right-handed faults.

Faults and Triaxial Stress. Faults are rarely oriented along directions that represent normal concepts of up and down, left and right, or forward and backward. Most fault motion has some component of obliquity that results from several directions of motion applying to any individual structural feature within a tectonic province. However, there are enough examples of deformation related to regional stress patterns to refer to deformation in terms of every-day dimensions. A classification of faults emerges that is based upon relative movements of fault blocks.

Folds

Folding in layered rocks consists of deformation of strata without faulting. Folds are formed by compression,

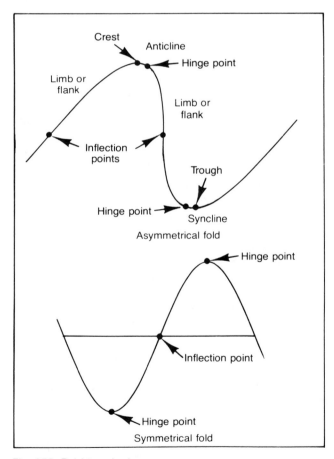

Fig. 255. Fold terminology

draping over basement blocks, and compaction. Faulting often accompanies folding and can precede it or follow it. The triaxial stress system for folding is representative for faulting since the two are intimately related.

Fold Terminology. Folds are manifest as different types according to morphology, origin, and type of internal deformation. General fold terminology is descriptive and relates to elements of individual features applicable to all types of folds.

1. An *anticline* is a fold with upward convexity (Fig. 255). Beds that make up an anticline dip in opposite directions away from the crest or high point of the fold.

2. A *syncline* is a concave upward (Fig. 255). Beds making up a syncline dip in opposite directions toward the trough, or low point, of the fold.

3. The *hinge point* occurs at the locus of maximum curvature of a fold (Fig. 255). The hinge point is applicable equally to anticlines and synclines. Hinge points connected on any particular bedding surface are joined along a hinge line.

4. The locus of all hinge lines of a fold is the *hinge surface* of the fold.

5. An *inflection point* occurs where bed curvature in one direction changes to bed curvature in the opposite direction (Fig. 255). Inflection occurs where concavity changes to convexity.

6. The *limbs,* or flanks of folds, are those portions adjacent to the inflection lines of folds (Fig. 255). They converge upon the inflection line synclines and diverge from it in anticlines.

7. A fold is *symmetrical* when its shape and mirror image are identical or when the fold is bilaterally symmetrical about the hinge surface (Fig. 255). Symmetrical folds are usually represented by a vertical hinge surface.

8. A fold that is not bilaterally symmetrical about the hinge surface is *asymmetrical* (Fig. 255). Its hinge surface is usually inclined.

9. A horizontal or nearly horizontal hinge surface is characteristic of a *recumbent fold* (Fig. 256). The beds in the bottom limb of a recumbent fold are upside down.

10. When the hinge surface of a fold is depressed below the horizontal, the fold is *overturned* (Fig. 256). The beds in the bottom limb of an overturned fold are upside down and depressed below the horizontal.

11. *Concentric (parallel) folds* comprise strata parallel to each other that maintain thickness throughout all dimensions of the fold and slip against each other as they fold (Figs. 257 and 258).

12. With *nonparallel folds,* the concentricity of folded beds cannot be maintained, and thinning along the fold

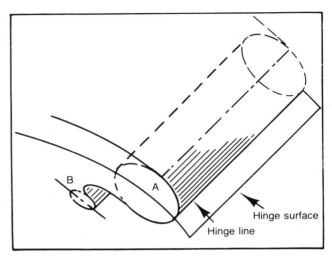

Fig. 256. Hinge surfaces of cylindrical folds (Recumbent Fold-A; overturned Fold-B)

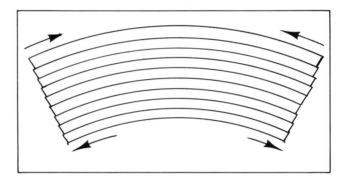

Fig. 257. Parallel fold bedding slippage

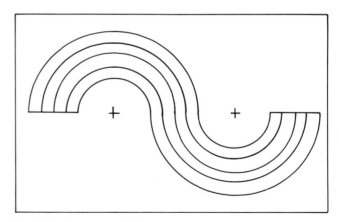

Fig. 258. Concentric folds

limbs occurs (Fig. 259). Flowage away from the limbs of nonparallel folds can occur as folding takes place.

13. Folds that have the same geometric form, but in which shearing or flowage has occurred, are *similar folds* (Fig. 260). Thickness of individual folded beds is not constant. Fold axes are usually thicker than fold limbs.

14. Sedimentary sequences subject to folding often comprise layers if variable thickness and competence.

The individual layers in such a sequence fold according to their ability to do so and produce varying degrees of deformation within an individual fold (Fig. 261). Folds produced in this way are not concentric, and because they are different are known as *disharmonic folds*.

Folds and Triaxial Stress. Inasmuch as folds usually form in association with faults, the stress patterns for both are identical and depend upon the type of deformation involved.

Compressional folds are often formed in association with thrusting where σ_1 and σ_2 are horizontal and σ_3 is vertical (Fig. 262). Compression is not exclusively an expression of thrusting so that local compressional folds

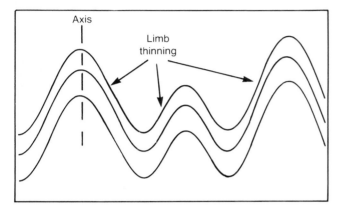

Fig. 259. Nonparallel fold

Fig. 260. Similar fold

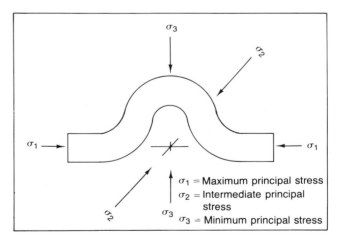

Fig. 261. Disharmonic fold

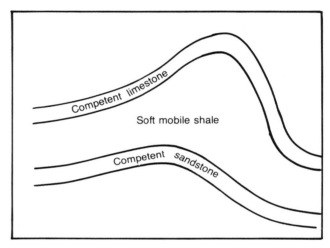

Fig. 262. Triaxial stress and compressional folding

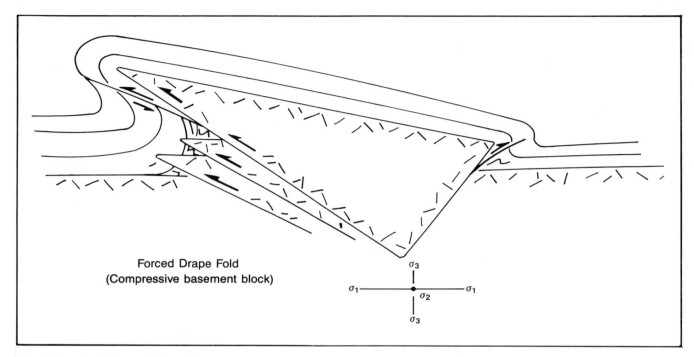

Fig. 263. Forced drape fold.

can form regardless of the orientation of the primary stresses.

Upthrusting, overthrusting, or vertical movement of basement blocks creates horizontal and vertical stress and the draping of sediments over the blocks (Fig. 263). In these instances, σ_1, is vertical and σ_2 and σ_3 are horizontal, or σ_1 and σ_2 are horizontal and σ_3 is vertical.

Strike and Dip

A stratum of rock dips when its attitude departs from the horizontal with which it describes an angle. Dipping beds can exceed 90° of departure from the horizontal when they are overturned. The line of intersection between a dipping bed and the horizontal is the strike of the bed. Strike is always at a right angle to the direction of dip and has a direction expressed as an azimuth or bearing. Strike and dip are represented on geologic maps by the symbol shown in Figure 264. The strike shown has an azimuth of 45°, or a bearing of north 45° east, with north toward the top of the page. Dip of the bed is represented as 30° to the southeast.

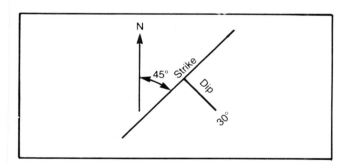

Fig. 264. Strike and dip symbol

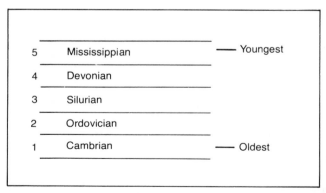

5	Mississippian	—— Youngest
4	Devonian	
3	Silurian	
2	Ordovician	
1	Cambrian	—— Oldest

Fig. 265. Superposition of rock strata

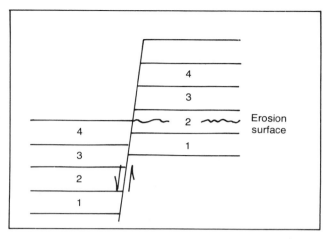

Fig. 266. Cross-section of faulted sequence before erosion

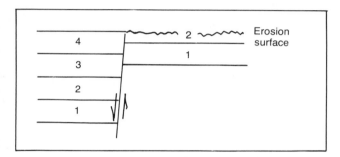

Fig. 267. Cross-section of faulted sequence after erosion

Superposition

Rocks are laid down in order of superposition, with the oldest on the bottom and the youngest on top (Fig. 265). Unless a rock sequence is overturned, the order of superposition is retained. If a sequence of rocks is faulted, one block moves up relative to the other (Fig. 266). Erosion of the upthrown block to the horizontal level of the downthrown block results in older beds adjacent to younger beds at the level of the eroded surface (Fig. 267).

The rule from this relationship states that when older beds are faulted into position adjacent to younger beds at the surface, the older beds are represented as the upthrown block of the fault.

Tilted Sequence. When a normal rock sequence is tilted and eroded to a horizontal surface, dip of the sequence is from the oldest beds toward the youngest (Fig. 268).

Faulted Tilted Sequence. A tilted sequence that has been faulted and eroded to a horizontal plane causes the formation contacts between the individual beds appear to move in the direction of dip (Fig. 269).

In the diagrams shown, *a* represents the position of the contact between beds 2 and 3 before the tilted sequence is eroded. The position of the contact between 2 and 3 is shown at *b* after erosion of the tilted sequence. Distance, *d*, represents the apparent amount of movement of the contact between 2 and 3 in the direction of the dip toward *b* after erosion to the erosion surface. The contact between 2 and 3 has not actually moved but has apparently moved because of the dip of the sequence.

A plan view of a tilted sequence illustrates the concept of the contacts between beds moving in the direction of

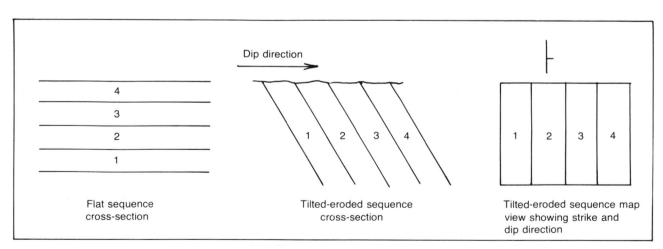

Fig. 268. Erosion of tilted sequence and dip direction

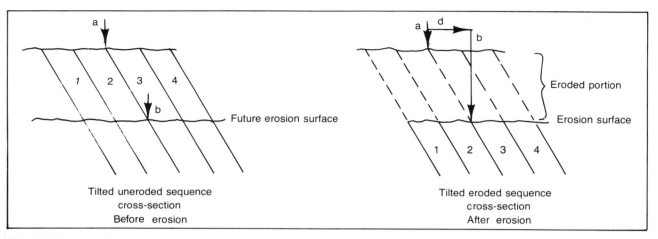

Fig. 269. *Faulted tilted sequence*

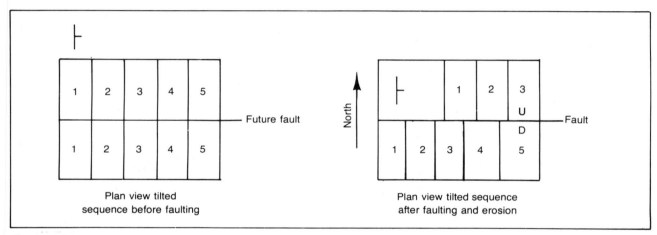

Fig. 270. *Faulted tilted sequence*

dip on the eroded upthrown block of a fault (Fig. 270). Here, the strike of the sequence is north-south and the dip is toward the east.

Contacts have apparently moved in the direction of dip on the block on the north side of the fault. This indicates that the north block was faulted up, and after erosion to the horizontal level of the south block illustrated the eastward movement of the contacts.

Eroded Anticline. An eroded anticline is represented on a horizontal surface by parallel beds that have a north-south strike (Fig. 271). The structural limbs of the anticline dip at 180° to each other away from the vertical hinge surface of the structure. An eroded syncline is represented in the opposite way with the structural limbs dipping toward the hinge surface.

A *plunging anticline* appears on a horizontal eroded

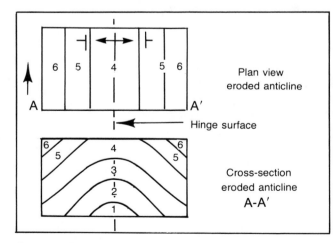

Fig. 271. *Eroded anticline*

surface as a horseshoe or parabola (Fig. 272). The structure plunges toward the closed end of a parabola since the beds always dip from oldest to youngest.

The diagrams show that the eroded anticline plunges to the north in the plan view. There is no way to determine whether the anticline is plunging from information shown on the cross-section.

Determination of the upthrown block on a *faulted plunging anticline* uses the principle of migration of bedding contacts in the direction of dip on the upthrown side (Fig. 273). Determinations of relative movement along the fault surface can be made from the plan and cross-section views shown in the diagrams.

It is evident from the plan view that the contacts between 3 and 4 have moved in the direction of dip (northeast and southwest) along the fault on the east block. This

means that the east block is the upthrown side of the fault. The same relationship is shown in the cross-section. It is not evident that the structure is a plunging anticline on the cross-section, however.

Block Diagrams

Three-dimensional block diagrams utilize the concepts developed in the previous section (Fig. 274). The following completed block diagram illustrates how structural relations on geologic maps and cross-sections can be determined.

Note that on this diagram, older beds dip under

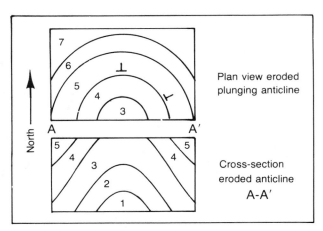

Fig. 272. Eroded plunging anticline

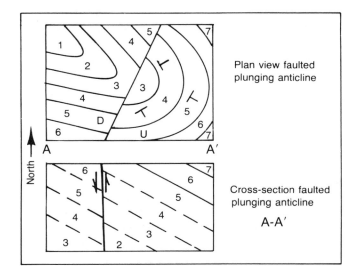

Fig. 273. Faulted plunging anticline

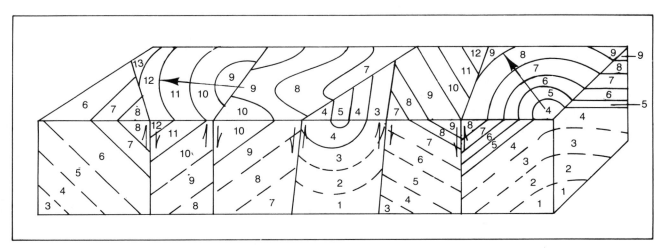

Fig. 274. Three-dimensional block diagram

younger beds, bedding contacts move in the direction of dip on upthrown fault blocks, and structures plunge toward the youngest beds. This diagram has been completed on the front and side from information given on the top of the block.

Descriptive Geometry

Geology is a three-dimensional science to which the fourth dimension of time has been added. This means that the geological processes of erosion, deposition, and deformation proceed through the abstraction of immense segments of geologic time. They relate to each other and modify each other. Geology and its processes are the physical three-dimensional manifestations of that time.

The concepts of geologic parameters are totally three-dimensional in the sense of distribution and construction. A sand bar, a river channel, a large anticline, or the evolution of life forms are three-dimensionally manifested because they all have length, width, and depth, which are, of course, functions of geologic time. At any particular instant in geologic time, however, geologic features appear to stand still momentarily and allow us to observe them, the products of their mechanical, developmental processes brought to the point of our observation by the kinematics of time.

Structural geology is three-dimensional because structure of any sort involves dimension. The world around us is three-dimensional, and we take much of it for granted because our perspectives are based upon datum values we are familiar with. It is sometimes more difficult to visualize perspectives with which we are less familiar. The perspective of structural geology and its datum values utilizes all of the things we daily take for granted and places them into a system of reference. For these reasons a brief discourse in the fundamentals of descriptive geometry is significant in the process of three-dimensional visualization of structural concepts.

Descriptive geometry utilizes three primary views of an object in space (Figs. 275 and 276). The most important are the top and front views, which establish direction, height, and depth. The side view completes the picture but is not essential since all necessary information is available on the top and front views. Our discussion will utilize the top and front views as basic information.

Top and front views of an object in space do not always provide all information about the object, however. Nor in most cases would the addition of the side view. It is clear that auxiliary views are often needed to provide the required information. Our objective is to illustrate the use of simple auxiliary views in the determination of certain parameters that apply to an object in space.

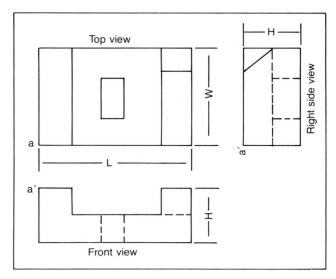

Fig. 275. Top, front, and side views of object in space

Consider the top and front views of line *ab* (Fig. 277). Line *ab* has a bearing that is determined from the top view. It is evident from the front view that *b* is higher than *a* by the amount of distance *d*, and that the line *ab* plunges or departs from the horizontal, from *b* toward *a*. However, because line *ab* is oblique to the top and front views, an auxiliary view is needed to find the true length and amount of plunge of the line. To view line *ab* in its true length and to view its plunge, the line must be observed from a vantage point normal to the line. This requires that a section parallel to the line be drawn to facilitate the required observation. So we have the following problem: to find the bearing, true length, and plunge or slope of line *ab*. To solve the problem, we consruct an auxiliary view of *ab* in a vertical plane containing line *ab* (Fig. 277).

The results of our construction show us that line *ab* has a bearing of north 40° west or an azimuth of 320, and from measurement on the diagram, a true length of 60 mm and a plunge of 33|° to the northwest from *b* to *a*.

We have discussed the concept of strike and dip in our treatment of faults and deformed beds. These concepts are important in the determinations of structural types, or more specifically, the attitudes of faults and sections of dipping strata. Strike and dip of a plane can be measured by descriptive geometry using the top and front views of the plane to construct auxiliary views (Figs. 277, 278).

The strike of plane *abc* is determined by drawing a horizontal plane *h,h*, which appears as a line, through points in the front view. Points *a',x'* are projected up to points *a,x* in the top view and connected by a line *ax* that represents the intersection of the horizontal plane with

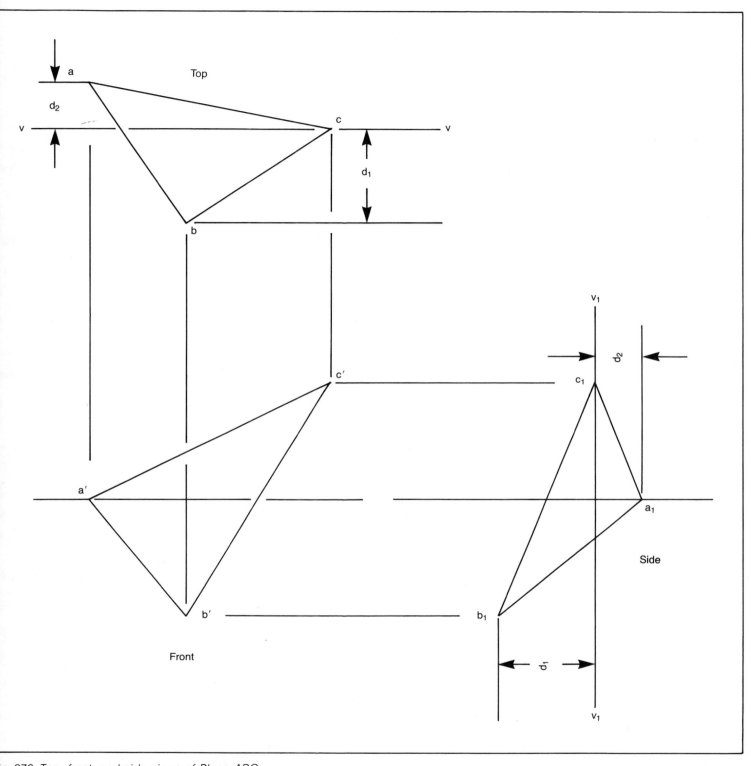

ig. 276. *Top, front, and side views of Plane ABC*

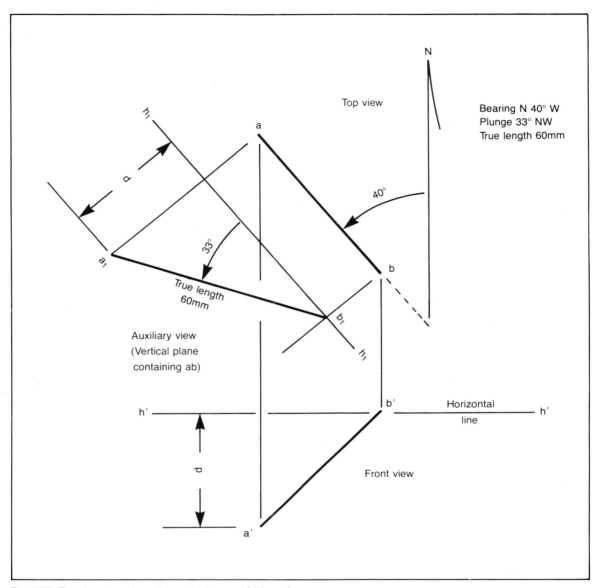

Fig. 277. Bearing, true length, and plunge of Line ab

plane *abc*. This line lies on the horizontal plane, is therefore horizontal, and represents the strike of plane *abc*, since it also lies on the plane *abc*. The strike has direction that can be measured from the top view.

To obtain the dip of plane *abc* an auxiliary view showing the plane on edge or as a line is constructed. This view shows the strike as a point and is constructed normal to the strike. The angle of the edge view of the plane *abc* with *h,h* in the auxiliary view, represents the dip of the plane. Since point *b* is the lowest part of plane *abc*,

the dip is normal to the strike toward the low part of the plane.

Descriptive geometry is useful in the determination of strike and dip in actual situations, where the elevations of three points on a fault plane or dipping bed are known. Differences in elevation between the highest and lowest points in the inclined surface are determined. By establishing a proportion between elevation and the length of the line between the highest and lowest points, a point with the same elevation as the third point on the surface

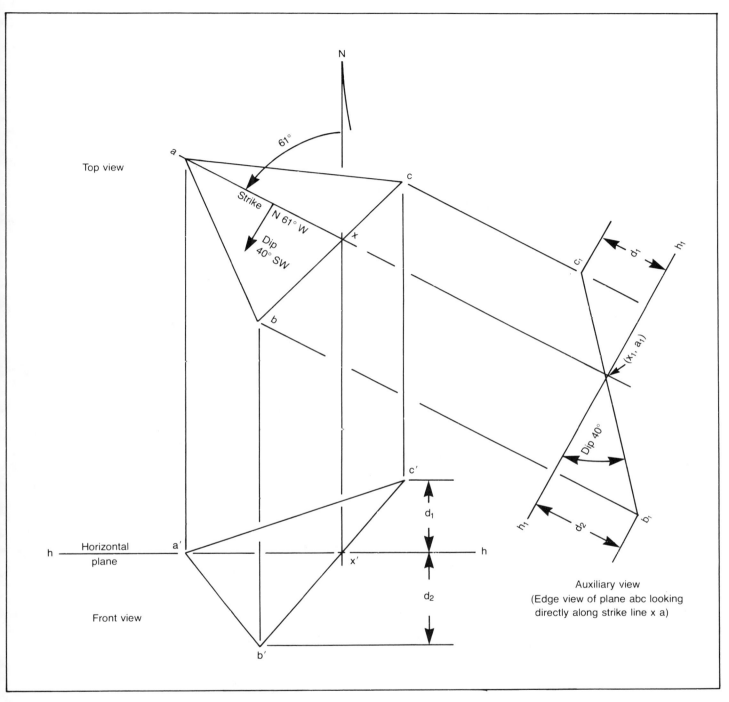

Fig. 278. Strike and dip of Plane abc

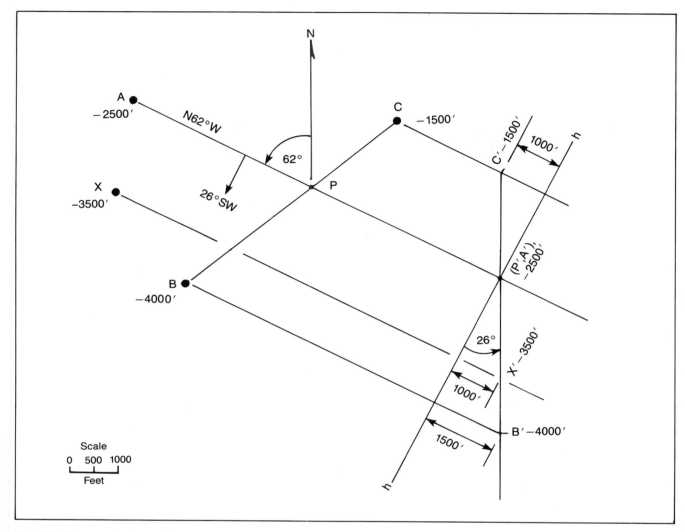

Fig. 279. Three point determination of strike and dip of a plane.

can be plotted on the line connecting the highest and lowest points. Connecting that point with the third point produces the strike line, since both points are at the same elevation. The bearing of the strike line can then be measured.

An auxiliary view of the edge of the plane, showing the strike as a point, will show the dip of the inclined surface (Fig. 278). The elevations of the high and low points are plotted above and below the h,h surface plotted through the strike point using the scale shown in the map. By connecting the high and low points with a straight line through the strike point, the edge view of the plane is shown and the dip can be measured from it.

This method of determining strike and dip involves the solving of this problem known as a three-point problem. It is widely used to determine attitudes of planes in geological problems (Fig. 279).

Three wells drilled in the development of a field (Fig. 279) encountered the top of a reservoir horizon in wells A, B, and C at 2500 ft, 4000 ft and 1500 ft (below sea level respectively). Strike and dip of the reservoir are required as is the depth of the reservoir in a well to be drilled at X. The reservoir dips in the general direction of B from C, which are at −4000 ft and −1500 ft respectively. A straight line connection B and C can be divided into five equal horizontal increments, each representing 500 ft vertically. A distance of two horizontal increments from C to B at P represents 1000 vertical feet lower than C or −2500 ft, which is the same elevation as A. A line connecting A and P lies at −2500 ft and represents the strike of the reservoir described by ABC at north 62° west. An auxiliary edge view of ABC illustrates the strike PA as a point at −2500 ft. A horizontal line h-h is drawn through PA perpendicular to lines of projection from A, B, and C parallel to the strike. By using the map scale to measure 1500 ft below h-h (at −2500 ft) B is determined at −4000 ft. The same scale is used to find A at −1500 ft or 1000 ft above h-h.

A straight line connection C, PA, and B shows ABC as a line or on edge. The angle between h-h and the edge-view line of ABC provides the 26° dip of ABC to the southwest.

Projection of X to the auxiliary edge view of ABC defines its depth below h-h. Use of the map scale measures that distance as 1000 ft that when added to −2500 ft establishes the top of the reservoir at location X as −3500 ft.

The results of this construction illustrate that the reservoir described by ABC strikes north 62° west, dips 26° to the southwest and will be encountered at −3500 ft at well location X.

Descriptive geometry can be useful in determination of strike and true dip from two apparent dips that may come from surface, well, or seismic date (Figure 280).

Two apparent dips have bearing and plunge of north 60° east; 19° northeast, and south 50° east; 22° southeast respectively. True strike and dip are required. Both apparent dips are plotted with their bearings northeast and southeast from a common point A to B and C respectively. An auxiliary view normal to AB shows its true length, the 19° northeast plunge, and distance d along the projection of B. An additional auxiliary view illustrates A, C, the 22° southeast plunge and the position of distance d (derived from the auxiliary view of AB), perpendicular to h-h and its intersection with AC at X. Projection of X established its position in the top view of plane ABX, illustrates XB as a point, distance d, and the true dip of the plane. The results indicate that ABX strikes north 11° east and dips 25° east.

Stereographic Projection

Another way of representing three dimensions and solving problems makes use of stereographic projection and plotting of structural data on the stereo net. Stereographic projection involves projection of points on a hemisphere below a horizontal surface through the hori-

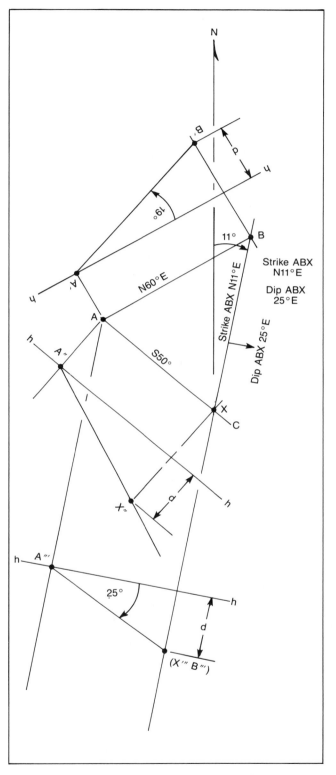

Fig. 280. True dip constructed from two apparent dips.

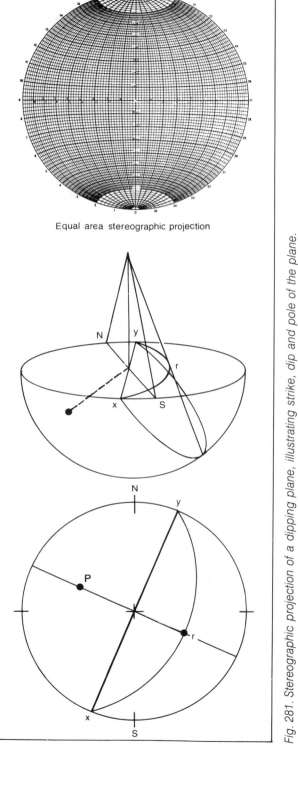

Equal area stereographic projection

Fig. 281. Stereographic projection of a dipping plane, illustrating strike, dip and pole of the plane.

zontal surface to a point above the center of the hemisphere (Fig. 281). The projection of the hemisphere to the horizontal surface presents a net consisting of meridional great circles and latitudinal small circles that intersect to form the net. It is important to understand that the net itself represents the projection to the horizontal of the meridians and parallels on the hemisphere below. To observe a stereo net is to look down upon the projection of a bowl to a horizontal surface.

The stereo net is used with an overlay sheet of transparent paper on which the north and south positions of the net are plotted. A thumbtack penetrating the center of the net from below and taped to the net allows the overlay to rotate freely and maintain its position over the net.

Plotting of structural values on the overlay is rapid and straightforward. To plot a strike of north 30° east and a dip of 40° southeast, the overlay is oriented north-south, and a mark is made at the margin, or primitive, of the limiting circle of the net at a point 30° east of north and 30° west of south (Fig. 282). A straight line representing the strike and connecting the two points is drawn through the center of the net, and the net is then rotated to place the strike line in the north-south position. The dip is plotted by counting 40° toward the center from the primitive along the east-west equator and drawing a great circle through the resultant point to the north and south ends of the strike line. The pole of the dipping surface plunges the complement of the 40° dip, or 50° in the opposite direction, at an angle of 90° to the dipping surface (Figure 283). Rotation of the overlay to its position of north and south places the strike of the surface at north 30° and the great circle connecting the strike ends shows a dip of 40° southeast.

Use of the stereonet can rapidly determine the true dip from two apparent dips (Fig. 284). Bearing and plunge of each (north 60° east, 19° northeast, south 50° east, 22° southeast from Figure 279) are plotted on the overlay of the stereonet. The overlay is rotated until both plunge values fall on the same great circle that is drawn between the north and south poles along the appropriate meridian. The dip of 25° east-southeast is measured along the equator and the overlay is rotated to true north to read the strike of north 11° east along the primitive. Both correspond to the results of the construction (Fig. 279).

Problems involving two dipping beds separated by an angular unconformity can be stereographically solved to determine the attitude of the older bed prior to tilting, erosion, and subsequent deposition and tilting (Fig. 285). The older bed strikes north 35° east and dips 55° southeast. The younger bed strikes north 20° west and dips 30°

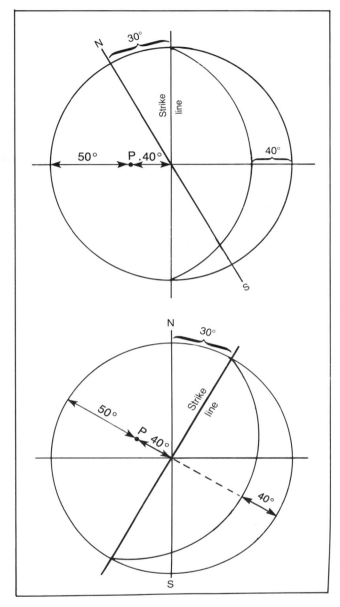

Fig. 282. Stereographic plot of strike, dip and pole of a dipping plane

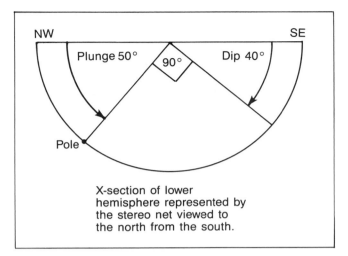

Fig. 283. Cross-section of lower hemisphere represented by the stereonet viewed to the northeast from the southwest.

northeast. Both strikes and dips are plotted on the overlay that is rotated to place the strike of the younger bed in a north-south position. Since the younger bed dips 30°, that amount of dip must be removed to restore it to the horizontal. During the restoration of the younger bed to the horizontal, the older bed will be rotated 30° around the strike of the younger bed along the small circles of parallels of latitude. A half dozen migrated points of the older

bed, when placed on the same great circle will produce its pre-unconformity position striking north 59° east and dipping 43° southeast.

Any dipping surface can be plotted on a stereo net provided strike and dip information are given. It is evident that intersecting faults or both flanks of a structure can be plotted in the same way.

Stress and the Stereonet. The triaxial stress system for any type of structure is easily plotted on the stereonet. Stress relations for the three types of faults and their strikes, dips and poles present characteristic patterns (Fig. 286).

Normal fault planes dip at about 60° and are oriented with σ_1 vertical bisecting the conjugate fault pair, σ_2 horizontal at the ends of the strikes of the faults representing their intersection, and σ_3 horizontal located in the primitive at 90° to σ_2 (Fig. 287). This presumes the intersection of the faults σ_2 to be horizontal. Slip direction of faults in the triaxial system occurs along the great circles of the fault planes 90° to σ_2. The slip directions, fault poles, and σ_1 and σ_3 lie on the same great circle normal to σ_2. This relationship is always true regardless of fault attitudes.

Vertical fault planes intersecting at 60° at the center of the stereonet represent a *strike-slip fault* conjugate set (Fig. 288). The triaxial stress system is represented by σ_1 horizontal bisecting the conjugate fault pair, σ_2 vertical at their intersection, and σ_3 horizontal at the primitive. Fault slip directions and poles, and σ_1 and σ_3 lie on the same great circle designated by the primitive.

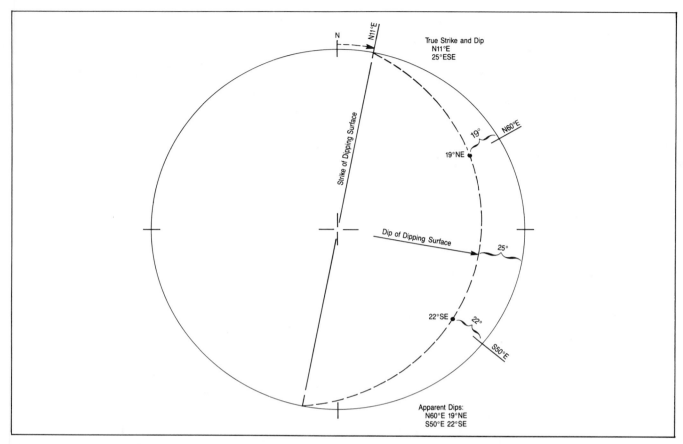

Fig. 284. Stereonet method of obtaining true strike and dip of a surface using two apparent dips on the surface.

Reverse faults dipping at 30° or less are shown on the stereo net with σ_1 horizontal bisecting the conjugate set, σ_2 horizontal representing the intersection of the thrusts, and σ_3 vertical (Fig. 289). The relationship of fault slip directions and poles, and σ_1 and σ_3 on the same great circle is manifest again.

The stereogram for normal faults shows the intersection of the faults as horizontal and represented by the strike line common to both faults along the line of σ_2. If the line of intersection departs from the horizontal, σ_2 assumes a position inside the primitive and indicates a plunge (Fig. 290). Rotation of σ_2 below the horizontal is made about σ_3 as the rotational axis, which does not change position. Rotation demands that σ_1 rotate to maintain its position normal to σ_2. In the diagram shown, the plunge of σ_2 is to the north, causing σ_1 to migrate to the south from the center of the net by an amount equal to the rotation. No longer does σ_2 represent the strike of the two faults although it still represents the intersection of

their planes. The positions of the slip directions remain normal to σ_2.

Structural information is often at a minimum but when properly used can be greatly expanded. Fault 1, striking north 60° east, dipping 40° northwest, with a slip direction of south 80° west provides sufficient data to determine the same information for the conjugate fault and the orientations of the primary stress axes (Figure 291).

A plot on an overlay illustrates strike, dip, slip direction and pole of Fault 1. A point 90° from the slip direction along the great circle of the fault dip locates σ_2. The great circle which includes the fault pole (Pf_1) and its slip direction (SD_1) indicates σ_1, 30° from the slip direction, the slip direction (SD_2) of the conjugate fault (Fault 2) 60° from it, and σ_3 90° from it along the circle. A great circle, which includes σ_2 and the slip direction (SD_2) of the conjugate fault (Fault 2), determines the strike and dip of the conjugate fault. The pole of the conjugate fault (Pf_2) can be plotted to show its position on the great circle

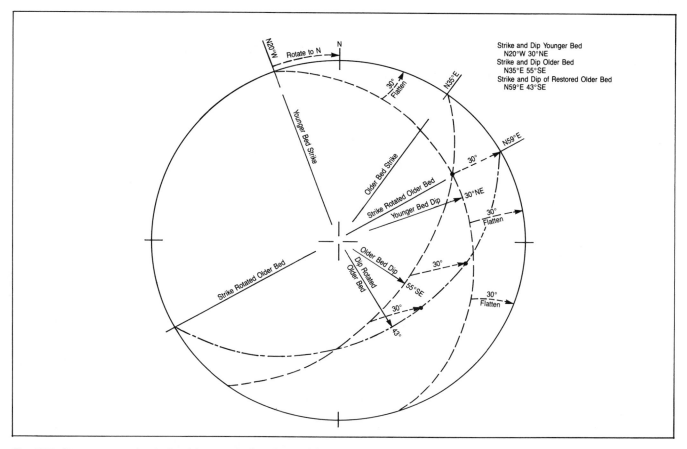

Fig. 285. Stereonet method of solving unconformity problem.

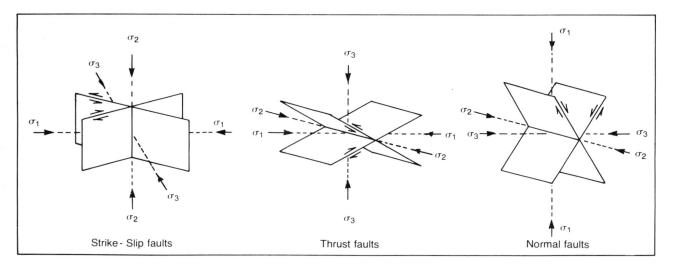

Fig. 286. Fault classification

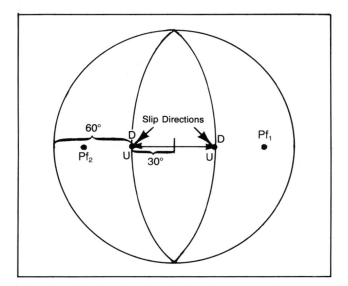

Fig. 287. Stereogram of normal faults

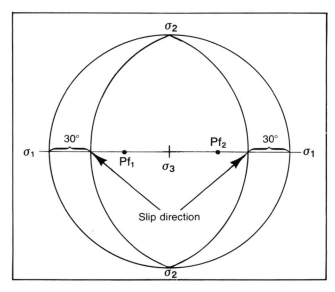

Fig. 289. Stereogram of thrust faults

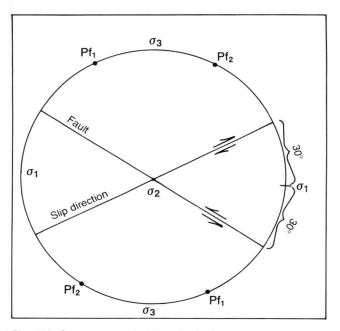

Fig. 288. Stereogram of strike-slip faults

which includes the slip directions of both faults, the pole of the first fault, and σ_1, and σ_3. By placing σ_2 and, σ_3 on the same great circle and σ_2 and σ_1 on another great circle, their 90° or separation can be confirmed.

Structural Analysis. Descriptive geometry and the stereo net are important in structural analysis, which requires recognition of planes, lines, intersections, and stress systems in space, and their relationships to each other. Recognition of stress patterns facilitates predictions of locations of structural trends and provides information on their deformational histories as well as the orientations of potential prospects. Structural analysis is important in complex exploration plays where several periods of deformation and deposition are involved.

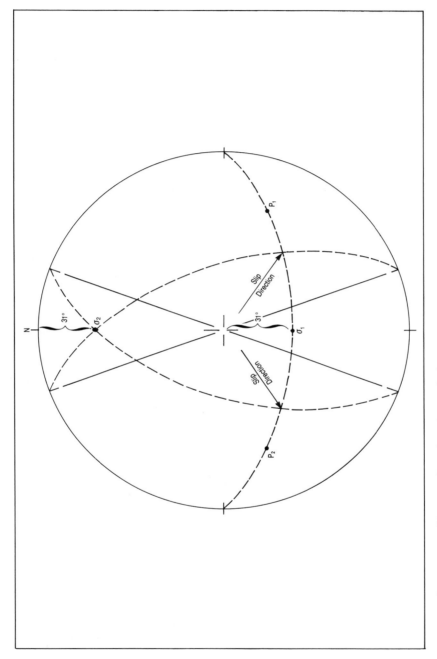

Fig. 290. Stereogram of intersection of conjugate faults

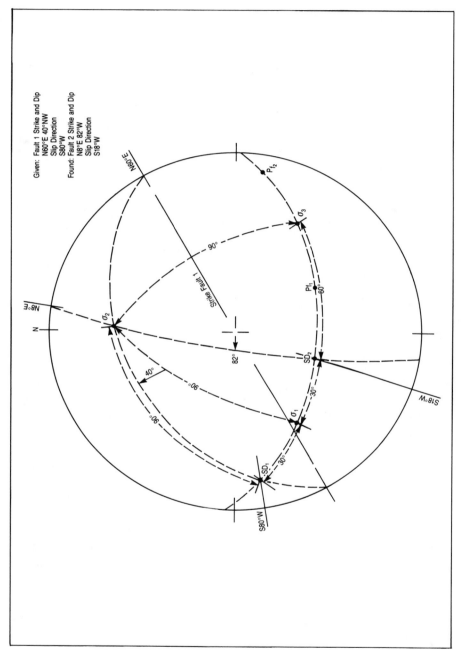

Given: Fault 1 Strike and Dip
N60°E 40°NW
Slip Direction
S80°W
Found: Fault 2 Strike and Dip
N8°E 82°W
Slip Direction
S18°W

Fig. 291. Finding strike and dip of conjugate fault given only strike, dip and slip direction of the fault.

10   Petroleum Traps

Hydrocarbon Traps

A trap is any combination of physical factors that promotes accumulation and retention of petroleum in one location. Petroleum traps can be structural, stratigraphic, or a combination of the two.

Structural Traps

Deformation of rocks creates structural traps by folding or faulting. Many styles of traps result from deformation. Some styles are very complex because complicated stresses are involved in their formation (Figs. 292 and 293). However, other styles are relatively simple because their patterns reflect uncomplicated stress regimes.

Anticline. An anticline is a simple fold that traps petroleum in its crest (Figs. 294 and 295). Anticlines form in a variety of ways, ranging from compression to gravity-induced fault block activity. As structural traps, anticlines are some of the most common productive structures. Producing anticlines can be symmetrical or asymmetrical structures (Figs. 295, 296).

Many producing anticlines are faulted according to the stress patterns that formed them (Figs. 297 and 298). Most anticlines are associated with some type of faulting in their deformational histories, whether they are normal or thrust-fault related. Faulted anticlines are usually asymmetrical.

Overthrust anticlines produce abundantly from fields in overthrust belt provinces in the United States amd Canada (Figs. 299, 300, and 301). In some cases, producing beds in overthrust anticlines are thrust against source beds of different ages not normally associated with the reservoirs. Overthrust anticlines are normally asymmetrical and complexly faulted. (Figs. 302, 303, and 304).

Typical overthrust belts do not significantly involve crystalline basement rocks in potentially productive structures. However, compressive foreland blocks detach the crystalline basement and often thrust it over prospective sediments (Fig. 305).

Faulting and tilting of crustal blocks produces a variety of trapping mechanisms. A single fault can provide a seal to prevent petroleum migration. Multiply faulted blocks create traps which are often connected, but can conversely be entirely independent of each other. Petroleum production usually comes from the fault-sealed upthrown block. Enough petroleum comes from downthrown blocks to consider them important exploration objectives (Fig. 306).

Fault Traps. Simple fault traps involve up-dip closure of a reservoir bed against a sealing fault (Fig. 308). A plunging anticline faulted up-plunge is a good example of a fault trap involving partial closure.

Where a reservoir is multiply faulted, a complex fault trap occurs (Figs. 307, 308, and 309). Multiply faulted anticlines form complex fault traps.

Subthrust traps occur in overthrust belts, where the deformed reservoir forms a trap below a thrust fault (Figs. 310, 311 and 312). Some subthrust traps are complexly faulted, and others are quite simple and relatively undeformed.

Salt-related Traps. Salt structure formation can produce several different types of traps described by varying deformational intensities. Most salt structures are faulted in response to the degree of deformation. Traps related to salt deformation do not usually produce from the salt

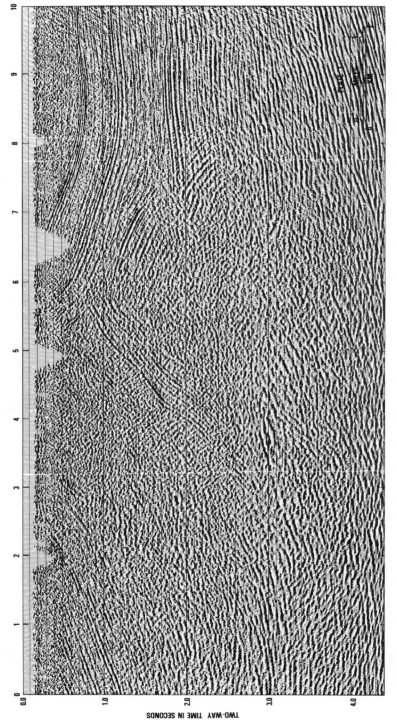

Fig. 292. Seismic section of complex strike-slip fault and fold deformation in southern Oklahoma. From Harding, et al., 1983. Permission to publish by AAPG.

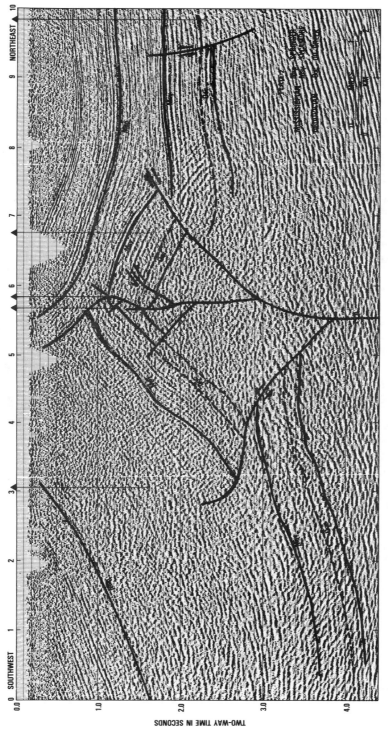

SOUTHWEST   NORTHEAST

TWO-WAY TIME IN SECONDS

Fig. 293. Interpretation of complex strike-slip fault and fold system in southern Oklahoma. From Harding, et al., 1983. Permission to publish by AAPG.

itself. They produce from beds deformed by the salt tectonics instead.

Low-relief peripheral salt ridges form by salt migration down-dip from the up-dip limit of a salt wedge (Fig. 313). Faulting in the overlying sediments occurs as a slight rollover produces a productive anticline in the beds overlying the salt.

Salt flowage from a relatively thin marginal salt source into a salt pillow produces a low-amplitude productive anticline, or low-relief salt pillow, in the overlying sediments (Fig. 314). Faulting can be an accompanying deformational feature.

Migration of salt from a somewhat thicker section into a salt pillow produces an intermediate-amplitude anticline in the overlying sediments (Fig. 315). Faulting can accompany deformation of the upper beds.

The final step in high-amplitude salt anticline deformation involves a thicker salt section and considerable faulting of overlying beds (Fig. 316). Separation of overlying beds can occur as well.

Salt dome traps are important in several areas of the world. Among the most notable is the Gulf Coast of the United States. Here, traps of many types have produced hydrocarbons from Tertiary sediments deformed by salt dome activity (Figs. 317, 318 and 319). Inasmuch as gravity, sediment loading, and salt buoyancy are the important deformational factors, vertical tectonics are the rule.

Faulting and folding related to upward and lateral salt movement produce a variety of traps. Salt overhangs and reservoir beds impinging against impermeable salt form other types of traps. Most salt movement accompanies sedimentation. This permits development of multiple source and reservoir beds throughout the depositional history of the sediments and the deformational history of the salt structure.

Stratigraphic Traps

When a trap is formed by depositional and sedimentary factors, it is considered a stratigraphic trap. Here, the porosity, permeability, and geometry of the depositional feature control the dimensions of the trap. Deformation is not a primary consideration in stratigraphic traps, which exist because of their intrinsic depositional and geometric characteristics.

Sand Body Traps. Inasmuch as there are numerous types of finite sand bodies, many different trap types are possible. Some trap types are modifications of each other because of changes in depositional environments. However, there are several well-defined types of sand bodies that form traps in many places (Tables 22 and 23).

Channel sands form traps enclosed in shale in many locations in the mid-continent and Rocky Mountains (Figs. 320, 321 and 322). These deposits have limited lateral extent (Figs. 323 and 324) but may persist linearly for several miles of production.

Delta reservoir traps consist mainly of channels enclosed by floodplain and swamp deposits. Traps of this type are well-developed in the Gulf Coast. Delta deposits are common on divergent continental margins where rapid sediment progradation occurs (Fig. 326).

Beach or barrier bar traps are relatively narrow linear deposits that become traps when enclosed by shale (Figs. 328, 329 and 330). These occur in the geologic column of the Gulf Coast where they are being formed under present conditions as well.

Reef Traps. Reef traps are important producing features in many localities (Fig. 331). Porosity in reefs can be excellent and provide highly prospective reservoir characteristics. A reef enclosed in a shale source unit that provides a hydrocarbon seal is a highly desirable exploration target (Figs. 332, 333 and 334).

Changes in porosity and permeability can be caused by a variety of depositional and diagenetic factors. Sandstone loses its permeability as it grades laterally into shale (Fig. 335). Hard, finely crystalline limestone becomes significantly more permeable when dolomized.

A trap often develops where a sand loses permeability or pinches out entirely whether by non-deposition or erosion. Traps in dolomite are often sealed by up-dip contact with impermeable limestone. Wherever the permeability of a reservoir rock of any lithology becomes reduced up-dip, the possibility of a trap is manifest. This can occur as a result of change in facies, erosion, or non-deposition.

Combination Traps

A depositional sedimentary body that has been deformed forms a combination stratigraphic-structural trap if both stratigraphy and structure contribute to the trapping mechanism. Since there can be many combinations of structure and stratigraphy, many different combination traps are possible (Fig. 336).

The geometry of a channel deposit is linear. It can trap hydrocarbons laterally as it shales out into the associated floodplain. However, in the absence of a trapping mechanism the channel deposit can become a trap if faulted across its longitudinal dimension.

Folding a beach deposit will develop a trap in the same way as faulting a channel. The intrinsic limiting dimensions of a beach provide trapping in one direction. Folding the beach sand provides trapping in another dimension and effectively closes the feature.

At Hibernia, on the eastern shelf of Newfoundland, faulting and tilting deform Lower Cretaceous deltaic reservoir sediments (Fig. 337 and 338). Over 20,000 barrels of oil per day have been tested from this significant discovery.

Elimination of up-dip porosity by formation of asphalt, where an oil-bearing bed crops out, can form an asphalt seal trap. Tilting of the oil-bearing bed causes the oil to migrate up-dip to the outcrop where it is chemically altered to form the asphalt seal.

Unconformity Traps

An unconformity represents a period of erosion or non-deposition. Several types of unconformities are possible. If they involve the proper combination of source, reservoir and sealing beds, they can be productive of oil and gas.

Angularity in an unconformable situation occurs when a sequence of horizontal beds is tilted, eroded, and subsequently overlain by younger flat-lying beds (Figs. 339, 340 and 341). Tilted permeable beds surrounded by conformable and tilted impermeable beds and angularly overlain by flat-lying impermeable beds, have potential as reservoirs. Hydrocarbons in the reservoirs move up-dip until they can move no farther against the flat-lying impermeable beds above the unconformable surface.

Deformation is responsible for the tilting of the reservoir beds in an angular unconformity. However, without the overlying sealing beds, there would be no trap.

A disconformity is an unconformity which has no angular relationships between the flat-lying beds below and above the erosional surface. The surface of the unconformity can be irregular, however. Trapping is accomplished by differences in permeability with sealing beds above the erosional surface, which may represent paleotopography (Figs. 342, 343 and 344).

Disconformities are often so flat that it is difficult to determine that they exist without the help of paleontology.

Sediments overlying a crystalline igneous or metamorphic terrain create a nonconformity (Fig. 345). Sometimes the crystalline terrain is sufficiently weathered or fractured to provide permeability for hydrocarbon accumulation against the overlying sediments.

Structural Geology and Petroleum

It is evident from the foregoing that structural geology has direct application to how hydrocarbons migrate and accumulate and the types of structures and traps into which they move. Temporal effects established the timing of structural events and add a fourth dimension of kinetic development of geologic processes. Inasmuch as geologic processes are spatially three-dimensional, their analyses must be similarly accomplished as they occur. Therefore, structural geology is important in the total evaluation of any exploration province because depositional and deformational histories are so intimately related.

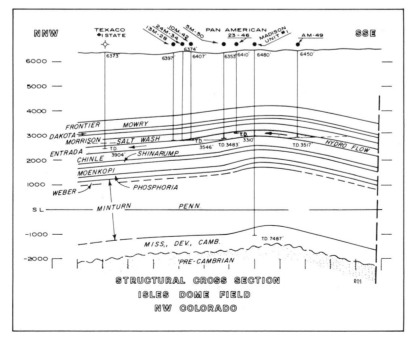

Fig. 294. Anticlinal, Isles Dome Field, Colorado. From Stone, 1975. Permission to publish by the Rocky Mountain Association of Geologists.

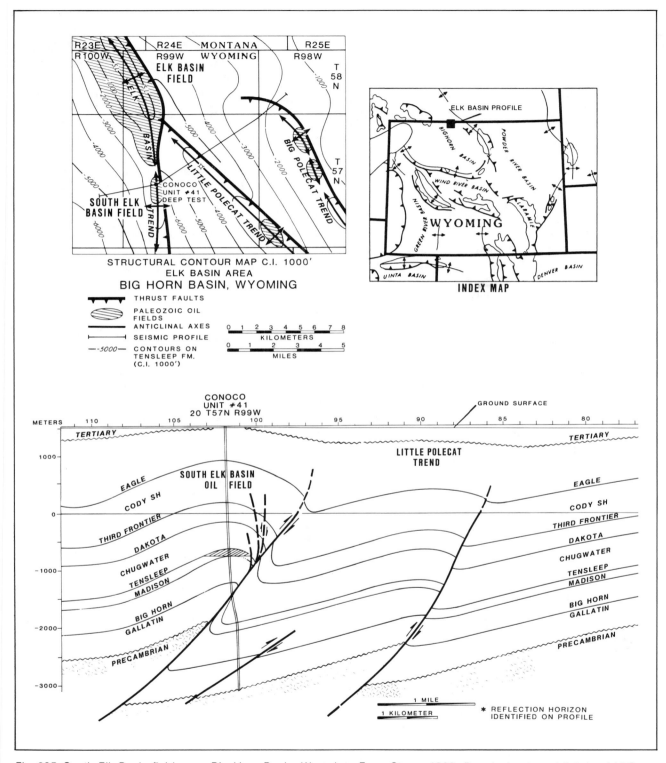

Fig. 295. South Elk Basin field area, Big Horn Basin, Wyoming. From Stone, 1983. Permission to publish by AAPG.

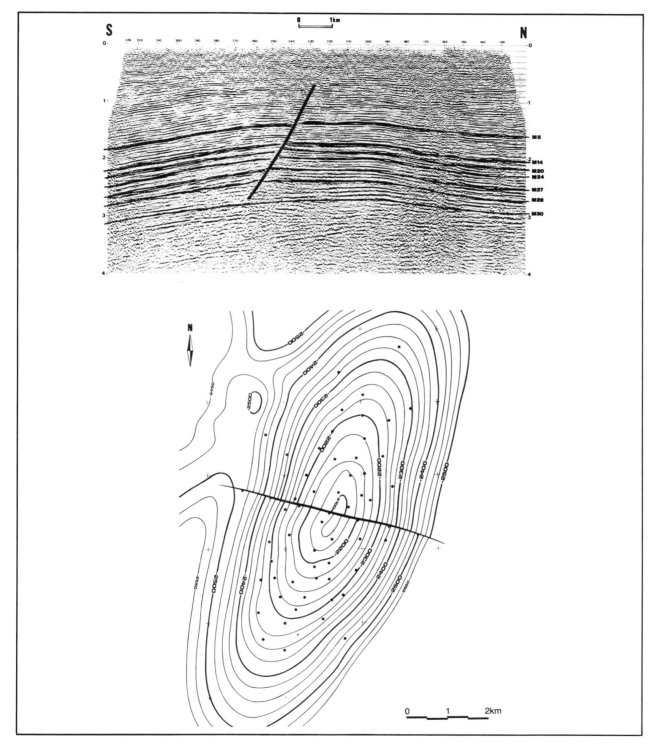

Fig. 296. Structure map and South-North seismic section of faulted, symmetrical anticline, Mandil Field, E. Kalimantan, In-donesia. From Verdier, et al., 1980. Permission to publish by AAPG.

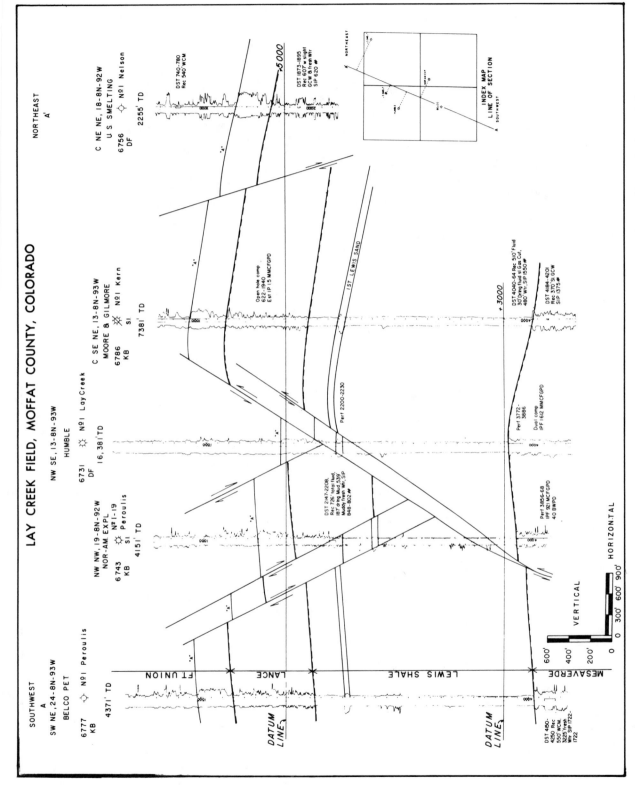

Fig. 297. Faulted anticline, Lay Creek Field, Colorado. From Duffy, 1975. Permission to publish by the Rocky Mountain Association of Geologists.

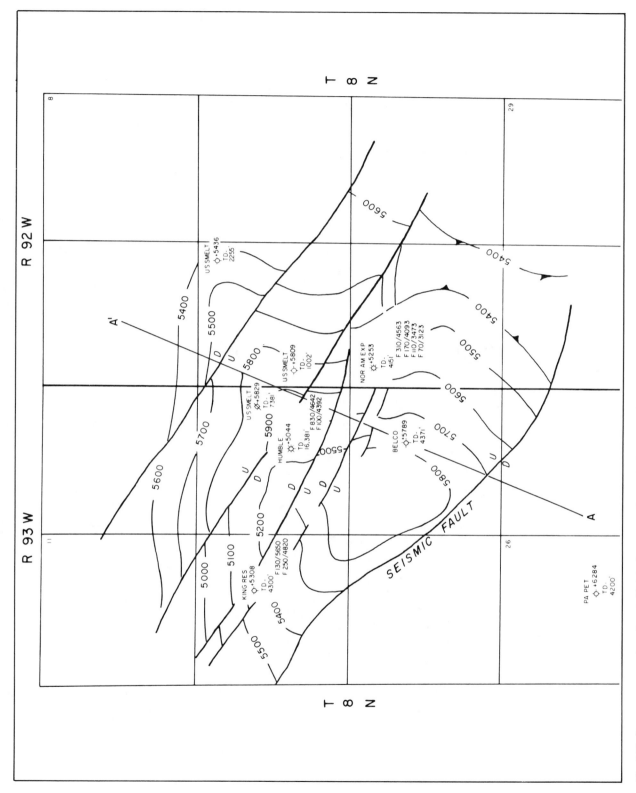

Fig. 298. Structure map, Lay Creek Field, Colorado. From Duffy, 1975. Permission to publish by the Rocky Mountain Association of Geologists.

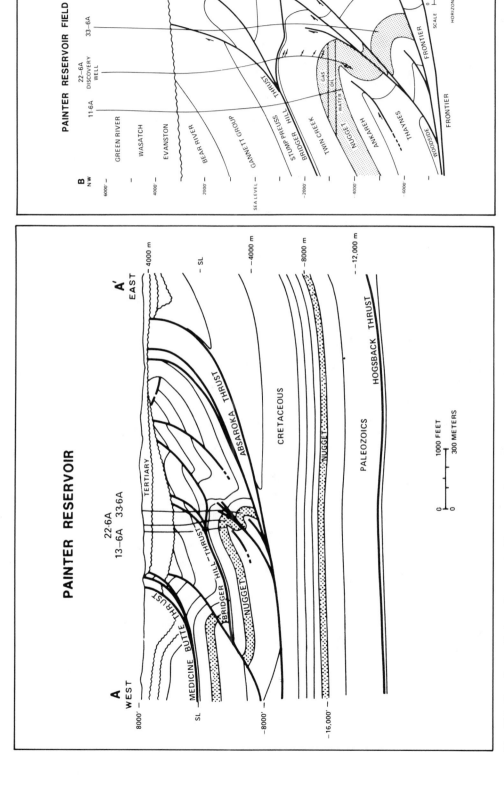

Fig. 299. Overthrust, Painter Reservoir, Wyoming. From Lamb, 1980. Permission to publish by AAPG. | *Fig. 300. Overthrust, painter Reservoir, Wyoming. From Lamb, 1980. Permission to publish by AAPG.*

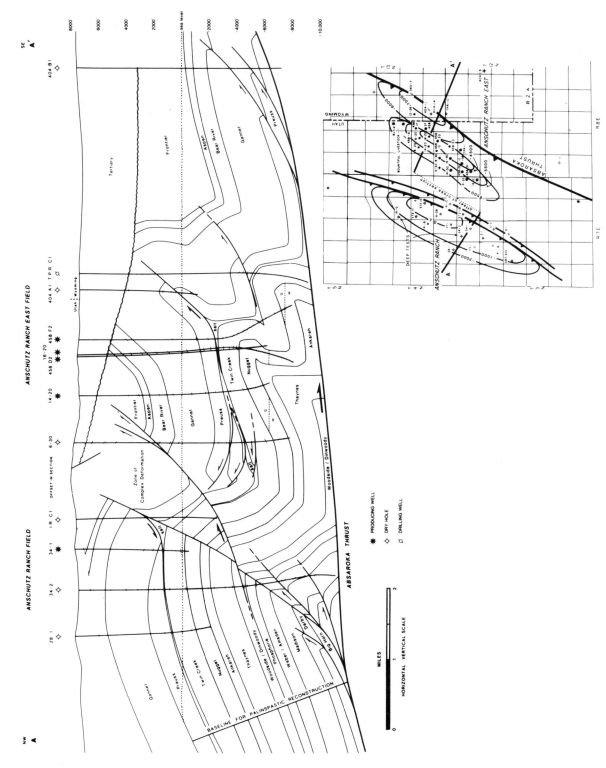

Fig. 301. Anschutz Ranch and Anschutz Ranch East Fields Overthrust Belt, Utah and Wyoming. From West and Lewis, 1982. Permission to publish by RMAG.

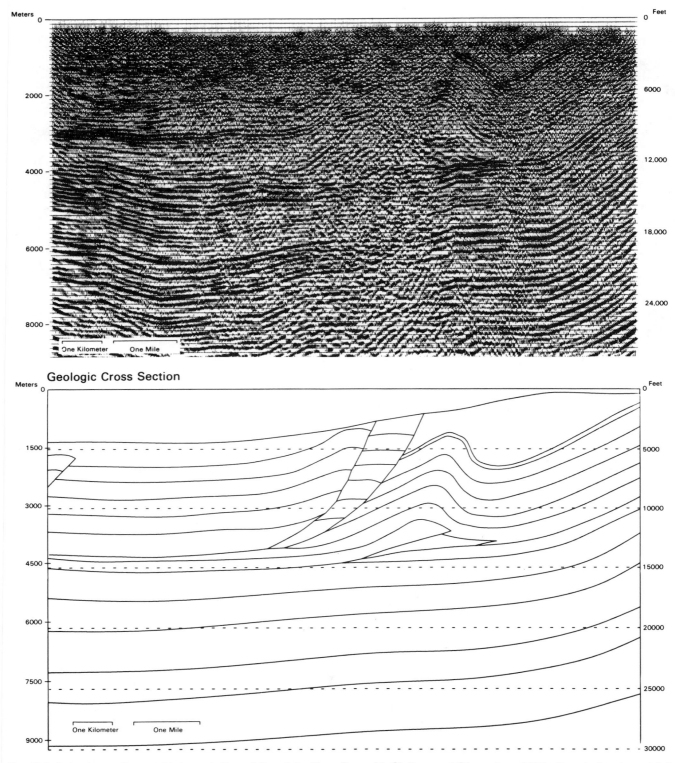

Fig. 302. Seismic section and interpretation of thrust faulting. From McClellan and Storrusten, 1983. Permission to publish by AAPG.

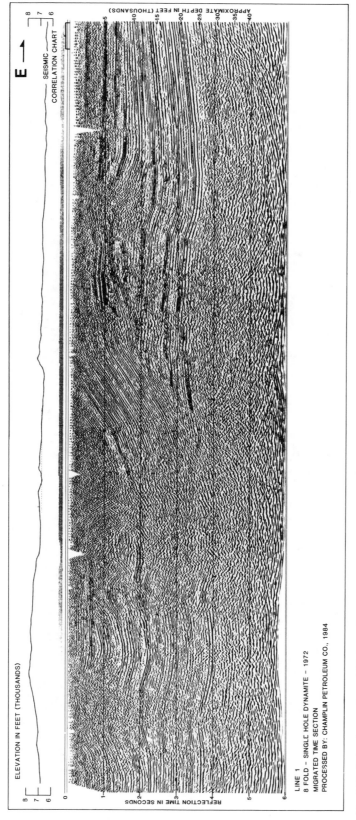

Fig. 303. Seismic section of Whitney Canyon and Ryckman Creek field overthrust belt, Utah and Wyoming. From Williams and Dixon, 1985. Permission to publish by RMAG.

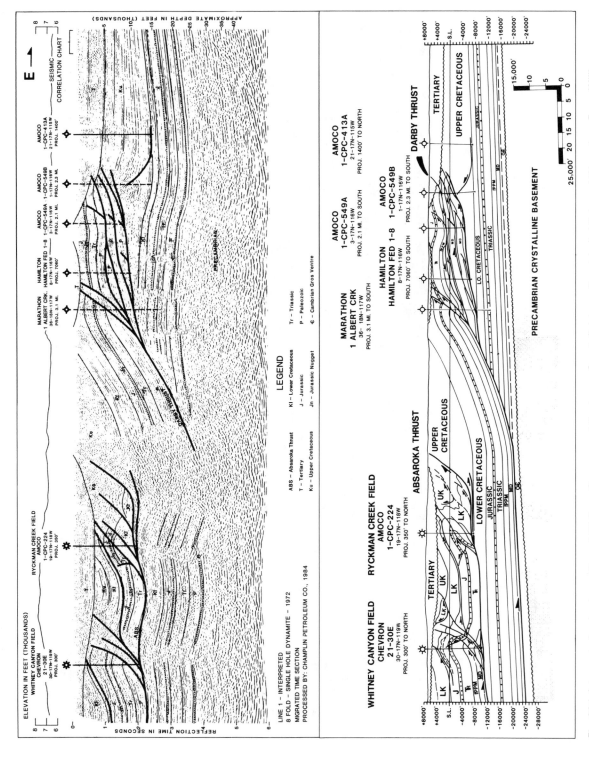

Fig. 304. Interpreted seismic section of Whitney Canyon and Ryckman Creek field overthrust belt, Utah and Wyoming. From Williams and Dixon, 1985. Permission to publish by RMAG.

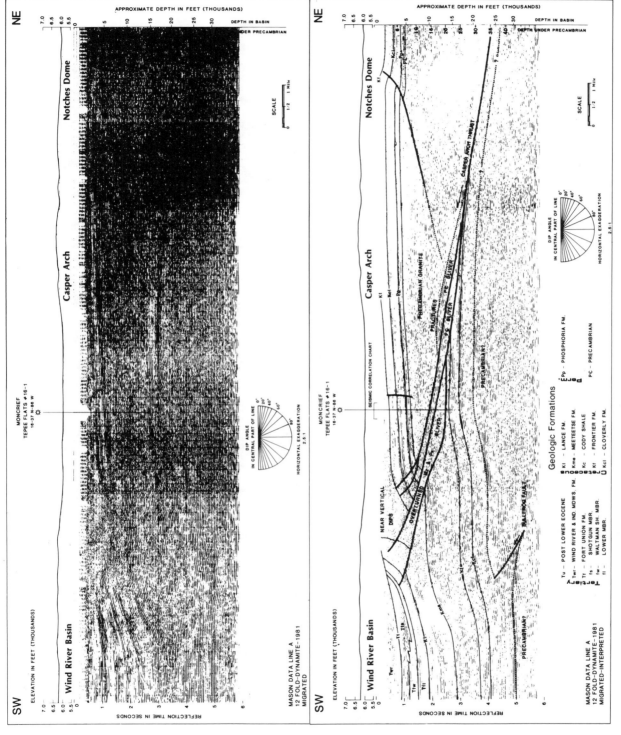

Fig. 305. Seismic section of overthrust crystalline basement in foreland compressive block in Wyoming. From Ray and Berg, 1985. Permission to publish by RMAG and Denver Geophysical Society.

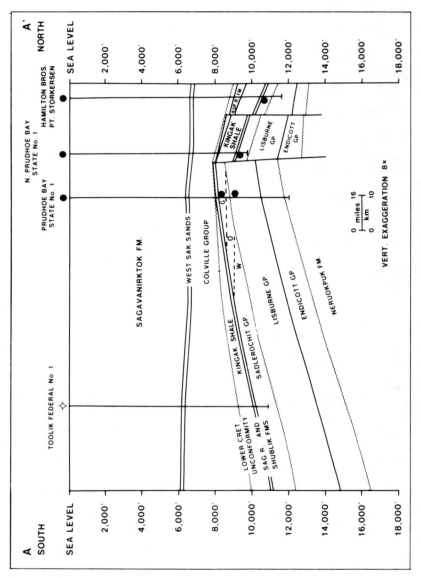

Fig. 306. North-South cross section of Prudhoe Bay, Alaska illustrating production from up and down thrown fault sealed blocks. From Jamison, et al., 1980. Permission to publish by AAPG.

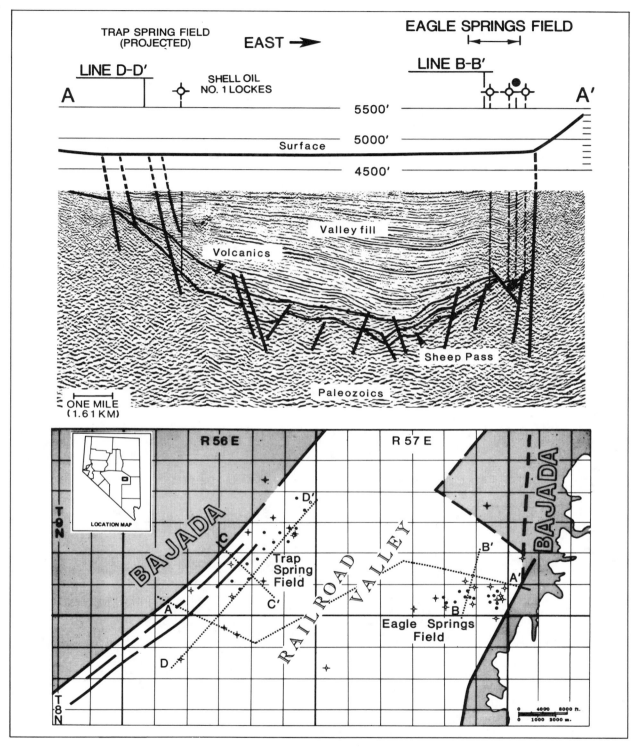

Fig. 307. Railroad Valley, Nevada. Upper: Interpretation of seismic section A-A', migrated. Lower: Location of seismic sections. From Vreeland and Berrong, 1979. Permission to publish by RMAG and Utah Geological Association.

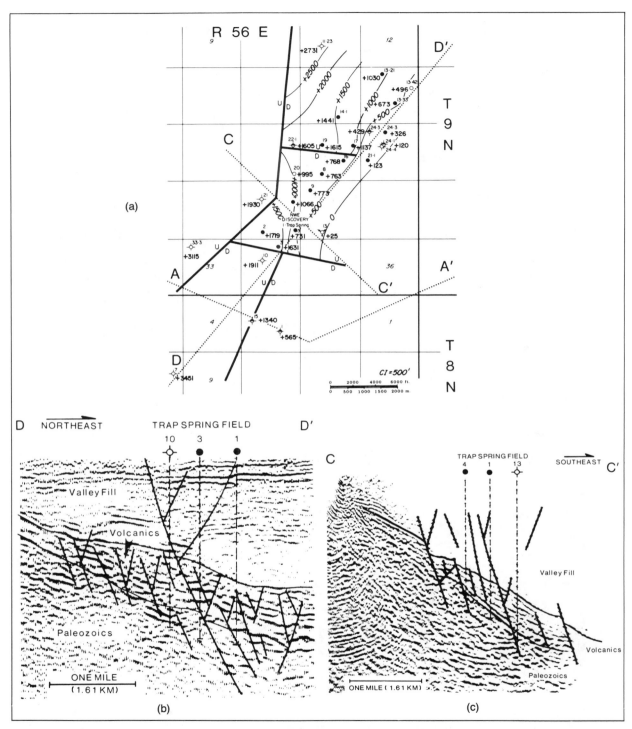

Fig. 308. Trap Spring Field. (a): Generalized structure map showing approximate location of seismic sections. Datum is the top of the volcanics. (b): Northeast, southwest seismic section. From Vreeland and Berrong, 1979. Permission to publish by RMAG and Utah Geological Association.

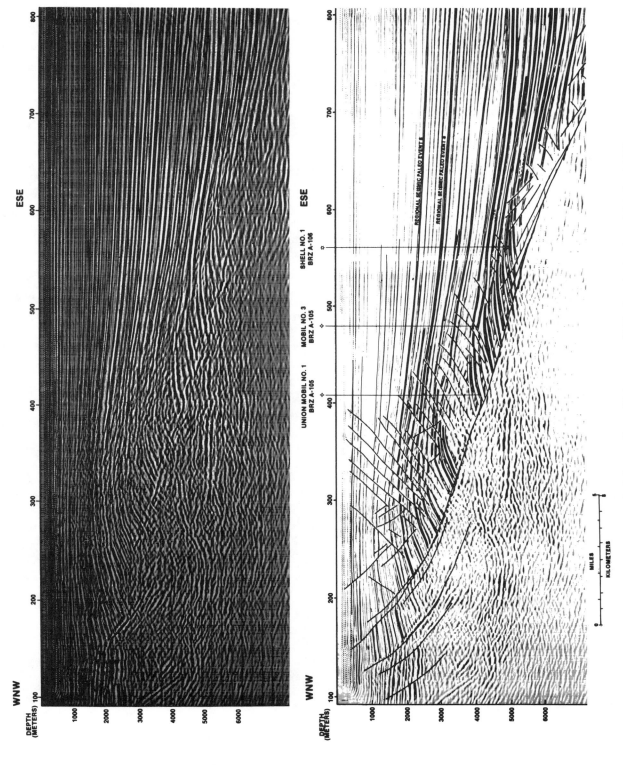

Fig. 309. Complex growth fault blocks, Texas Gulf Coast. From Christensen, 1983. Permission to publish by AAPG.

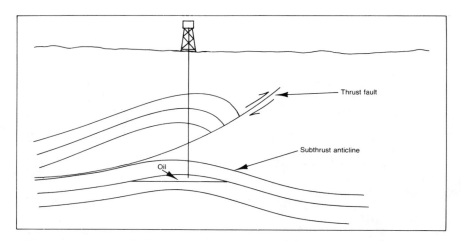

Fig. 310. Subthrust trap, Overthrust Belt, Wyoming

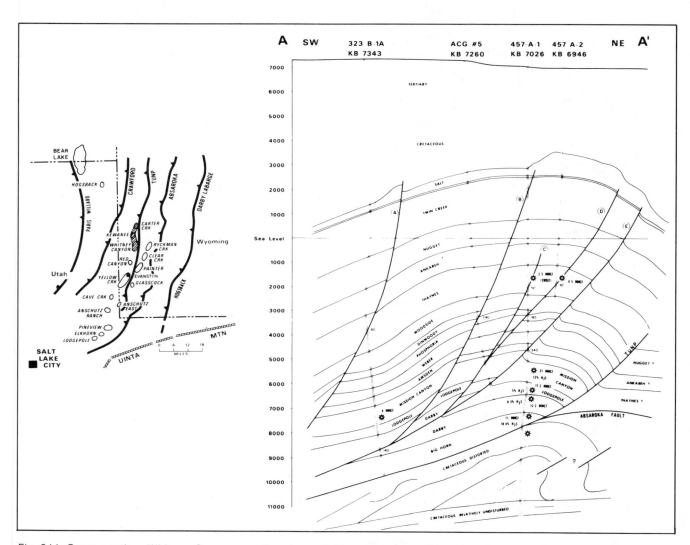

Fig. 311. Cross-section, Whitney Canyon overthrust gas field, southwestern Wyoming Overthrust Belt. From Bishop, 1982. Permission to publish by RMAG.

240

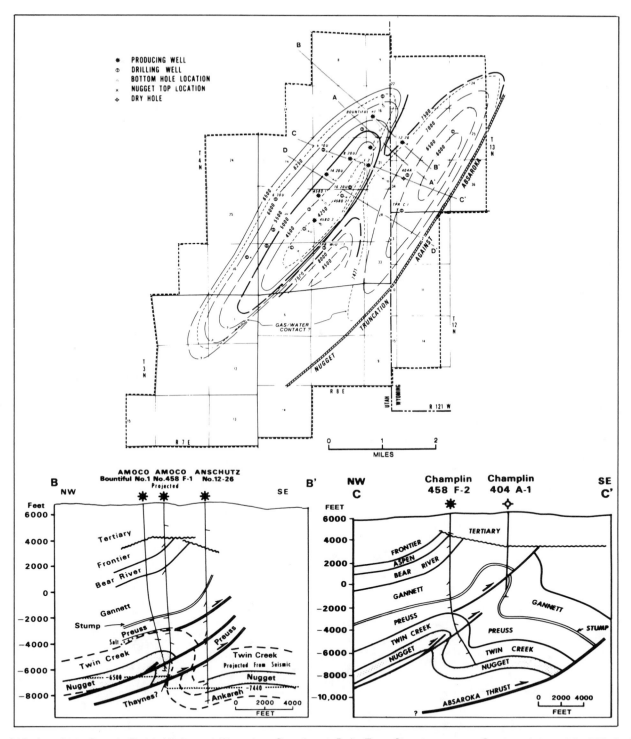

Fig. 312. Anschutz Ranch Field, Utah and Wyoming Overthrust Belt. Top: Structure map. Contour interval is 500 ft. Left: Geologic cross section B-B' showing subthrust production. Right: Geologic cross section C-C'. From Lelek, 1982. Permission to publish by RMAG.

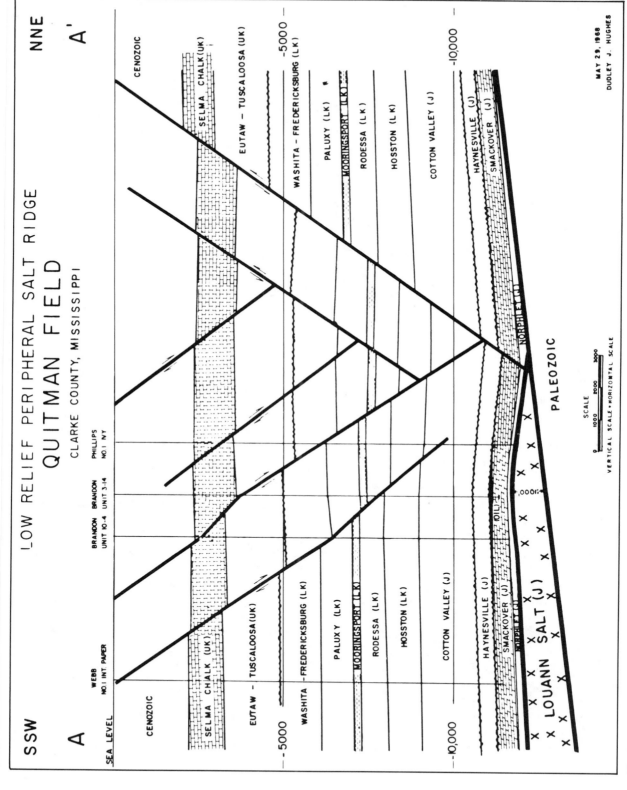

Fig. 313. Quitman Field, low relief salt ridge, Mississippi. From Hughes, 1968. Permission to publish by the Gulf Coast Association of Geological Societies.

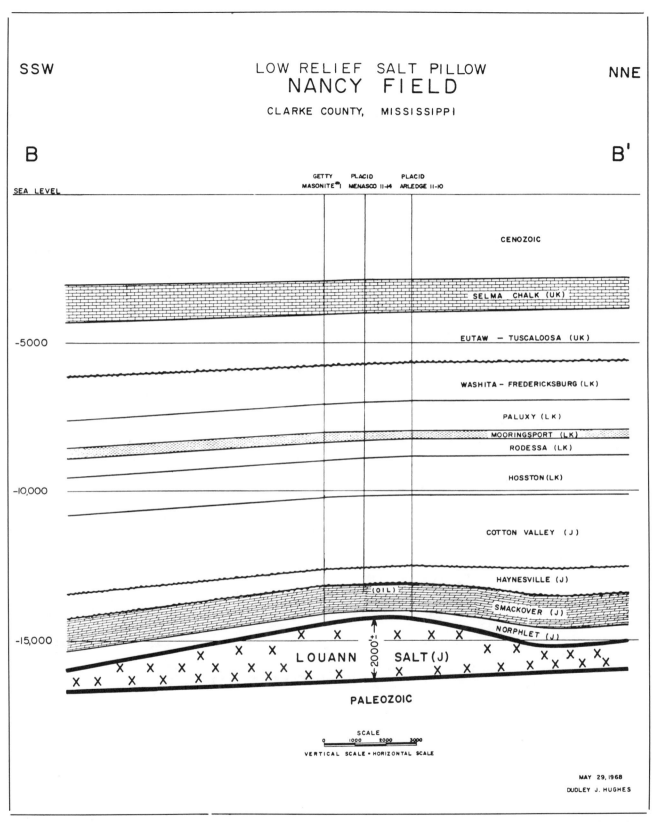

Fig. 314. Nancy Field, low relief salt pillow, Mississippi. From Hughes, 1968. Permission to publish by the Gulf Coast Association of Geological Societies.

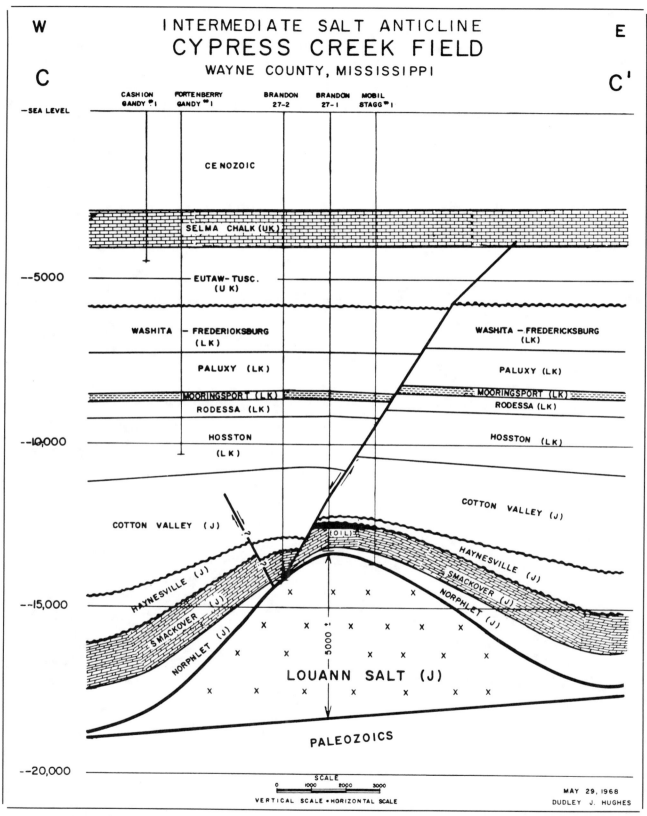

Fig. 315. Cypress Creek Field, intermediate salt anticline, Mississippi. From Hughes, 1968. Permission to publish by the Gulf Coast Association of Geological Societies.

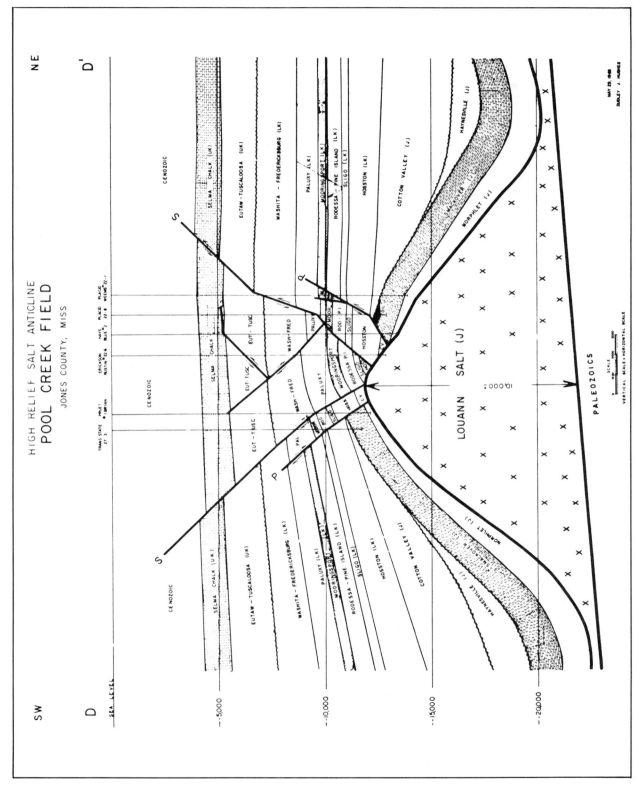

Fig. 316. Pool Creek Field, high relief salt anticline, Mississippi. From Hughes, 1968. Permission to publish by the Gulf Coast Association of Geological Societies.

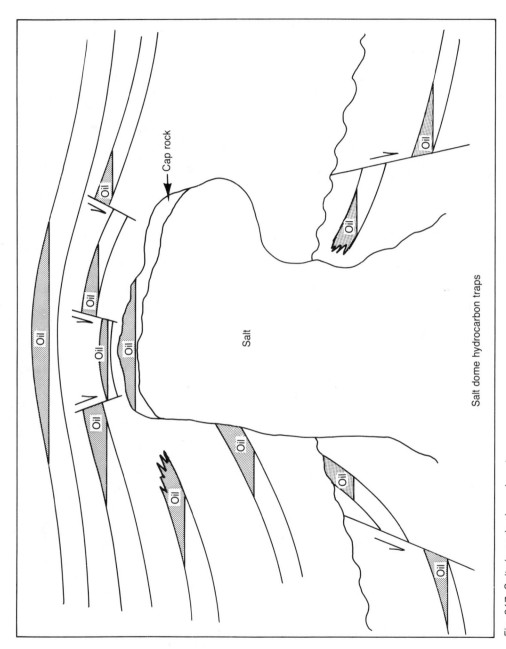

Fig. 317. Salt dome hydrocarbon traps

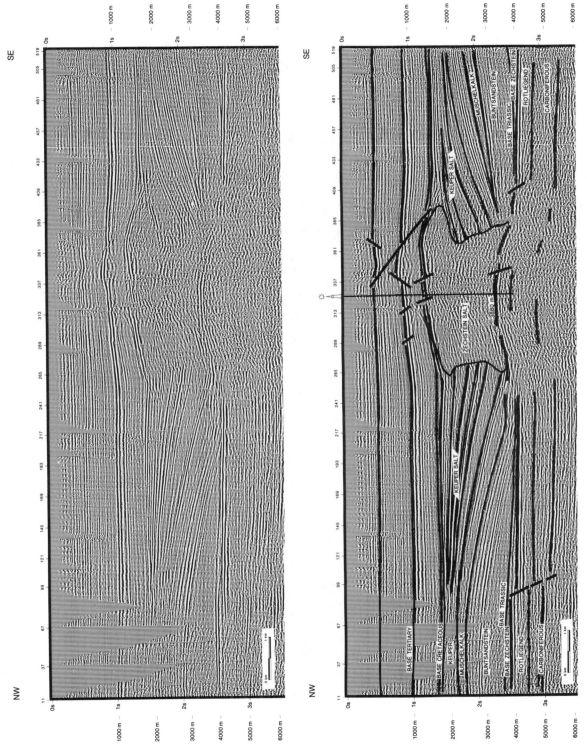

Fig. 318. Salt plug, Siegelsum, northwest Germany. From Lukic, et al., 1983. Permission to publish by AAPG.

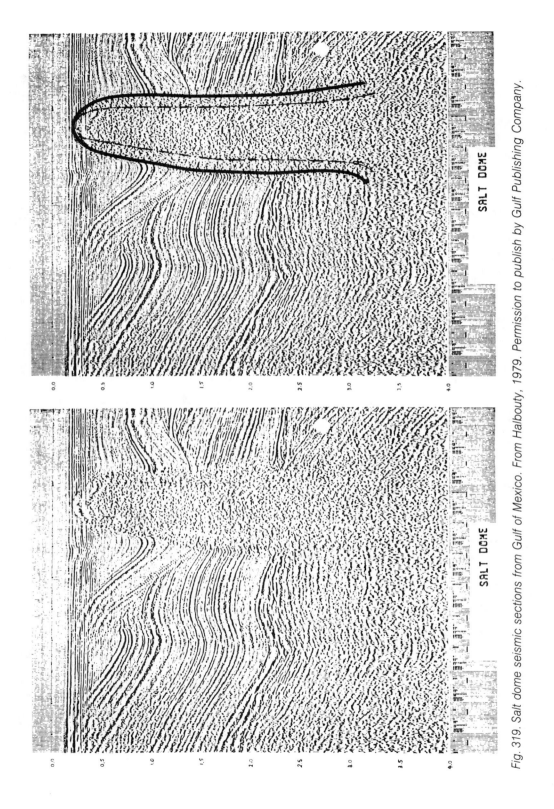

Table 22
Characteristics of Sediment Bodies

SEDIMENT BODY	GENERAL LITHOLOGY	THICK-NESS (FEET)	SHAPE, HORIZONTAL DIMENSIONS	DISTRIBUTION, TREND	RELATIONSHIP TO ADJACENT OR ENCLOSING FACIES	LITHOLOGY, COMPOSITION, TEXTURE, FAUNA	BOUNDING CONTACTS	OVERALL VERTICAL GRAIN-SIZE CHANGE	STRATIFICATION	CROSS-STRATIFICATION: CONTACTS, SET THICKNESSES	NATURE OF LAMINAE	SHAPE OF SETS	RIPPLES	DEFORMATIONAL AND ORGANIC SEDIMENTARY STRUCTURES	MISCELLANEOUS REMARKS	REFERENCES
SUBAERIAL MIGRATED DUNE SANDS	sands, no muds; homogeneous	10 to 1000	elongate, or sheets up to 1000's of sq. miles in area	downwind from source of sand	commonly the end stage of a regressive sequence	well sorted sands, pebbles & clasts rare	variable	not systematic	conspicuous high-angle cross bedding	erosional, horiz. or sloping; sets up to 10's of feet thick	lee dips 25°-34°, commonly tangential to lower boundary	tabular; sometimes enormous troughs	high indices; crests often parallel to dip of lee beds	slumps not uncommon; vertebrate tracks		McKee (1966)
ALLUVIAL SANDS (DELTAIC DEPOSITS)	sands, muds, some gravels	usually 30-80, sometimes 200-300	continuous bodies, usually ½ to 5 mi. wide, 10's to 100's of miles long	make large angles with shoreline trends	lower contacts erosional; lateral contacts erosional or indeterminate	pebbles and clasts common, proportion of mud variable	base erosional, top usually transitional	upward decrease	many beds lenticular; abundant cross bedding	erosional, planar or concave up, usually ½-2 feet thick	maximum dips usually 20°-25°, inclined or tangential to lower boundary	trough	short-crested; linguoid; microtrough in cross-section; abundant	slumps common; burrows uncommon		Harms (1966), Hewitt & Morgan (1965), Fisk (1944), Potter (1967)
DISTRIBUTARY CHANNEL FILLS (DELTAIC DEPOSITS)	sands, muds	up to 200	continuous sinuous bodies, usually ~1 mi. wide		commonly enclosed in nonmarine or blackish muds									slumps, burrows not uncommon		Frazier (1967), Brown (1969)
DELTA-FRONT SHEET SANDS (DELTAIC DEPOSITS)	sands	20-80	sheets		underlain by marine pro-delta muds; overlain by marsh muds	similar to barrier sand bodies →								↑		
REWORKED TRANSGRESSIVE SANDS (SHALLOW MARINE SANDS)	sands	1-40	sheets		underlain by or adjacent to deltaic deposits	well sorted, may contain coarse sand lag	both sharp	not systematic	high-angle cross bedding; orientation diverse	erosional, planar; ½-2 feet thick	maximum dips 20°-25°, tangential to lower boundary	wedge or tabular	not conspicuous, microtroughs present	slumps, burrows uncommon	lateral facies changes may provide proximity indicators	Frazier (1967), MacKenzie (1965)
REGRESSIVE SHORELINE SANDS (BARRIER-ISLAND SANDS) (SHALLOW MARINE SANDS)	sands, rare muds	20-60	elongate or sheets, up to miles wide, 10's of miles long	parallel to shoreline where elongate	transitional downward and seaward into muds, landward into lagoonal or deltaic deposits	well sorted, pebbles & clasts rare, marine fauna	base transitional, top sharp	upward increase, but middle may have coarsest beds	upper & lower: subhorizontal stratification with low-angle truncations, esp. near base; sets < 1 foot thick. middle: high-angle cross bedding, ½-2 feet thick, tangential laminae; cross-laminae dip obliquely shoreward; local scours				most abundant near base; symm., long-crested	load structures & burrows common at base; uncommon		Bernard et al (1962), Weimer (1966), McCubbin & Brady (1969)
OFFSHORE BARS (SHALLOW MARINE SANDS)	sands with mud partings	several to 10's	elliptical lenses, less than a few sq. miles in size	scattered, orientation variable	enclosed in and intertongues laterally with marine muds and silts	pebbles, clasts, glauconite, phosphate, marine fauna	sharp, or narrowly transitional	not systematic	low-angle cross bedding	erosional, planar; ~1 ft. thick	most dips < 10°; laminae parallel to lower set boundary	wedge?	common, some symm., long-crested	burrows abundant only in marginal facies	gradual outward decrease in sand/clay may provide proximity indicator	Exum & Harms (1968)
STRIKE-VALLEY SANDS (SHALLOW MARINE SANDS)	fine to coarse sands and muds, heterogeneous	10-50	elongate, up to several miles wide, 10's of miles long	parallel to pre-unconformity paleo strike	fills erosional strike valleys; intertongues with marine muds seaward; onlaps landward				tabular units with high-angle cross bedding dipping parallel to sand body elongation	erosional, planar; ½-5 feet thick	max. dips 25°-30°, tangential to lower boundary	tabular; sets straight & continuous for 100's of feet	common locally, esp. at toes of x-sets; some are long-crested wave ripples	burrowing common	paleogeologic and paleotopographic maps effective in exploration	McCubbin (1969)
PROXIMAL TURBIDITES (DEEP-WATER SANDS)	interbedded sands, silts and muds	100's to 1000's	fans or sheets up to 1000's of sq. miles in area	high flanks of deep basins near sand source	may be middle part of regressive sequence from deep to shallow-water deposits	graded bdg; displaced shallow-water fauna; proximal turbidites	variable	variable	parallel stratified or structureless; may have large mud-lined scours			trough-shaped sets found rarely	asymmetric ripples, both short and long-crested; found at tops of individual beds	burrows uncommon, bedding uncommon		Walker (1966, 1967)
DISTAL TURBIDITES (DEEP-WATER SANDS)	interbedded sands, silts and muds	100's to 1000's	sheets up to 1000's of sq. miles in area	sumps of deep basins	sands interbedded with deep-water muds	marine fauna, if any			1-3' continuous beds, parallel or ripple stratified			trough-shaped sets found rarely		plane tracks and trails often present	cf. distal beds, proximal beds are thicker, coarser grained, less well graded, less regular, more deformed, and more porous and permeable	Walker (1966, 1967)

From MacKenzie, 1972.

249

Table 23
Selected Giant Primary Sandstone Stratigraphic Traps

Name	Location	Depth (ft)	Age	Year of Discovery	Ultimate Recoverable Reserves (million bbl)	Depositional Type
Greater Red Wash	Utah, U.S.A.	5500	Eocene	1951	135	Lacustrine delta
Coalinga East	California, U.S.A.	8000	Eocene	1938	520	Shallow marine(?)
Pembina	Alberta, Canada	5000	L. Cret.	1953	1800	Shallow marine
Bell Creek	Montana, U.S.A.	4500	E. Cret.	1967	114	Barrier bar and Delta front Barrier bar
Cut Bank	Montana, U.S.A.	3000	E. Cret.	1926	200	Alluvial valley
Burbank	Oklahoma, U.S.A.	3000	Penn.	1920	500	Shoreline Alluvial valley
Bradford (partly structural)	Pennsylvania, U.S.A.	1500	L. Dev.	1871	660	Shallow marine Turbidite
Mitsue	Alberta, Canada	5700	M. Dev.	1964	300	Deltaic
Nipisi	Alberta, Canada	5500	M. Dev.	1965	200	Deltaic (?)

From MacKenzie, 1972.

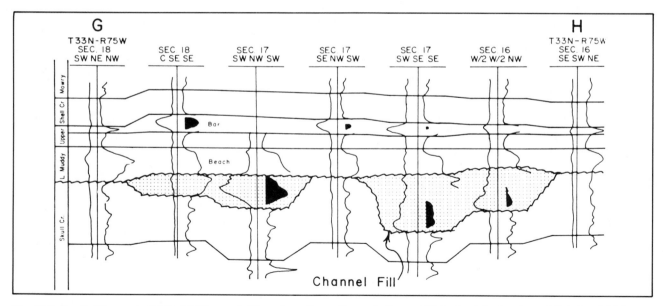

Fig. 320. South Glenrock Oil Field Channel, Wyoming. From Curry & Curry, 1972. Permission to publish by AAPG.

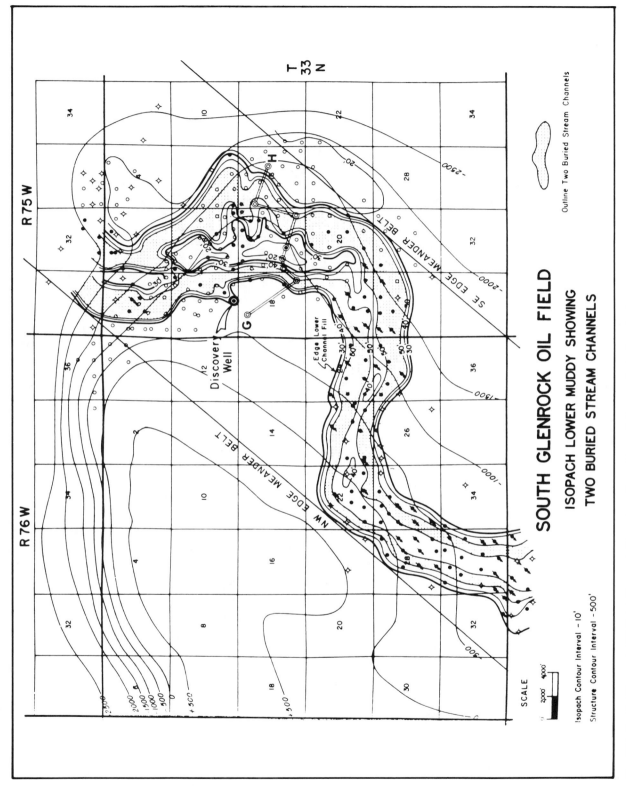

Fig. 321. South Glenrock Oil Field Channel, Wyoming. From Curry & Curry, 1972. Permission to publish by AAPG.

Time Section

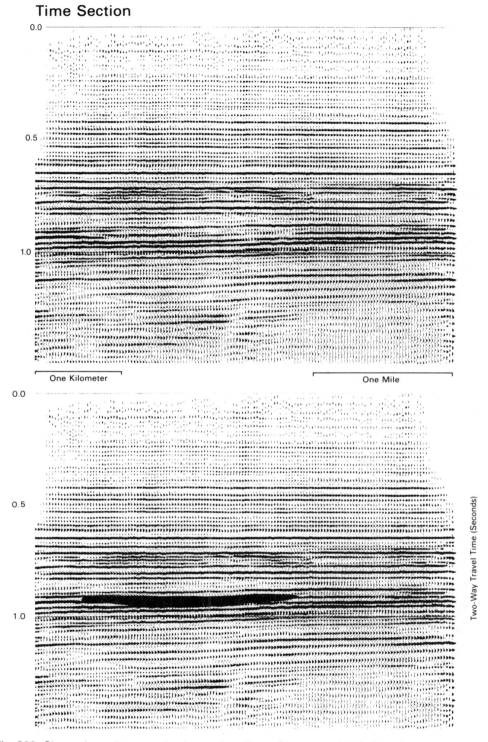

Fig. 322. Channel sandstone seismic section. From Chapman, 1983. Permission to publish by AAPG.

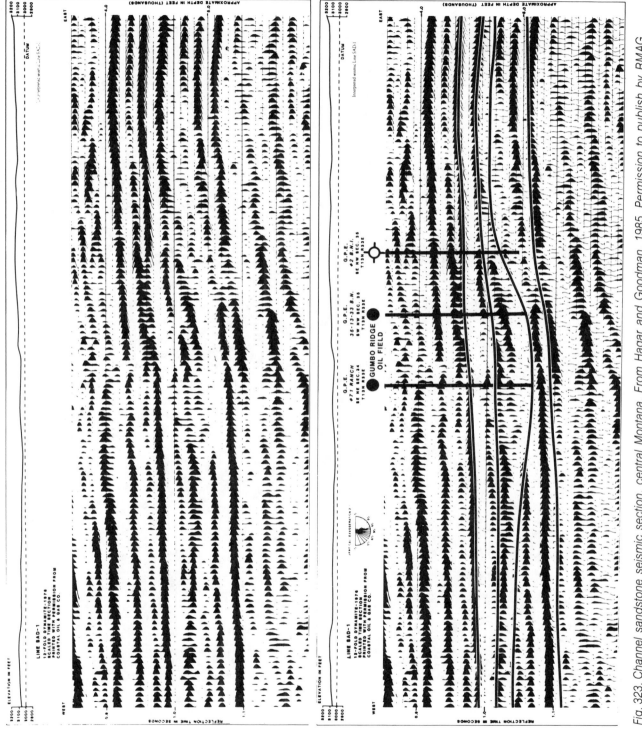

Fig. 323. Channel sandstone seismic section, central Montana. From Hagar and Goodman, 1985. Permission to publish by RMAG and Denver Geophysical Society.

WEST

EAST

GUMBO RIDGE

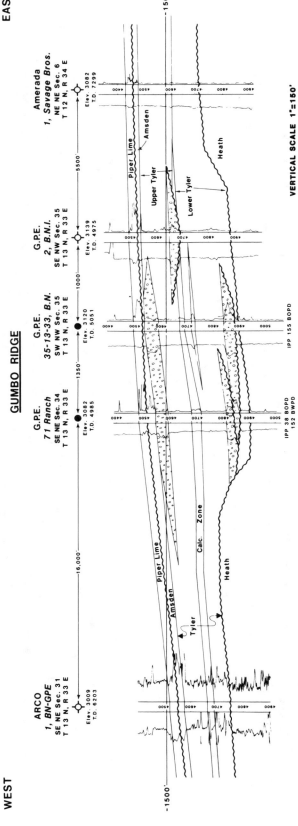

Fig. 324. Electric log section of channel sandstone system, central Montana. From Hagar and Goodman, 1985. Permission to publish by RMAG and Denver Geophysical Society.

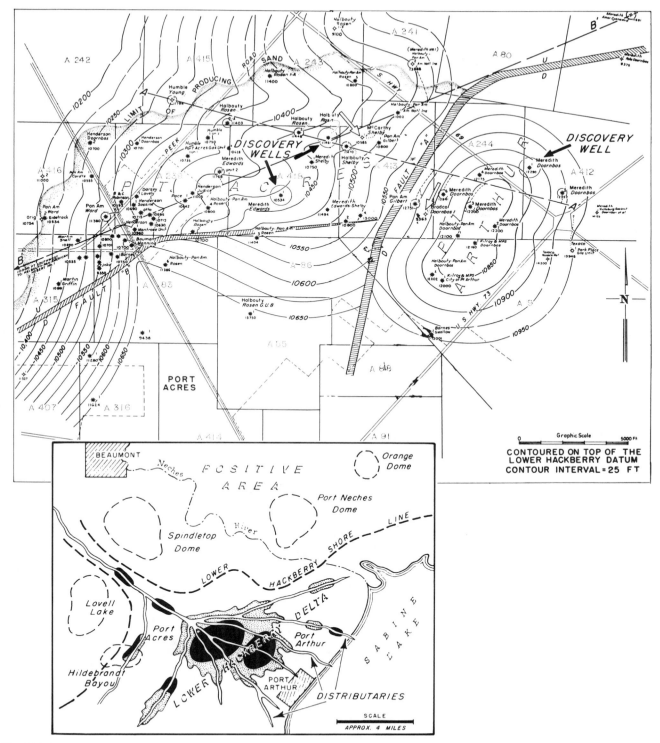

Fig. 325. Index map and structure map of Port Acres, Port Arthur Lower Hackberry delta fields. Halbouty and Barber, 1972. Permission to publish by AAPG.

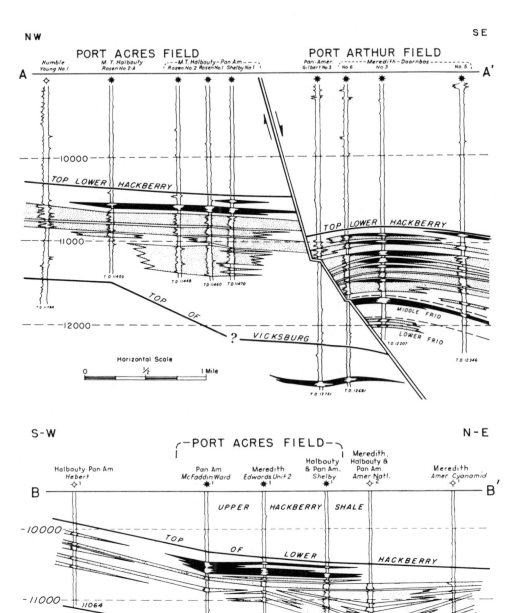

Fig. 326. Structure cross-sections across Port Acres, Port Arthur Lower Hackberry delta fields. Halbouty and Barber, 1972. Permission to publish by AAPG.

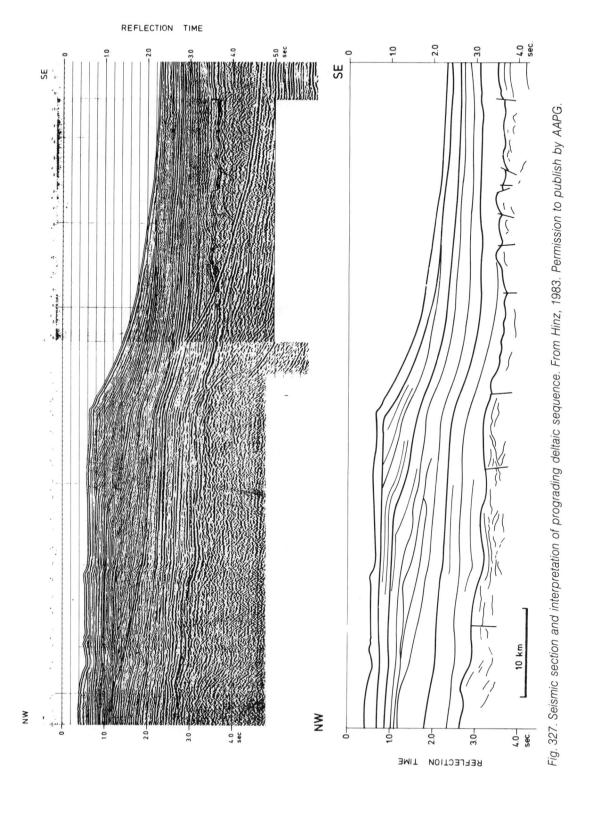

Fig. 327. Seismic section and interpretation of prograding deltaic sequence. From Hinz, 1983. Permission to publish by AAPG.

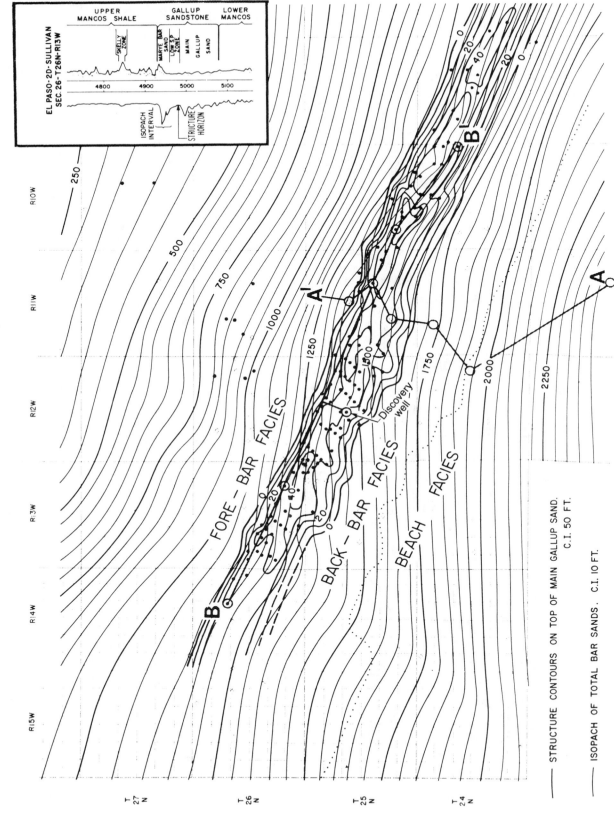

Fig. 328. Bisti Bar Oil Field, New Mexico. From Sabins, 1972. Permission to publish by AAPG.

STRUCTURE CONTOURS ON TOP OF MAIN GALLUP SAND. C.I. 50 FT.

ISOPACH OF TOTAL BAR SANDS. C.I. 10 FT.

258

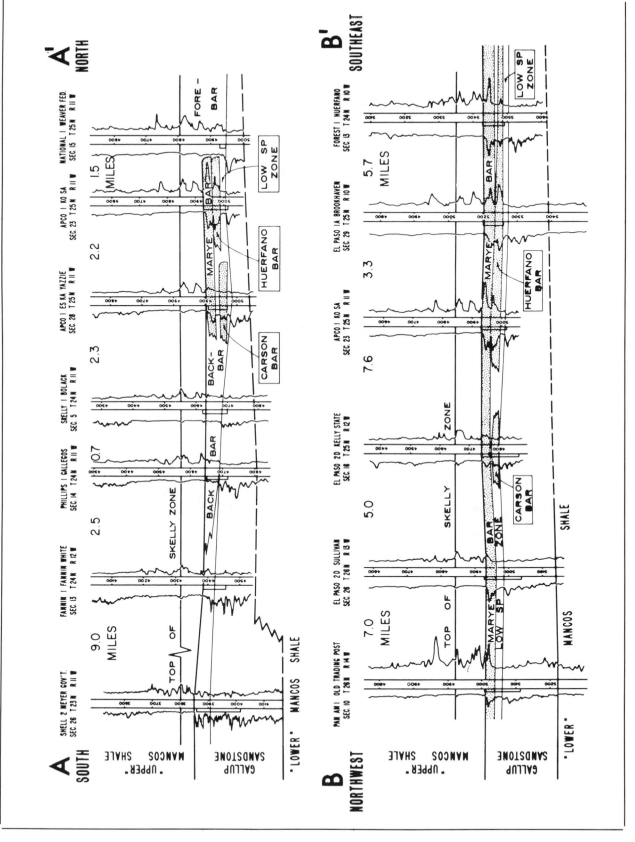

Fig. 329. Bisti Bar Oil Field, New Mexico. From Sabins, 1972. Permission to publish by AAPG.

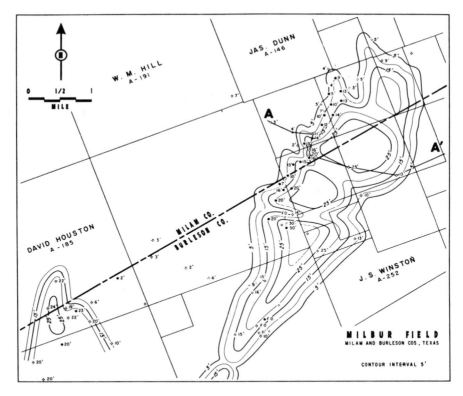

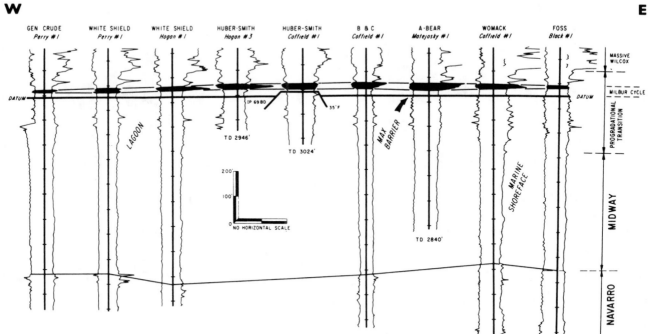

Fig. 330. Top: Isopach map of "Milbur" barrier bar sandstone, Milbur field, Texas. Bottom: Electric-log stratigraphic cross-section A-A', Milbur field area. From Chuber, 1972. Permission to publish by AAPG.

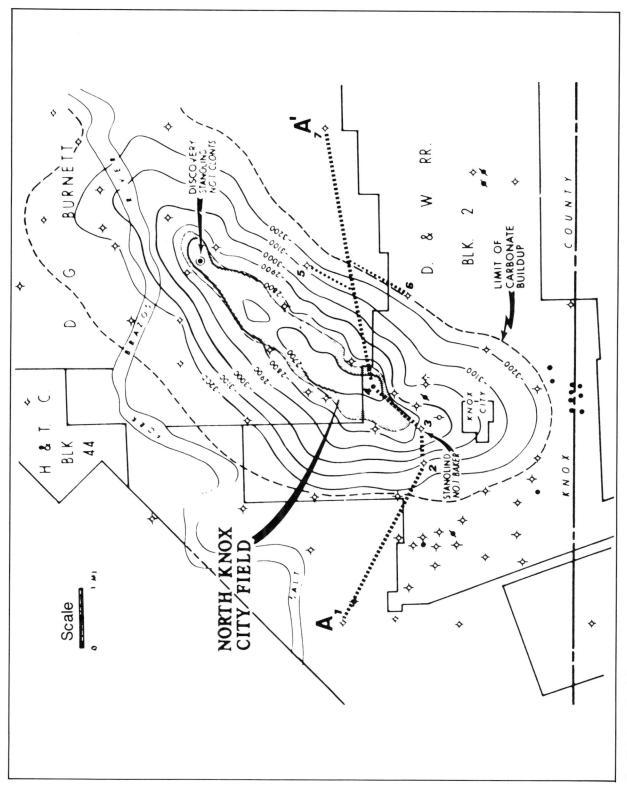

Fig. 331. Knox Reef Oil Field, Texas. From Harwell and Rector, 1972. Permission to publish by AAPG.

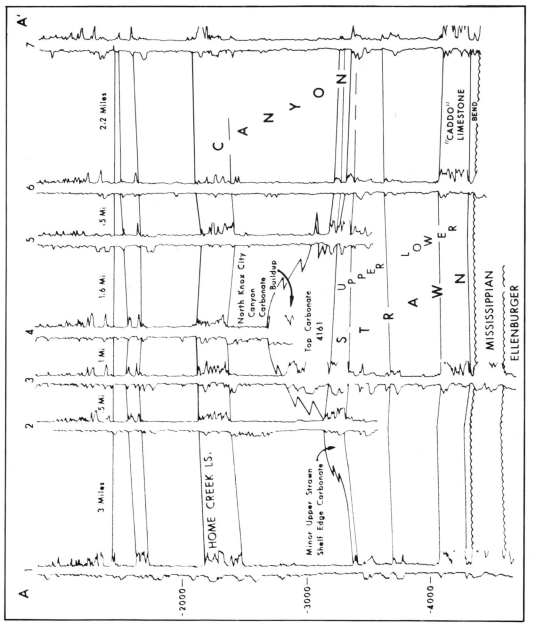

Fig. 332. Knox Reef Oil Field, Texas. From Harwell and Rector, 1972. Permission to publish by AAPG.

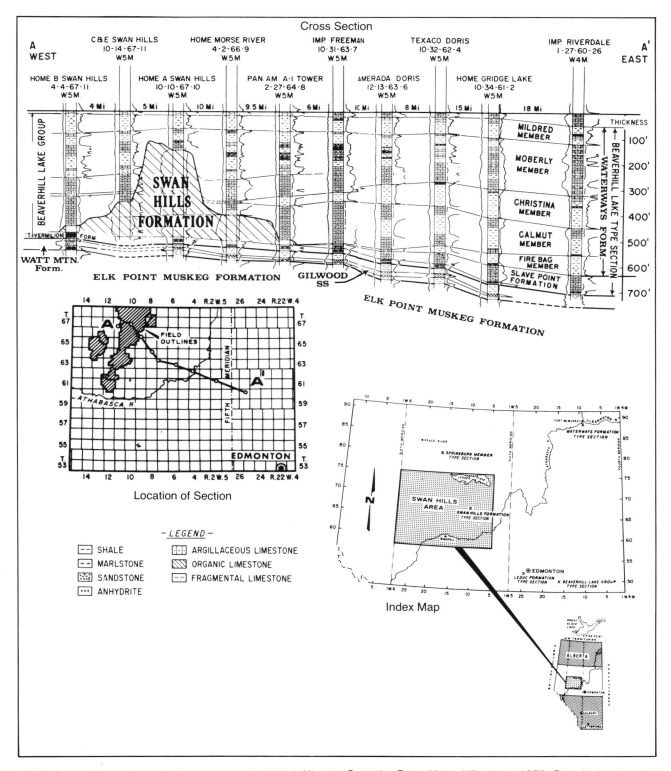

Fig. 333. Swan Hills reef producing area, west central Alberta, Canada. From Hemphill, et al., 1970. Permission to publish by AAPG.

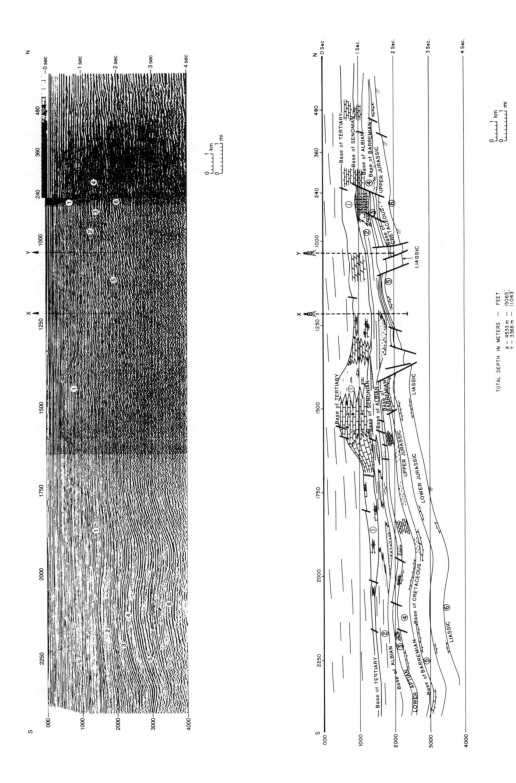

Fig. 334. Reef complex offshore Mesapotamia basin, France. Top: Seismic section of reef complex. Bottom: Interpretation of reef complex seismic section. From Cunnelle and Marco, 1983. Permission to publish by AAPG.

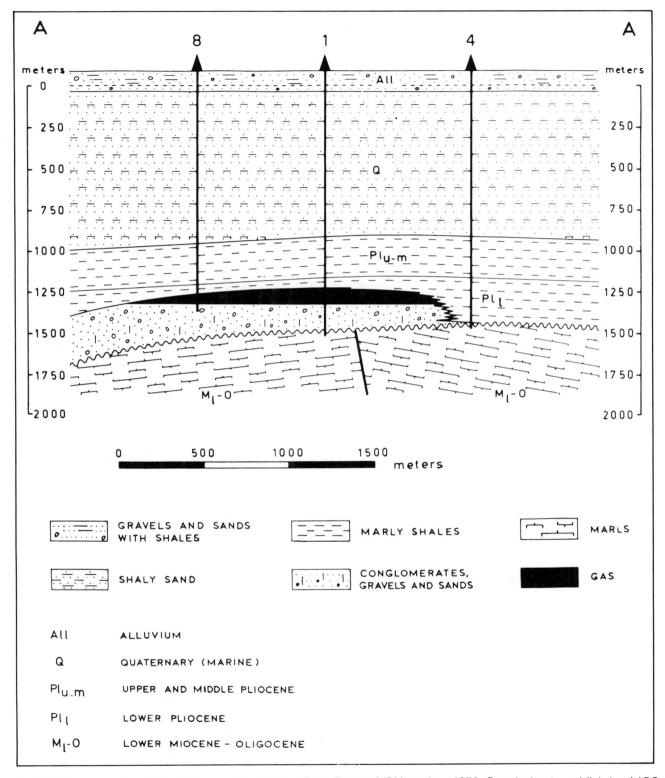

Fig. 335. *Sergnano Gas Field, Italy, porosity change. From Rocco & D'Agostino, 1972. Permission to publish by AAPG.*

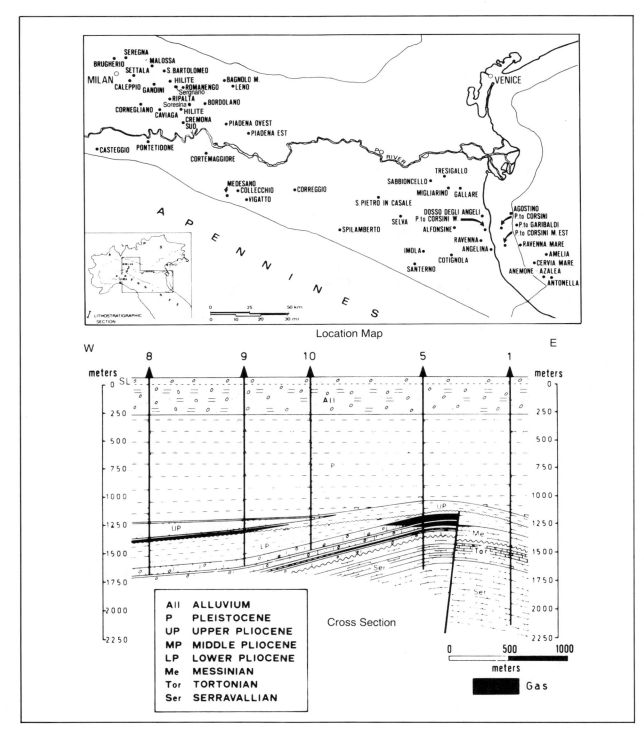

Fig. 336. Soresina gas field, northern Italy. Top: Index map. Bottom: Cross-section of a folded depositional feature plus a structural trap. From Rocco and D'Agostino, 1972. Permission to publish by AAPG.

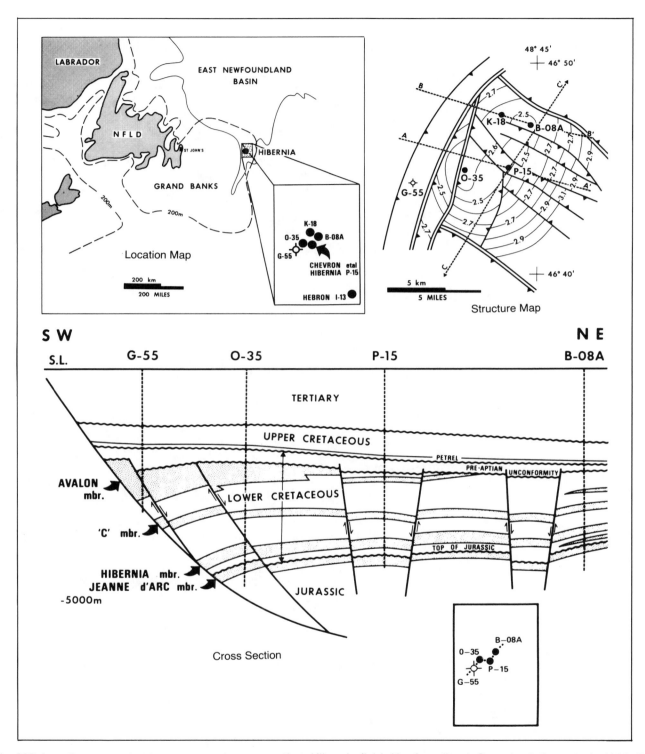

Fig. 337. Location map, structure map, and cross section, Hibernia field, Newfoundland, Canada. Arthur, et al., 1982. Permission to publish by AAPG.

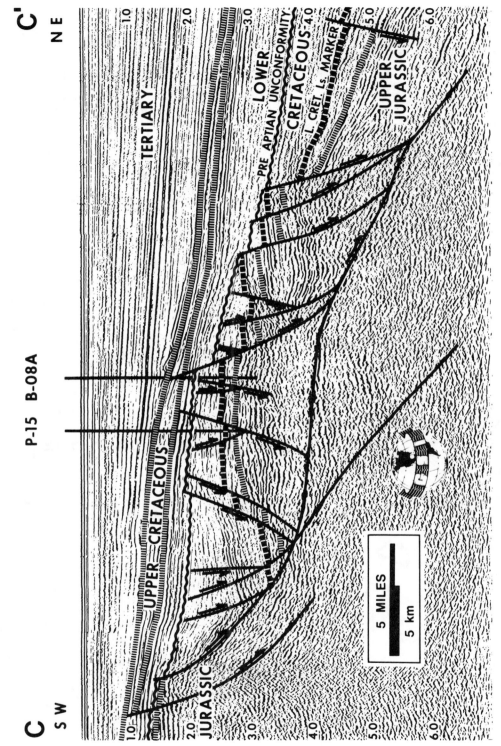

Fig. 338. Seismic section across the Hibernia structure. From Arthur, et al., 1982. Permission to publish by AAPG.

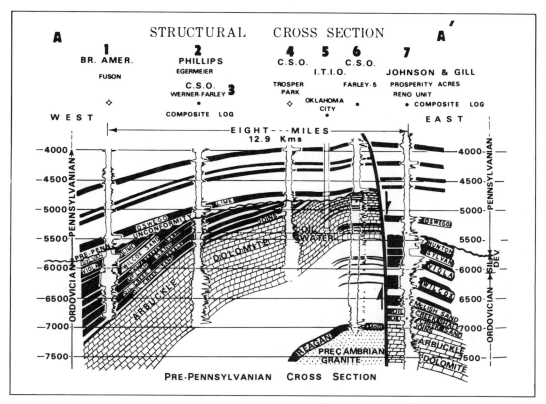

Fig. 339. Oklahoma City Field, angular unconformity. From Gatewood, 1970. Permission to publish by AAPG.

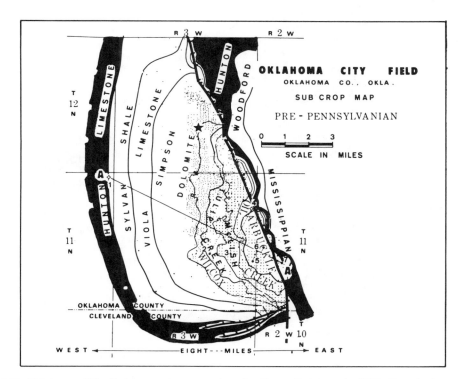

Fig. 340. Oklahoma City Field; angular unconformity. From Gatewood, 1970. Permission to publish by AAPG.

269

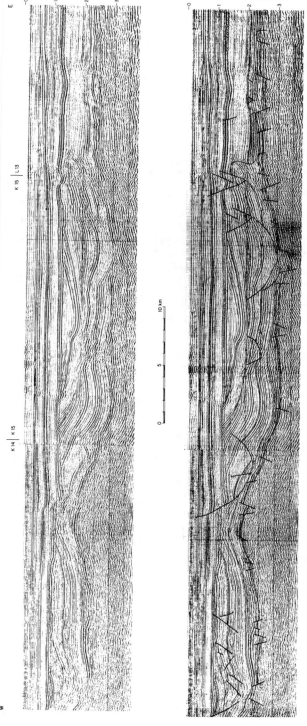

Fig. 341. Seismic section and interpretation of angular unconformity, western North Sea. From Ziegler, 1983. Permission to publish by AAPG.

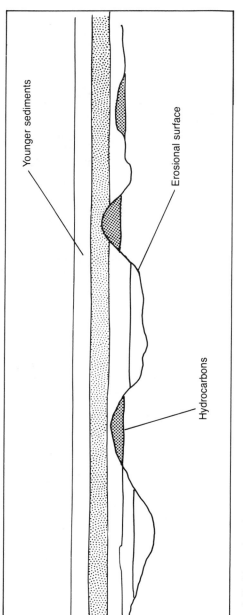

Fig. 342. Disconformity trap

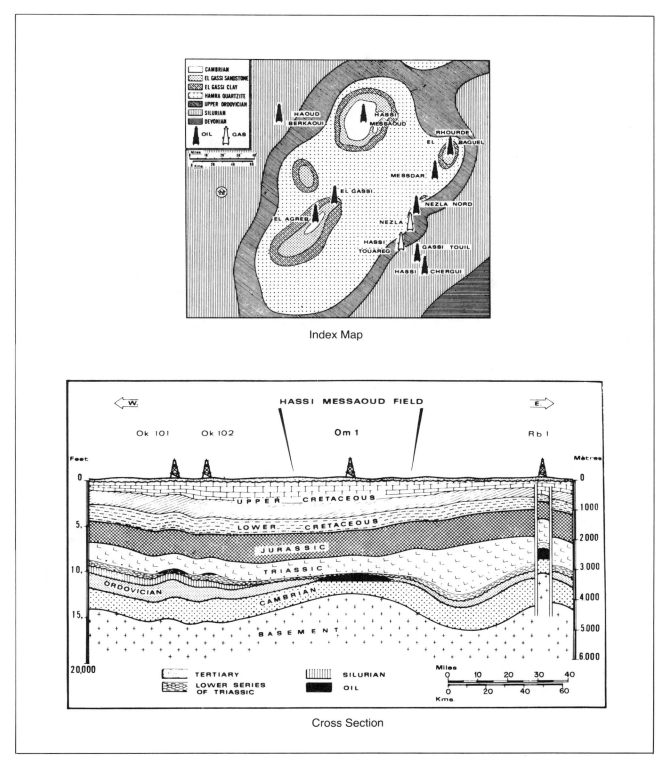

Index Map

Cross Section

Fig. 343. Hassi Messaoud disconformity oilfield, Algeria. From Bolduchi and Pommier, 1970. Permission to publish by AAPG.

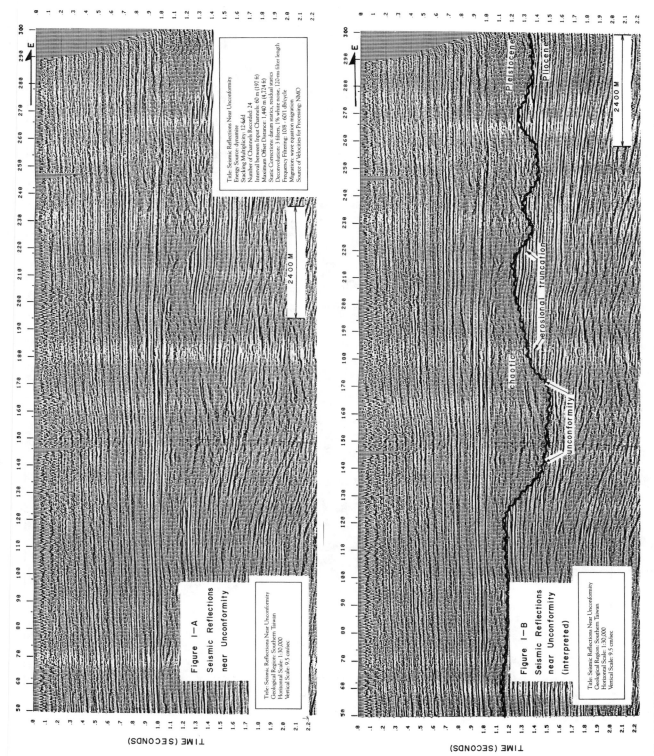

Fig. 344. Seismic section and interpretation of a discomformity, southern Taiwan. From Chang and Pan, 1983. Permission to publish by AAPG.

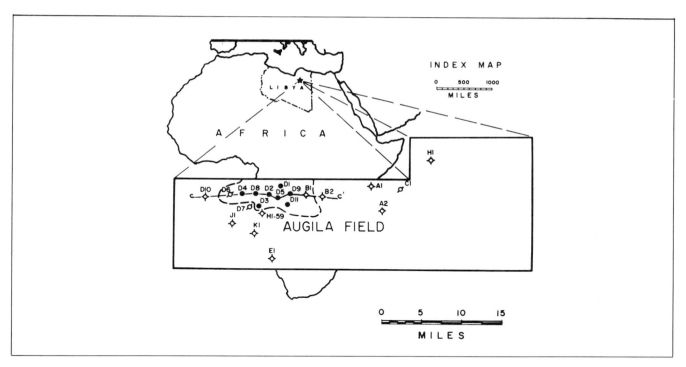

Fig. 345. Location map of Augila Field, Libya. From Williams, 1972. Permission to publish by AAPG.

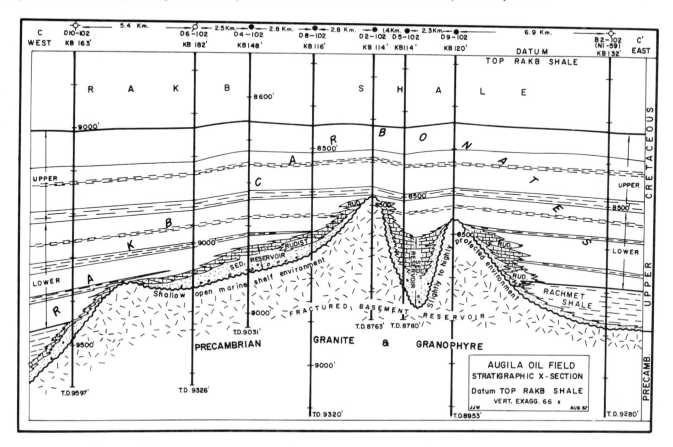

Fig. 346. Augila Field; nonconformity trap, Libya. From Williams, 1972. Permission to publish by AAPG.

11   Petroleum and Reservoirs

Origin, Migration, and Accumulation of Petroleum

This is one of the more problematical subjects in the discussion of petroleum geology since there is only limited agreement about how petroleum forms, how it migrates, and how it accumulates. However, since there is oil in structures that include reservoirs far away from what appear to be potential source beds, the fact that oil does form, migrate, and accumulate is indeed a reality.

Theories concerning oil and gas formation (catagenesis) involve organic and inorganic considerations. The extensive arguments over which is the source have been challenged with little ultimate resolution. However, present evidence brings most weight to bear on the side of the organic origin of petroleum because hydrocarbon compounds are present in organic material derived from plant and animal life. For our purposes, the origin of petroleum will be considered to be from organic sources.

Source rock analysts and geochemists are not in complete agreement on the types of organic material. Some suggest that only plant material is involved. Others conclude that animal and plant material contribute to petroleum generation. Of those who are partial to plant material, some differentiate between fibrous, structured plants, i.e., trees, shrubs and grass, to generate coal and nonfibrous unstructured algae to generate oil. Others, however, indicate that some coal generative environments can form significant amounts of oil as well as gas.

Additional problems concern theories of how petroleum migrates, the driving mechanisms for migration, and the distances that it migrates. It seems evident that oil and gas migrate, but the manner in which they move and the distances involved are less than clear and probably manifold. They are probably a function of reservoir and contained-fluid characteristics, diagenesis and pressure as well as petroleum generation considerations.

Accumulations of petroleum occur in places from which migrating fluids cannot escape. How migrating petroleum responds to changes in reservoir and trap considerations in the site of accumulation is difficult to assess. In addition, the amount of time required for accumulation is probably sufficiently variable to preclude the establishment of definite parameters.

Origin of Petroleum

Current information dealing with the organic origin of petroleum indicates that hydrocarbons are products of altered organic material derived from microscopic plant and animal life. Microscopic plants and animals are carried in great volume by streams and rivers to lakes or the sea, where they are deposited under deltaic, lacustrine and marine conditions with finely divided clastic sediments. The marine, deltaic and lacustrine environments produce their own microscopic plants and animals, which are deposited with the organic materials introduced by the streams and rivers. As deposition of the organic material takes place in the marine, deltaic and lacustrine environments, burial and protection by clay and silt accompany it. This prevents decomposition of the organic material and allows it to accumulate.

The amount of burial is a function of how much sediment is discharged by streams and rivers into lakes and the sea as well as the amount of time involved in the depositional process. Thick accumulations of silt, clay, and organic material can produce large volumes of petroleum if there is enough time for the alteration process to occur. Some research indicates that terrestrial organic material generates coal and gas, and marine organic material forms oil. However, there are positive indications that the coal-forming environment of the Upper Cretaceous-Eocene fluvio-deltaic Latrobe Group generates oil in the Australian Gippsland basin. Distance from terrestrial sediment and organic sources may also have significance as to petroleum type.

Conversion of the organic material is called catagenesis. It is assisted by pressure caused by burial, temperature and thermal alteration and degradation. These factors result from depth, some bacterial action in a closed nonoxidizing chemical system, radioactivity and catalysis. Temperature, as thermogenic activity, appears to be the most important criterion with assistance from other factors as applicable. Accumulation of organic and clastic material on a sea or lake bottom is accompanied by bacterial action. If there is abundant oxygen, aerobic bacteria act upon the organic matter and destroy it. However, the aerobic destruction of organic matter is greatly reduced or eliminated if enough low permeability sediment is deposited to stop the circulation of oxygen-bearing water. As aerobic bacterial action ceases with the decrease in available dissolved oxygen, anaerobic bacterial action involving oxygen from dissolved sulfates begins and a reducing environment develops. Anaerobic activity occurs within the upper 40 feet or so of the sediments, below which, bacterial action ceases. Biogenic gas is the result of bacterial action which is not involved in thermogenic activity.

The organic origin of petroleum is strongly suggested by the great quantities of organic compounds continuously being deposited in sedimentary basins around the world at the present time. Plant and animal remains contain abundant carbon and hydrogen, which are the fundamental elements in petroleum.

It is well documented that shale and some carbonates contain organic material that bears hydrocarbons of types similar to those in petroleum. Shale and carbonate rocks of this type are not reservoir rocks and could be considered ultimately to be source beds. The hydrocarbons are of the same type as those found in living plants and animals and consist of asphalt, kerogen, and liquid forms. Recent studies show large concentrations of organic material in some hypersaline environments. Evaporites produced under certain hypersaline conditions may ultimately be proved as petroleum source beds where no clastic source rocks were deposited.

Diagenesis of source rocks eliminates some of the contained organic materials but allows retention of residual amounts that are found in some quantities in most nonreservoir rocks. Greater concentrations of these materials

can cause some rocks to act as source beds when affected by the various hydrocarbon conversion processes.

Variations in the compositions of different crude oils are probably due to chemical variation in the compositions of the organic material that produced them. Crude oil high in asphalt probably came from organic material that was high in protein. Paraffin oil, on the other hand, was probably derived from fatty organic material.

1. Normal heat flow within the earth's crust produces an average geothermal gradient of approximately 1.5°F for each 100 feet of depth. Maturation studies on various crude oil types indicate that temperatures required to produce oil occur between the approximate depths of 5,000 and 20,000 feet under average heat-flow conditions. Temperatures below 20,000 feet are excessive for oil generation and most often produce gas. Above 5,000 feet the crust is too cool to generate oil or gas. Inasmuch as most localities have either above or below average heat flow, petroleum generating depths are substantially variable from place to place.

In some instances, the geothermal gradient is altered by geological considerations. In these places oil can be generated at lesser depths because of high heat flow or at greater depths because of low heat flow. Erosion or addition of overburden following petroleum generation can result in lesser or greater than optimum generating and emplacement depths as well.

2. Like temperature, pressure is a function of depth and increases one pound per square inch for each foot of depth. Pressure is caused by the weight of sedimentary overburden, which can be considerable.

3. Bacterial action is considered important in the conversion of organic material to petroleum at shallow depths. It is involved in the process of breaking down the original material into hydrocarbon compounds, which eventually become biogenic gas.

4. The best source rocks are considered to be organically rich, black-colored shales deposited in a nonoxidizing, quiet marine environment. Shale thickness is probably an important factor in the amount of petroleum generated. Carbonate source rocks generate petroleum where no shale is present in Iran and southern Florida. Highly saline evaporite environments often contain large organic concentrations and may act as source beds in some places. Some coal environments are drawing increased attention as having oil source potential.

5. Most petroleum is generated and occurs in sedimentary rocks and is organic in origin. Some petroleum occurrences in igneous and metamorphic rocks may have inorganic or altered organic origins and represent other than common sedimentary and diagenetic processes.

Hot water solutions are known to acquire dissolved

minerals and transport hydrocarbons as they pass through mineral-rich or organic-rich sediments. When they encounter igneous rocks, bitumen can be deposited in fractures, intergranular spaces and/or fault zones. Much organic material can be lost this way but some can be retained.

High temperatures associated with igneous intrusion can mobilize petroleum from surrounding organic-rich sediments. Cooling of the intrusions permits occurrence and preservation of the petroleum in them. Such occurrences have been observed in western Russia.

Mobile petroleum, including gas and crude oil, migrate into reservoirs to be preserved. Most often the reservoirs are sedimentary rocks. Crystalline reservoirs, which comprise igneous and metamorphic rocks, can have excellent fractured and/or weathered porosity and permeability and provide significant reservoir potential.

Many metamorphic rock suites resulted from alteration of sediments, which included some source beds. The organic material in the sediments, subjected to heat of metamorphism, was converted to petroleum. Some of the petroleum was retained in the resulting metamorphic rocks but most was dissipated and lost. Metamorphic rocks, like igneous rocks, can provide excellent reservoir potential with fractured and weathered porosity and permeability. Several fields in the mid-continent of the United States and others in Indonesia produce from metamorphic rock reservoirs.

Petroleum formation from igneous processes may be affected as inorganic alteration of mineralogic carbon by elevated temperature and mineral solutions. Generation of petroleum in this manner is considered unusual and does not contribute significantly to reserves. Igneous rocks, once cooled, fractured and weathered, can provide important reservoir potential, which currently produces in the Rocky Mountains of the United States, northern South America, and North Africa.

Generation of Crude Oil

Organic material in shale averages approximately one percent of the shale rock volume. Clay mineral constituents comprise the remaining 99 percent (Fig. 347). Some shale has much greater concentrations and some much lower.

Kerogen is an insoluble, high molecular weight, polymeric compound which comprises about 90 percent of the organic material in shale. The remaining ten percent comprises bitumens of varying composition, which according to some researchers is thermally altered kerogen.

Living organisms and their metabolic and biochemical

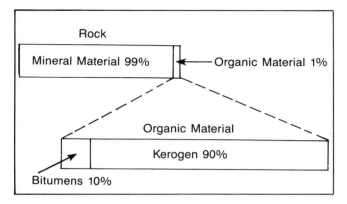

Fig. 347. Percentages of mineral and organic material and bitumen and kerogen and rock. From Barker, 1979. Permission to publish by AAPG.

processes produce many types of organic materials, of which some are chemically stable and persist long after the death of the organisms. Other organic materials are much less stable, and some are eliminated very quickly. Because of a considerable variation in organic material stability, processes that act upon sediments affect the organic material at correspondingly variable rates.

Original plant and/or animal organic material is deposited with sedimentary sequences. It can be deposited under no oxygen anoxic, quiet water, high-preservation conditions or under oxidizing conditions where it is not preserved. Under some circumstances of rapid deposition, oxidation in upper layers may occur until available oxygen is depleted as overburden increases and anoxic conditions develop. As alteration occurs kerogen is developed by the increasing temperature in the closed system. Carbon content increases in the kerogen, which is progressively more condensed, and graphite is produced. If hydrogen increases, crude oil forms before gas that forms later (Fig. 348).

Kerogen, as the major organic constituent in rocks, is a primary factor in forming bitumens that increase and migrate to accumulate as crude oil. Thermal conversion of kerogen to bitumens is the important process of crude oil formation. Thermal alteration increases the carbon content of the migratable hydrocarbons, which leave the unmigratable residual kerogen components behind.

Maturation of kerogen is a function of increased burial and temperature and is accompanied by chemical changes. As kerogen thermally matures and increases in carbon content, it changes from an immature light greenish-yellow color to an overmature black, which is representative of a progressively higher coal rank (Fig. 349).

Several criteria are available to determine thermal ma-

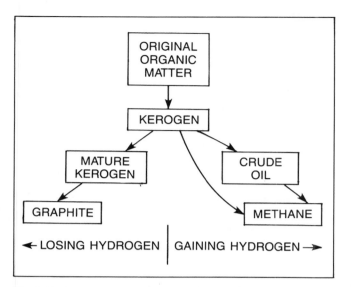

Fig. 348. Alteration of organic material to hydrogen-poor and hydrogen-rich compounds. From Barker, 1979. Permission to publish by AAPG.

turity of the organic material and therefore that of the sediments (Fig. 349). A significant method measures *vitrinite reflectance* or the ability of vitrinite, a progressively coaly, shiny, woody material, to reflect incident light. An increasing ability to reflect light indicates progressively greater thermal maturity.

In addition to coal rank and vitrinite reflectance, thermal alteration index, carbon percent in kerogen and fixed carbon ratio illustrate additional thermal maturity parameters that correlate with degrees of sediment diagenesis—metamorphism and petroleum development (catagenesis).

Generation of Natural Gas

Natural gas comprises biogenic gas and thermogenic gas with differences contingent upon conditions of origin (Figs. 350 and 351).

Biogenic gas forms at low temperatures at overburden depths of less than 3000 ft under anaerobic or

Fixed Carbon Ratio	Coal Rank	Vitrinite Reflectance (R_0) %	Petroleum Generation and Destruction	Thermal Alteration Index	Weight % Carbon In Kerogen	Metamorphism and Hydrocarbons
	Peat	0.2		1 Yellow	65	Unmetamorphosed (Biogenic Gas)
50	Lignite	0.3			70	
		0.4			75	
	Bituminous	0.5	Oil	2 Orange		Mature (Oil)
		0.6	Peak Oil Generation		80	
60						
70		1.0	Wet Gas Dry Gas Peak Wet Gas Gen.		85	
		1.2	Peak Dry Gas Gen.	3 Brown		
		1.35	Oil Limit		90	
80						
	Anthracite	2.0	Wet Gas Limit	4 Black		Over Mature (Gas)
90						
		3.0	Dry Gas Limit			
		4.0		5 Black		Slate
100	Graphite	5.0			95	Phyllite

Fig. 349. Sediment thermal maturity indicators.

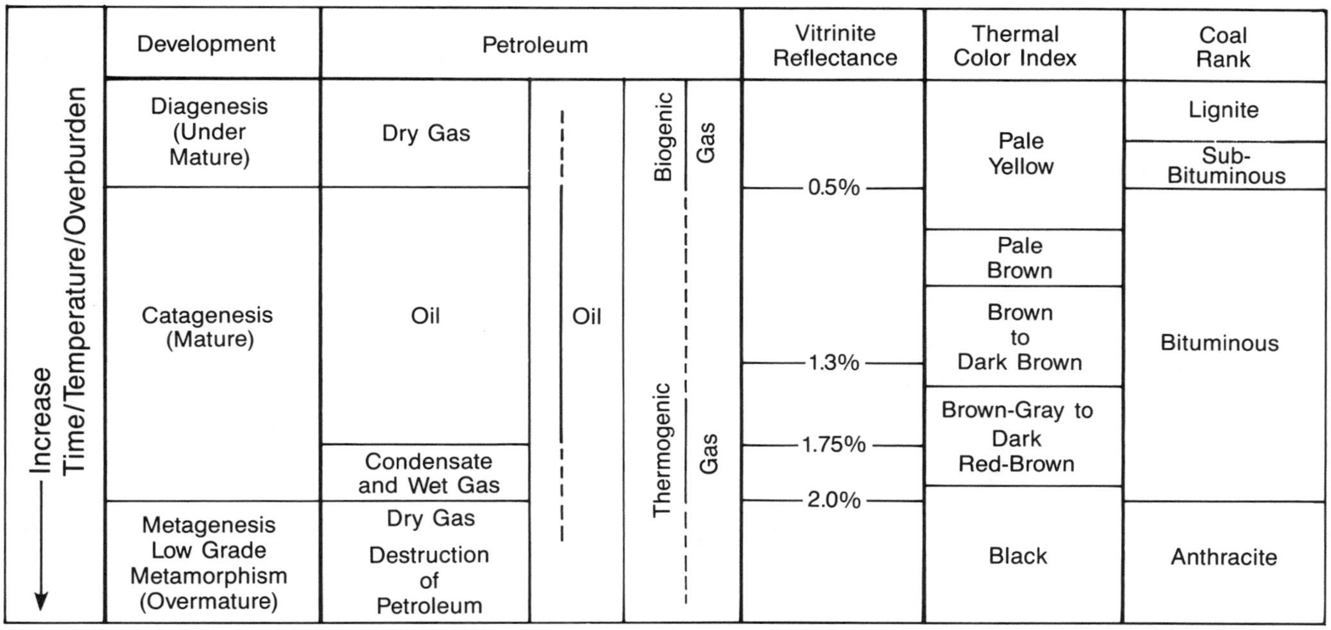

Development	Petroleum				Vitrinite Reflectance	Thermal Color Index	Coal Rank
Diagenesis (Under Mature)	Dry Gas		Biogenic Gas			Pale Yellow	Lignite
					— 0.5% —		Sub-Bituminous
Catagenesis (Mature)	Oil	Oil	Thermogenic Gas			Pale Brown	Bituminous
						Brown to Dark Brown	
					— 1.3% —		
	Condensate and Wet Gas				— 1.75% —	Brown-Gray to Dark Red-Brown	
					— 2.0% —		
Metagenesis Low Grade Metamorphism (Overmature)	Dry Gas Destruction of Petroleum					Black	Anthracite

(Left axis label: Increase Time/Temperature/Overburden)

Fig. 350. Sediment thermal maturity indicators.

ditions often associated with high rates of marine sediment accumulation. Oxygen in the sediments is consumed or eliminated early and before reduction of sulfates in the system. Methane, the most common of natural gas constituents, forms after the sulfates are eliminated by hydrogen reduction of carbon dioxide. Carbon dioxide, from organic material, is chemically produced at slightly elevated temperatures or metabolically by anaerobic biota. Anaerobic oxidation of carbon dioxide produces methane. It is quantitatively affected by sediment overburden, amounts of organic materials, quality of the anoxic environment that affect quantities of sulfates and carbon dioxide, reduced temperatures and sufficient space for bacterial formation.

Since biogenic gas forms early under conditions of rapid sedimentation and limited overburden, it can occur in a variety of environments including, very notably, contemporary deltas of the Nile, Niger, Orinoco, Mississippi and Amazon rivers. Current estimates suggest that approximately 20 percent of the world's known natural gas is biogenic.

Representative worldwide accumulations of biogenic gas occur in rocks as old as Cretaceous and as young as Pleistocene (Table 24). Depths of occurrence are, comparatively, very shallow and are suggestive of low temperatures. Biogenic gas is produced in a number of North American locations (Fig. 352).

Thermogenic gas forms at significantly higher temperatures and overburden pressures than does biogenic gas. It forms during the catagenetic middle and low-grade metagenetic later phases of sediment and rock diagenesis from alteration of predominant kerogen and other minor organic constituents by molecularly dissociative thermal cracking and degradation. Thermogenic gas contains methane and significantly larger amounts of heavier hydrocarbons than biogenic gas. As time and temperature increase, progressively lighter hydrocarbons form as wet gas and condensate in the later stages of thermogenesis. Thermal methane replaces biogenic methane during catagenesis. Early oil-wet gas and condensate accompany progressive catagenesis followed by low grade metamorphism (metagenesis) and thermogenic methane.

Thermal, temporal and metamorphic criteria eventually eliminate all petroleum generation as metamorphism becomes progressively higher grade, and mineral constituents in the sediment remobilize and consolidate. All but the lowest grades of incipient metamorphism are hostile to petroleum generation.

Migration

The concept of petroleum migration remains a reality notwithstanding limited agreement about mechanisms and distances. When petroleum moves from source beds to

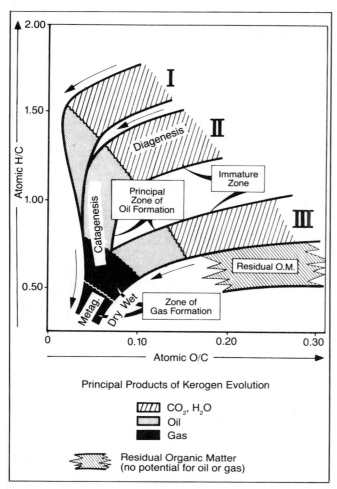

Principal Products of Kerogen Evolution

- ▨ CO_2, H_2O
- ░ Oil
- ■ Gas

〰 Residual Organic Matter (no potential for oil or gas)

Fig. 351. Evolution of kerogen in petroleum formation. From Tissot, 1984. Permission to publish by AAPG.

Table 24
Worldwide Accumulations of Biogenic Gas

Location	Reservoir Age	Depth (m)
United States		
Cook Inlet, Alaska	Tertiary	910–1,650
Offshore Gulf of Mexico	Pleistocene	460–2,800
Rocky Mountains (Colorado, Kansas, Montana, Nebraska, New Mexico, South Dakota)	Cretaceous and Tertiary	120–840
Illinois	Pleistocene	40
Japan	Tertiary	100–1,000
Italy	Tertiary	400–1,830
Germany	Tertiary	900–1,800
Canada	Cretaceous	300–1,000
USSR		
North Aral	Tertiary	320–350
Siberia	Cretaceous	700–1,300
Stravapol	Cretaceous and Tertiary	200–1,200
North Priaral	Tertiary	300–500
Trinidad	Tertiary	980–3,350

reservoir rocks, it does so by primary migration. If it moves within the reservoir after it has accumulated, it does so by secondary migration. How far petroleum migrates is not known. It is probable that some small amounts of petroleum generate within the reservoir rocks in which they accumulate. However, the bulk of it probably comes from source beds external to the reservoir.

That petroleum does migrate is suggested by the very common occurrence of active seeps where oil and gas come directly to the surface from reservoir rocks. Petroleum occurrence in now water-bearing reservoirs indicates that petroleum was previously present as a stable accumulation or had passed through earlier. Other evidence derives from the fact that fluid contacts are established in reservoirs that stratify various fluid phases, including water. Migration and equilibrium must be established for this to happen, particularly since traps rarely remain deformationally static.

Considerable petroleum movement may be the result of hydrodynamic pressure. Diagenesis can cause the expulsion of water from shale by compaction, which will cause migration of water and petroleum into porous beds. Water movement carries the petroleum from the source bed into the reservoir, where it establishes a position of equilibrium for the given hydrodynamic and structural conditions. Movement of cementing solution through a reservoir can similarly displace oil and gas to other locations as the initial pore space is occupied by mineral material. Movement of petroleum because of hydrodynamic pressure appears likely but is debated as to its importance relative to other factors.

Factors Affecting Migration

Some of the disagreement about migration concerns the role of water in movement of petroleum and movement involving separate phase transport with no water (Table 25). Several mechanisms are applicable under both systems, but are different in their specific physical characteristics.

Under most situations of migration, pressure gradient

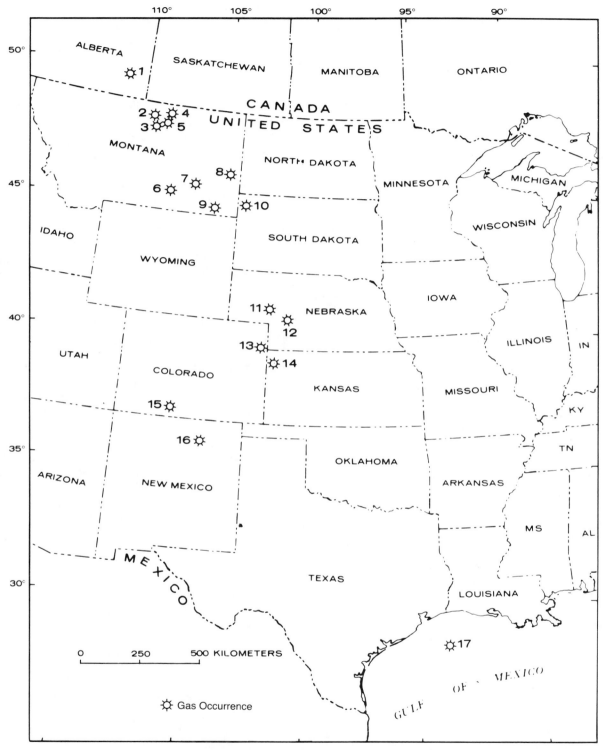

Fig. 352. Some biogenic gas locations in North America. From Rice and Claypool, 1981. Permission to publish by AAPG.

Table 25
Petroleum Migration Mechanisms and Carriers

Water-Controlled Transport		
Oil in water solution		Meteroric water
Gas in water solution		Compaction water
Other organics in water solution		Clay dehydration water
Micellar solution		Mineralization water
Emulsion		Pressure gradient
Diffusion		
Separate Phase Transport		
Oil		Diffusion
Gas	In	Pressure gradient
Gas in oil solution	Pore	
Oil in gas solution	System	

Adapted from Jones, 1978.

and/or volume change are prerequisite. Without these factors a gravity drainage system may be involved.

Migration involves properties of the rocks and the fluids including petroleum moving through the rocks (Tables 26 and 27). Kerogen properties contribute to migration factors as well (Table 28). All migration factors relate to mobility without which no accumulation can occur.

Multiple, potential migration factors inveigh against any paradigm applicable to all petroleum movement situations. Individual conditions control a specific migration from one specific source system to a specific reservoir system through specific conditions in transit.

As kerogen is thermally altered, the bitumen content in the source rock increases. When bitumen saturation reaches a certain level in the source rock, migration begins with or without water as applicable to the conditions.

Table 26
Rock Properties and Petroleum Migration

Porosity	Absorption
Permeability	Adsorption
Pore and grain size distribution	Interstitial Clay Minerals
Capillary Pressure	Hydrated Minerals
Wettability	Fracture Characteristics

Adapted from Jones, 1978.

Table 27
Fluid Properties and Migration

Pressure and Pressure Gradient	Viscosity
Temperature and Temperature Gradient	Saturation
Water Composition	Compressibility
Surface Tension	Density
Clay Surface Water	Buoyancy

Adapted from Jones, 1978.

Table 28
Kerogen Properties and Migration

Quantity and Distribution	Absorption Products
Generation Products	Maturation

Adapted from Jones, 1978.

So migration occurs as a direct result and concomittant of peak oil generation.

Some research suggests that oil may migrate in response to generation of carbon dioxide from kerogen. Carbon dioxide is used in secondary recovery, miscible flood programs and may be important in increasing primary oil mobility in initial migration.

If petroleum migrates and if it migrates for long distances, a closed system within which the oil remains is essential. Any breach of the system results in escape and loss of the petroleum. Therefore, permeable rocks through which petroleum migrates must retain their integrity through a process that seals them against the escape of the migrating petroleum.

Migration of petroleum will occur wherever there is an appropriate pressure gradient in a medium through which flow is possible (Table 29). The medium must be porous and permeable to permit the migration, which may be directed according to internal characteristics of the medium. So, a channel sandstone that grades to shale as it thins laterally has a migration axis along its length parallel to the flow direction of its depositional regime.

It seems that petroleum must migrate from geographically large source areas to smaller areas of accumulation with economically sufficient concentrations. Therefore it is not likely that petroleum in place, regardless of naturally generated amount, will be sufficiently concentrated to be economic (Figs. 353, 354, 355).

Table 29
Possible Migration Routes

Reservoir Quality Rocks with depositional porosity and permeability:	
Channel Sandstone	Bedding Plane
Beach and Bar Sandstone	Lamination
Dune Sandstone	Cross Bedding
Sand Lense	Parting
Weathered and Eroded Surfaces	
Unconformity	Solution-enhanced Fracture
Solution-Altered Rock	
Fractured Rock	
Oriented Fracture and Microfracture Fault	
Fractured and Fractured Crystalline Rock	
Intrusive Structure	
Salt, Shale, Metamorphic Diapir	
Sandstone Intrusion (dike)	

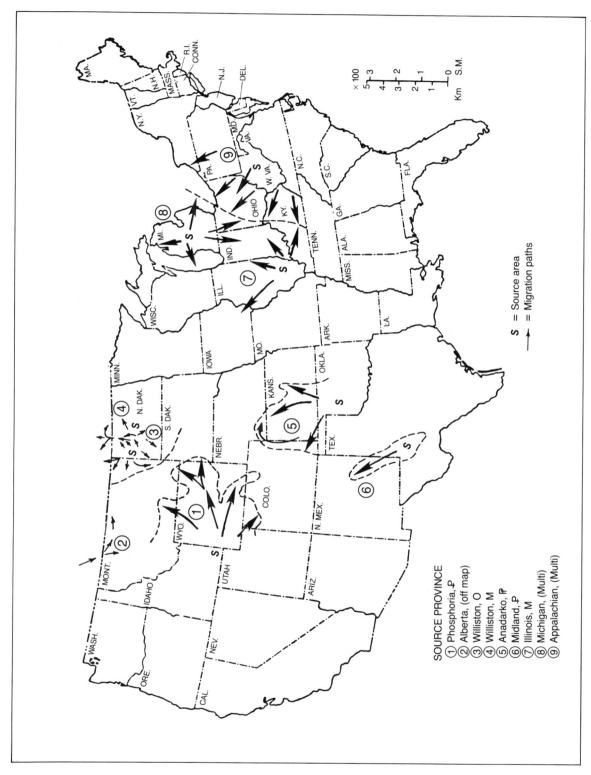

SOURCE PROVINCE
① Phosphoria, P
② Alberta, (off map)
③ Williston, O
④ Williston, M
⑤ Anadarko, P
⑥ Midland, P
⑦ Illinois, M
⑧ Michigan, (Multi)
⑨ Appalachian, (Multi)

S = Source area
→ = Migration paths

Fig. 353. Examples of long distance oil migration in the United States. From Momper, 1978. Reprinted by permission.

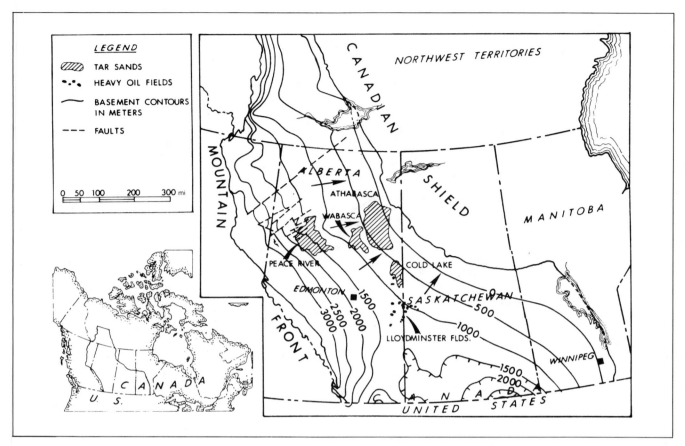

Fig. 354. Petroleum migration in the Alberta Basin, Canada. From Momper, 1978. Reprinted by permission.

Accumulation

Hydrocarbons accumulate and are stratified according to their fluid phases and the amounts of formation water. Gas is lightest and accumulates above oil, which overlies water. The quality of gas dissolved in oil depends on pressure, temperature, and hydrocarbon characteristics.

Hydrocarbons accumulate in the highest permeable portions of the reservoir because of hydrodynamics. This is why the highest area of an anticline is usually the best place to drill an exploratory well (Fig. 356).

Petroleum accumulations probably require long periods of time to form, particularly in reservoirs of low permeability. Mobility of fluids within a reservoir is enhanced by increasing permeability.

Composition of Hydrocarbons

Crude oil and gas are primarily of organic origin and contain many components typical of animal and plant material. Alteration of these materials produces compounds that form crude oil and are called hydrocarbons, which consist of hydrogen and carbon. Hydrocarbons come in many molecular forms and are classified according to the shapes of their molecules and the ratios of carbon and hydrogen. Specifically, they are classified by molecular type or pattern and by molecular weight.

The simplest molecular types are straight-chain molecular types (Fig. 358). They start with a single atom of carbon surrounded by four of hydrogen to form methane, a common natural gas.

Branched or isoparaffin hydrocarbons form a more complex molecular type than the normal paraffin straight chains. Isoparaffin hydrocarbons are branched and with the same chemical formula can be molecularly constructed in different ways. There are several variations of constructions of isoparaffin molecules (Fig. 359).

Closed paraffin molecules produce cycloparaffin hydrocarbons, or naphthenes. These molecules develop closed

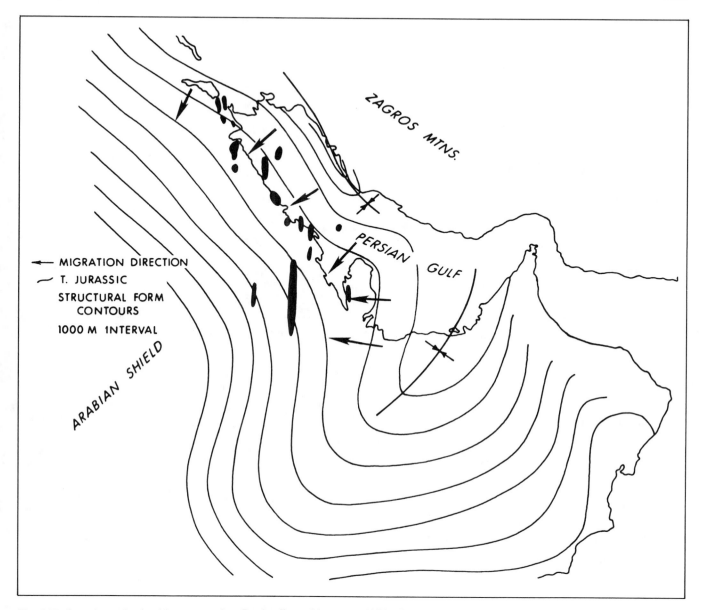

Fig. 355. Petroleum in the Mesopotamian Basin. From Momper, 1978. Reprinted by permission.

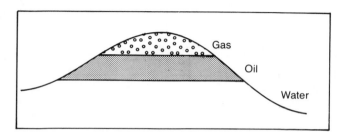

Fig. 356. Anticlinal petroleum accumulation

geometric patterns, some of which are fairly complex (Fig. 360). Closed molecule hydrocarbons with double bonds between alternating carbon atoms are called aromatic hydrocarbons (Fig. 361).

Crude oils vary in the amount of sulfur they contain (Fig. 357). High sulfur crude oils are extremely corrosive and damage drilling and production equipment. Sulfur is a by-product of crude oil refining where sulfur content is high. Removal of sulfur from crude oil is required by

Name, state, and country	Gravity API	Sulfur %	Viscosity SSU @ 100° F
Smackover, Ark., USA	20.5	2.30	270
Kern River, Cal., USA	10.7	1.23	6,000+
Kettleman, Cal., USA	37.5	0.32	
Loudon, Ill., USA	38.8	0.26	45
Rodessa, La., USA	42.8	0.28	
Oklahoma City, Ok., USA	37.3	0.11	
Bradford, Pa., USA	42.4	0.09	40
East Texas, USA	38.4	0.33	40
Leduc, Alberta, Canada	40.4	0.29	37.8
Boscan, Venezuela	9.5	5.25	
Poza Rica, Mexico	30.7	1.67	67.9
La Rosa, Venezuela	25.3	1.76	
Kirkuk, Iraq	36.6	1.93	42
Abqaiq, Saudi Arabia	36.5	1.36	
Seria, Brunei	36.0	0.05	

CHARACTERISTICS OF SOME CRUDE OILS.

Fig. 357. Characteristics of some crude oils. From Buthod, 1962. Copyright 1962, SPE-AIME.

government regulations prior to distribution of refinery-run products to the market place.

Properties of Crude Oil

Crude oil specific gravity is the ratio of its density to the density of water. It is expressed as API Gravity, which is defined by the following formula:

$$\text{Degrees API} = \frac{141.5}{\text{specific gravity @ 60°F}} - 131.5$$

High gravity oil has low viscosity and contains volatile fractions. Low gravity oil has high viscosity and contains fewer volatile fractions.

Crude oil viscosity is the property that controls its ability to flow under certain temperature conditions (Fig. 357).

The temperature at which a crude oil will congeal and no longer pour is its pour point. Some crude oils have very high pour points because of contained constituents. Paraffin-base oils in the Uinta Basin of Utah require spe-

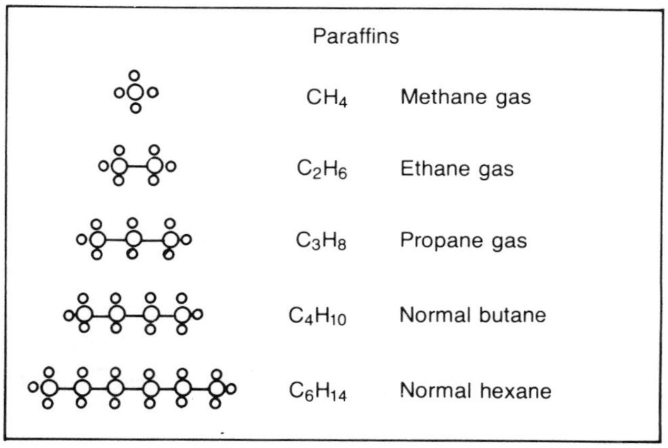

Fig. 358. Straight chain paraffin molecular structure

	CH_4	Methane gas
	C_2H_6	Ethane gas
	C_3H_8	Propane gas
	C_4H_{10}	Normal butane
	C_6H_{14}	Normal hexane

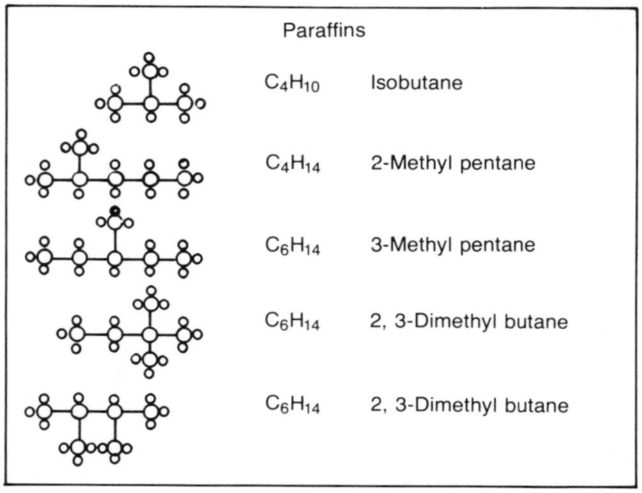

Fig. 359. Branched chain paraffin molecular structure

	C_4H_{10}	Isobutane
	C_4H_{14}	2-Methyl pentane
	C_6H_{14}	3-Methyl pentane
	C_6H_{14}	2, 3-Dimethyl butane
	C_6H_{14}	2, 3-Dimethyl butane

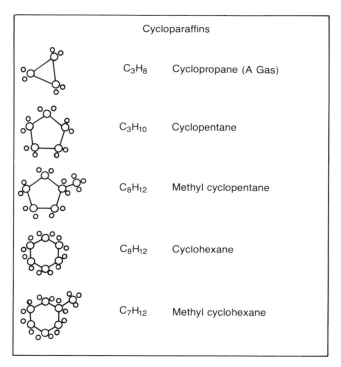

Fig. 360. *Cycloparaffin molecular structure*

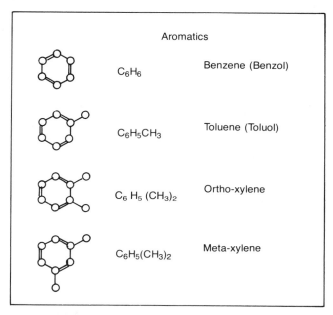

Fig. 361. *Aromatic molecular structure*

cial handling and transportation in heated tank trucks. Some Indonesian crude oils have pour points of 100°F.

Crude oil is composed of several compounds with relatively familiar names. These components result from the refining of the crude into its various fractions. Molecular size and molecular types of crude oil components are variable with different oil types (Fig. 362).

Properties of Natural Gas

Natural gas often consists of various components that occur in different ways within reservoirs. They are represented according to their occurrences:

1. Gas dissolved in crude oil is associated gas that is retained in solution by reservoir pressure. Associated gas most commonly occurs at a depth where pressure is sufficient to keep it in solution.

2. Gas in a reservoir containing no crude oil comprises nonassociated gas. Nonassociated gas may be the result of deep, high temperature hydrocarbon formation where liquid petroleum does not form. However, most nonassociated gas reservoirs currently being produced involve normal reservoir temperatures and depths perhaps because of erosion of overburden. Dry gas is nonassociated gas with no condensable hydrocarbons at normal oil and gas separation conditions.

3. Light hydrocarbons with a gas phase occur as gas condensate. Gas condensate contains variable amounts of condensable hydrocarbons.

COMPOSITION OF A CRUDE OIL			
Molecular size	*Wt. %*	*Molecular type*	*Wt. %*
Gasoline ($C_4 - C_{10}$	31	Paraffins	30
Kerosene ($C_{11} - C_{12}$)	10	Naphthenes	49
Gas Oil ($C_{13} - C_{20}$)	15	Aromatics	15
Lubricating Oil ($C_{20} - C_{40}$)	20	Asphaltics	6
Residuum (C_{40+})	24		100
	100		

Fig. 362. *Crude oil composition. After Hunt. Permission to publish by AAPG.*

AVERAGE COMPOSITION OF VARIOUS COMMERCIAL NATURAL GASES.							
Pool and location	SG (air = 1.0)	Methane	Ethane C_2H_6	Propane C_3H_8	Butane C_4H_{10}	Pentane and heavier	CO_2
United States							
Panhandle-Amarillo		91.3	3.2	1.7	.9	.56	.1
Hugoton, Kansas		74.3	5.8	3.5	1.5	.6+	
Carthage field,							
Texas	.616	92.54	4.7	1.3	.8	.6	
Velma, Oklahoma		82.41	6.34	4.91	2.16	1.18	
Canada							
Turner Valley, Alta.		92.6	4.1	2.5	0.7	0.13	

Fig. 363. Natural gas composition. From Levorsen, 1967. Permission to publish by W. H. Freeman.

Natural gas composition, like that of crude oil, is variable according to the contained components. Some components are more abundant than others, with methane being the most common constituent. Figure 363 illustrates natural compositions from several selected fields. Note that methane is the predominant constituent in each of the fields represented.

Oil and Gas Phases

Hydrocarbons can occur as liquid or gas phases that are subject to temperature and pressure (Fig. 364). Solubility of oil and gas is an important factor in their phase occurrences. The accompanying graph shows that for high temperature and low pressure hydrocarbons occur in the gas phase. Where temperature and pressure exceed the critical point, hydrocarbons are represented by a single phase that is not a gas or a liquid. The critical point is the temperature above which the hydrocarbon will not exist as two phases. The temperature and pressure at this point are called the critical temperature and critical pressure.

Reservoir Characteristics and Petroleum

It is not quite sufficient to say that to be a reservoir, a rock requires porosity and permeability. Reservoir behavior relative to oil and gas accumulation and production certainly involves porosity and permeability, but its performance is based upon several important engineering factors.

Porosity

Porosity represents the amount of void space in a rock and is measured as a percentage of the rock volume (Fig. 365). Connected porosity where void space has flow-through potential is called effective porosity. Noneffective porosity is isolated. Summation of effective and nonef-

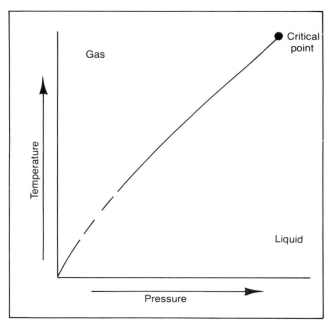

Fig. 364. Hydrocarbon vapor pressure vs. temperature (pure hydrocarbon component)

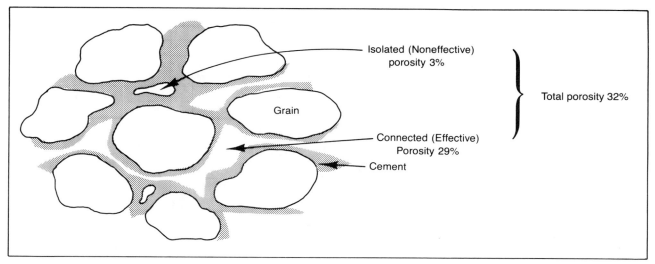

Fig. 365. Effective, noneffective, and total porosity

fective porosity produces total porosity, which represents all of the void space in a rock.

Pore space in rocks at the time of deposition is original, or primary porosity. It is usually a function of the amount of space between rock-forming grains. Original porosity is reduced by compaction and groundwater-related diagenetic processes.

Groundwater solution, recrystallization, and fracturing cause secondary porosity, which develops after sediments are deposited. Cavern formation in limestone is a good example of secondary porosity development.

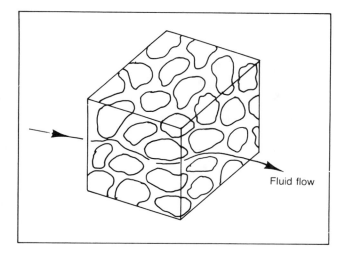

Fig. 366. Fluid flow through permeable sand

Permeability

A rock that contains connected porosity and allows the passage of fluids through it is permeable (Fig. 366). Some rocks are more permeable than others because their intergranular porosity, or fracture porosity, allows fluids to pass through them easily.

Permeability is measured in darcies (Fig. 367). A rock that has a permeability of 1 darcy permits 1 cc of fluid with a viscosity of 1 centipoise (viscosity of water at 68°F) to flow through one square centimeter of its surface for a distance of 1 centimeter in 1 second with a pressure drop of 14.7 pounds per square inch.

Permeability is usually expressed in millidarcies since few rocks have a permeability of 1 darcy. Intergranular material in a rock, such as clay minerals or cement, can reduce permeability and diminish its reservoir potential. It is evident, however, that mineral grains must be cemented to some degree to form coherent rock and that permeability will reduce to some extent in the process (Fig. 368).

Relative Permeability. When water, oil, and gas are flowing through permeable reservoirs, their rates of flow will be altered by the presence of the other fluids (Fig. 369). One of the fluids will flow through a rock at a certain rate by itself. However, in the presence of one or both of the other fluids, its rate of flow can be changed.

The flow rate of each fluid is affected by the amounts of the other fluids, how they reduce pore space, and to what extent they saturate the rock. Comparison of the flow rate of a single fluid through a rock with that same

$$Q = \left(\frac{k}{\mu}\right)\left(\frac{A(p_1 - p_2)}{B \quad L}\right)$$

where:

Q	= rate of flow (barrels per day)
k	= permeability (Darcys)
μ	= viscosity of the fluid (centipoises)
A	= cross sectional flow area (ft.²)
$p_1 - p_2$	= pressure drop (psi)
L	= length of flow path (ft.)
B	= formation volume factor (Bbl./stock tank Bbl.)

Fig. 367. Darcy equation

flow rate, at the same pressure drop, in the presence of another fluid determines the relative permeability of the system.

When rock pores decrease in size, the surface tension of fluids in the rock increases (Fig. 370).

If there are several fluids in the rock, each has a different surface tension, which exercises a pressure variation between them. This pressure is called capillary pressure (Fig. 371) and is often sufficient to prevent the flow of one fluid in the presence of another.

Reservoir Performance

Reservoirs produce oil and gas in relation to porosity and permeability. However, they are motivated to pro-duce by internal pressures, the types of fluids they contain, the phases of the fluids involved, and how the fluids interact with the ability of the rock to transmit the fluids.

When a reservoir is produced, its pressure is reduced, and the fluid phases are altered, which changes its ability to produce at initial flow rates. Ultimately, reservoirs must be placed on secondary and tertiary recovery programs. Production rates and cumulative totals are affected by several reservoir and fluid factors.

1. As reservoir pressure is reduced, gas in solution comes out of the oil. The pressure at which this occurs is the bubble point for the particular condition of the reservoir, its fluids, and its pressure regime.

2. The ratio of the volume of gas produced to the

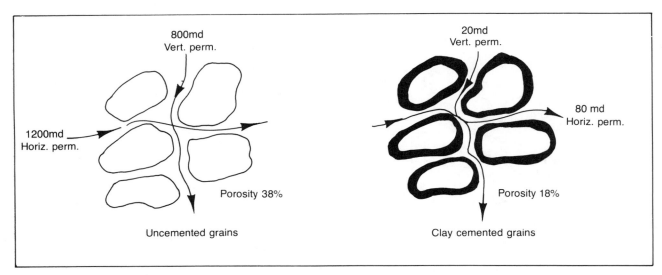

Fig. 368. Clay cement and porosity and permeability

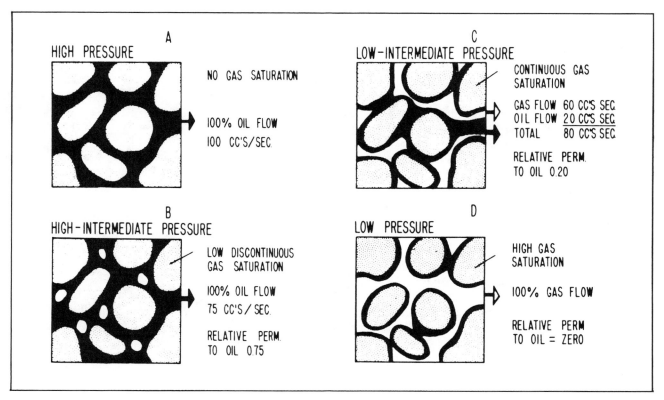

Fig. 369. Relative permeability. From Clark, 1969. Copyright 1969, SPE-AIME.

volume of oil produced is the gas-oil ratio. As the reservoir pressure declines, the gas-oil ratio increases until all gas is depleted.

3. Overburden, gas in solution, and meteoric water are several reasons for pressure to develop in reservoirs. Pressures are manifested as a function of the fluids in the

reservoir and are called reservoir fluid pressures. Initial fluid pressures are usually the highest, but they drop as the reservoir is produced.

4. Pressures representing reservoir fluids cause fluid drive production. Initially reservoir pressures are sufficient to force the fluids up the wellbore to the surface.

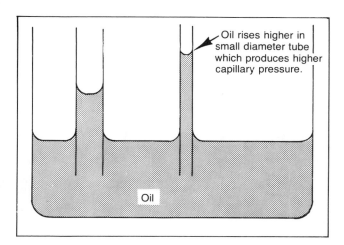

Fig. 370. Capillarity and tube size

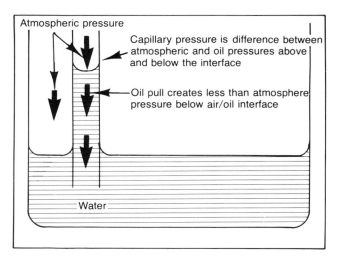

Fig. 371. Capillary pressure

When reservoir pressures drop because of production, gas dissolved in the oil expands and is released, forming a gas phase. As expansion occurs, fluids remaining in the reservoir are swept toward the wellbore, accompanying the characteristically rapid pressure decline (Figs. 372 and 373).

The pressure exerted upon the fluids in a reservoir by a gas cap above an oil leg can produce the oil as well as the gas. Inasmuch as the gas is under pressure, it is compressed and expands when the reservoir is produced (Figs. 374 and 375).

Pressure reduction in fields with gas cap drive is slower than with dissolved gas drive. Expected recoveries are higher as well.

Expansion of reservoir formation water is responsible for production of oil by a formation water drive (Fig. 376). Overburden pressure, tectonic deformation, and artesian pressure can increase the pressure of formation water to substantial levels. In good water drive reservoirs pressures decline slowly, and production rates are high (Fig. 377).

Comparisons of the three drive mechanisms show that water drive can be the most effective means of produc-tion. Oil recovery rates can reach seventy-five percent with a water drive.

Under normal circumstances, several drive mecha-nisms assist in the production of any one reservoir (Fig. 378). This is particularly important where one type of fluid drive declines but production is sustained by another type of drive with greater longevity. Figure 379 shows gas-oil ratio trends for various drive mechanisms as res-ervoir depletion occurs.

Gravitational segregation or gravity drainage is often classed as a fourth type of drive mechanism. In most reservoirs, gravity drive does not become a significant factor in oil recovery until reservoir pressure nears deple-tion. However, in some steeply dipping reservoirs, for example, the Hawkins Field in east Texas, gravity drain-age is a major recovery factor.

Reservoir Pressure

The preceding discussion alluded to various types of subsurface pressure resulting from a number of factors. For the most part, subsurface and therefore reservoir pressure comes from lithostatic, hydrostatic, or tectonic sources.

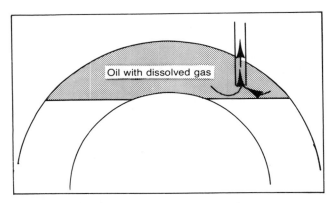

Fig. 372. Dissolved gas drive reservoir

DISSOLVED GAS DRIVE RESERVOIRS

Characteristics	Trend
1. Reservoir pressure	Declines rapidly and continuously
2. Surface gas-oil ratio	First low, then rises to maximum and then drops
3. Water production	None
4. Well behavior	Requires pumping at early stage
5. Expected oil recovery	5 to 30 per cent of original oil in place

Fig. 373. Dissolved gas drive reservoir performance

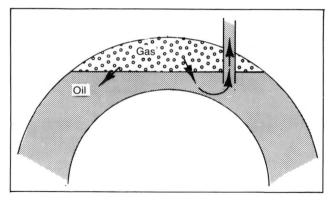

Fig. 374. Gas cap drive reservoir

GAS CAP DRIVE RESERVOIRS

Characteristics	Trend
1. Reservoir pressure	Falls slowly and continuously
2. Surface gas-oil ratio	Rises continuously in up-structure wells
3. Water production	Absent or negligible
4. Well behavior	Long flowing life depending upon size of gas cap
5. Expected oil recovery	20 to 40 per cent

Fig. 375. Gas cap drive reservoir performance

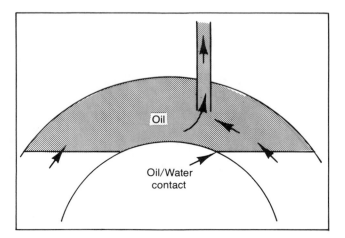

Fig. 376. Water drive reservoir

WATER DRIVE RESERVOIRS

Characteristics	Trend
1. Reservoir pressure	Remains high
2. Surface gas-oil ratio	Remains low
3. Water production	Starts early and increases to appreciable amounts
4. Well behavior	Flow until water production gets excessive
5. Expected oil recovery	35 to 75 per cent

Fig. 377. Water drive reservoir performance

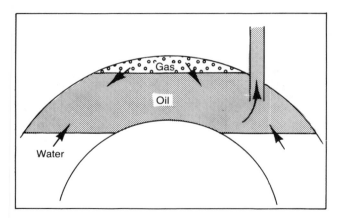

Fig. 378. Combination drive reservoir

Lithostatic pressure was mentioned in the section dealing with the origin of petroleum. It is derived from the weight of rock strata above rocks at any particular depth. Lithostatic pressure increases at the rate of one pound per square inch for each foot of depth and affects formation fluids, as well as the rocks themselves.

The weight of a column of fresh water increases at the rate of 0.433 pounds per square inch per foot of depth compared to 0.465 for sea water. Hydrostatic pressure is an important consideration where petroleum reservoirs have access to meteoric water and the buildup of a hydrostatic column.

Directed dynamic tectonic pressure can substantially increase formation pressure where the system remains closed and does not rupture. Gravity sliding, thrusting, and salt and shale tectonics can cause tectonic pressure.

High pressures normally develop where depositional loading of unstable, water-filled sediments occurs. Gravity translated to buoyancy and lateral sliding of unstable sediment masses commonly creates overpressured shaley zones in the Tertiary section of the Gulf of Mexico coastal area of Texas and Louisiana. Down-to-the-coast faulting due to gravity effects upon the unstable sediments creates pressures that are sometimes retained until the affected formations are drilled. Heavy drilling mud and special blow-out prevention precautions are necessary under such circumstances.

Lenticular sand bodies surrounded by strongly compacted dry shale can have abnormally low pressures because they do not compact as much as the shale. Impermeability of the shale precludes pressure equalization by meteoric water percolating downward from the surface.

Subnormal pressure can result from early sealing of a reservoir at the pressure compatible with the overburden and hydrostatic pressures at the time (Fig. 380). Additional deposition causes greater overburden but does not necessarily result in greater reservoir pressure, which creates subnormal pressure.

Geothermal Gradient. Crustal temperatures were discussed in the section dealing with the origin of petroleum. Temperature has bearing upon reservoir pressure, conditions, and the phases of the reservoir fluids, not to mention the actual generation of oil and gas. Temperatures normally increase about 1.5°F for each 100 feet in depth.

Reservoir Loss and Destruction

Geologic processes that are in constant operation form, preserve, and eliminate most things, including reservoirs.

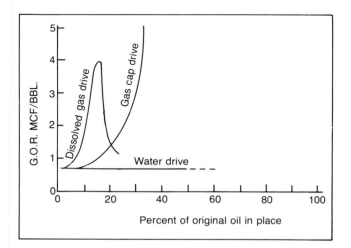

Fig. 379. Various drive performances

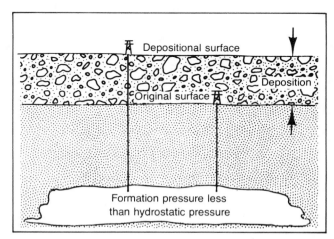

Fig. 380. Subnormal pressure occurrence

Reservoirs containing petroleum in trapping circumstances are routinely subject to change by a variety of processes.

1. Any petroleum-bearing trap is eventually subject to erosion and dissipation of the hydrocarbons.

2. Folding, faulting, fracturing, and intense manifestations of all of these can destroy any trap in a position to be affected by them. Extreme deformation accompanied by erosion can easily eliminate petroleum traps.

3. Exposure of hydrocarbons to weathering effectively eliminates them. Any time that deformation and erosion damage or eliminate a trap, weathering and oxidation complete the degradation of the petroleum.

4. Deep burial of a reservoir can eliminate the contained petroleum by increasing the temperatures beyond the normal tolerance levels. Oil in particular can be altered to gas by high temperatures increasing with depth.

Origin, migration, and accumulation of petroleum obviously involve numerous factors that are the primary considerations in the establishment of petroleum-related parameters. Hydrocarbons form from organic materials in complex ways that are not completely understood. They migrate in ways that are subject to debate. They accumulate under conditions that allow for reduced recoveries under some circumstances and greater recoveries under others.

Accumulation of petroleum in reservoirs prompted studies on how to recover it in terms of reservoir and petroleum characteristics and how they intimately relate to each other. Factors that affect reservoirs affect the fluids they contain and therefore the methods required for production.

Ultimately, petroleum accumulations can be eliminated by geologic processes. If they are not produced, they will be lost, as they have been since before the advent of Man and the petroleum industry.

12

Geological Considerations and Engineering Practices

Engineers routinely confront geologic problems in drilling, completion and recovery programs, because geologic factors determine reservoir geometry, size, and internal characteristics. Drilling rates, circulation parameters, hole conditions, and mud characteristics relate directly to internal formation factors. Exploration programs relate directly to reservoir geometry as well as internal reservoir characteristics. A summary of all geologic factors that relate to environments of reservoir deposition facilitates the determination of the types of engineering treatment used to affect oil and gas recovery. Cooperation between geologists and engineers improves the opportunities for maximum reservoir recovery if all variables are considered in any given recovery program. It is therefore beneficial for operating engineers to be familiar with geologic technology required for maximum oil or gas recovery.

The importance of geology in engineering operations involves relating rock characteristics to reservoir type, geometry, distribution, and quality. Accomplishment of this objective derives from study of rock types by sample and core examination, which establishes the lithology and depositional environment of the reservoir rock. Specific reservoir parameters including porosity, permeability, and thickness can also be determined by rock study. Information generated by rock studies establishes the size and shape of the reservoir and how its internal characteristics vary within its framework.

Reservoir porosity, permeability and quality studies must necessarily be augmented by well test information that relates to pressure and production performance. The ability of the reservoir to transmit fluids, maintain pressure, and produce petroleum under primary, second-ary, and tertiary recovery programs is contingent on its geological characteristics and the ability of the geologists and engineers to recognize and exploit these characteristics.

Oil and Gas Occurrences

Porous and permeable sedimentary rocks comprise the bulk of petroleum reservoirs. Sedimentary rocks are the only rock types that generate petroleum, although igneous and metamorphic rocks can serve as reservoirs.

Sedimentary reservoirs consist largely of clastics such as conglomerate and sandstone, or carbonates that make up reefs and banks. They have primary, depositional porosity and permeability and contain most of the world's petroleum. Fractured or weathered igneous, metamorphic, or impermeable sedimentary rocks serve as reservoirs much less frequently than depositionally and diagenetically porous and permeable sediments.

Occurrences and production of petroleum indicate that sandstone is more abundant, but carbonate rocks produce more petroleum (Fig. 381). The latter applies because of the substantial Middle East and Mexican production from carbonate reservoirs.

For reservoir rocks to be effective in accumulating oil and gas, they must be included in traps. Petroleum source rocks generate oil and gas that migrate into traps where they accumulate. The traps must contain porous and permeable reservoir rocks that are effectively sealed against petroleum escape by surrounding impermeable rocks. Localization of the petroleum is accomplished by the formation of a trap, which can be depositional or stratigraphic, deformational or structural, or a combination of

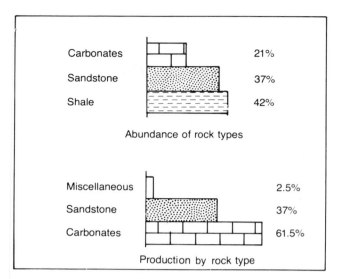

Carbonates   21%

Sandstone   37%

Shale   42%

Abundance of rock types

Miscellaneous   2.5%

Sandstone   37%

Carbonates   61.5%

Production by rock type

Fig. 381. Rock type abundance and production by rock type

the two where deposition and deformation combine. Regardless of how favorable a trap may appear, it will not contain petroleum unless it contains porous and permeable reservoir rocks.

Reservoir Rocks

For a rock to be an effective reservoir, it must contain adequate porosity and permeability that is properly sealed against erosion or tectonic destruction. Many reservoirs and traps have been generated and eliminated by erosion and deformation in the geologic past. Their occurrences in retrospect are presently manifest as dead oil, petroliferous odor, and light staining in cores and samples, or as dead-oil seeps and asphalt at surface exposures. Inasmuch as oil and gas are always being formed during the evolution of the earth, new accumulations will result. However, the millions of years required for generation and accumulation may be longer than we can afford to wait after we have exhausted our current supplies.

Geologic Factors and Reservoir Properties

Weathering, erosion, and deposition, which themselves are controlled by many factors, determine reservoir character from reservoir type and size to grain sizes, porosity and permeability, and type of cement. Diagenesis, which affects sediment subjected to the rock-forming process and generally reduces porosity and permeability, is an important additional factor.

Porosity and Permeability. A clastic rock with no in-

tergranular porosity or permeability can not act as a reservoir unless it is weathered or fractured. In that event, the intrinsic characteristics of the rock do not contribute to its reservoir capabilities and have no significance pertinent to its clastic origin.

Clastic rocks that retain depositional and diagenetic permeability are important as reservoirs because they intrinsically can transmit and contain petroleum. Porosity, as a function of rock fabric in clastic rocks, embraces a variety of factors with characteristics important to reservoir potential (Table 30) and concerns relations dealing with clastic granularity. All important clastic rock porosity parameters are related to granularity, which translates into reservoir potential.

Carbonate porosity is a function of many depositional factors and includes intergranular or interparticle porosity typical of the primary depositional environment (Table 31). There are other forms of carbonate porosity that relate to biological and diagenetic considerations and when eliminated or substantially reduced may be subsequently improved by fracturing or weathering.

Rock porosity is represented by void space and is expressed as a percentage of the total void space divided by the rock volume. Porosity of a rock determines how much fluid the rock can hold.

Permeability is the measure of how well a rock can transmit fluids through its pores. If the pores are not connected, the rock has no permeability regardless of the amount of porosity. Certain volcanic rocks contain abundant unconnected pores and although very porous are not permeable.

Table 30
Sandstone Reservoir Porosity

Porosity Factors	Characteristics
Sediment Primary Porosity	25–40%
Rock Porosity	10–30%
Primary Porosity	Intergranular
Ultimate Porosity	Intergranular; cemented; infilled
Pore Sizes	Diameter and throat sizes related to grain size and sorting
Pore Shapes	Related to grain shape, size and intergranular cement and clay
Pore size, shape, and distribution	Uniform in well-sorted homogeneous sandstone
Diagenetic influence	Porosity reduction by compaction, cementation, dissemination
Fracturing influence	Not a primary factor in sandstone
Permeability/porosity	Consistent and dependent upon grain size and sorting

After Choquette and Pray, 1970. Permission to publish by AAPG.

Table 31
Limestone Reservoir Porosity

Porosity Factors	Characteristics
Sediment Primary Porosity	40–70%
Rock Porosity	0–15%
Primary Porosity	Intercrystalline, interparticle
Ultimate Porosity	Variable-diagenetic-post diagenetic
Pore Sizes	Diameter and throat sizes not related to particle size and sorting
Pore Shapes	Variable; often not related to original particles
Pore Size, Shape and Distribution	Uniform to heterogeneous
Diagenetic Influence	Significant porosity reduction or enhancement
Fracturing Influence	Often significant
Permeability/Porosity	Variable; often independent of particle size and sorting

After Choquette and Pray, 1970. Permission to publish by AAPG.

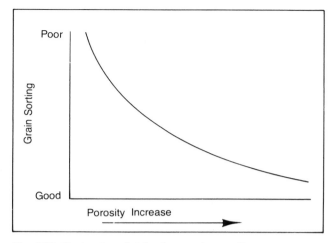

Fig. 382. Grain size distribution and porosity

Grain Size. The sizes of grains in clastic sedimentary rocks can be widely variable and relate directly to their mineral composition, types of weathering, distances and intensity of transporation, and ultimate environment of deposition. Large, uncemented grains usually indicate greater porosity and permeability than small grains, which provide lower porosity and permeability. Pore spaces between large grains are not as likely to be affected by permeability reduction caused by multiphase fluid flow, which reduces transmissibility through finer grained sediments.

Rocks containing grains of a single size are more porous and permeable than those comprising a variety of grain sizes (Fig. 382). This means that rocks consisting of well-sorted grains are better reservoirs than those that contain poorly sorted grains. It is intuitive that a clean sandstone, containing well-sorted grains, will have better porosity and permeability than a pebbly mudstone containing grains that range from very coarse to very fine (Fig. 383).

Grain Shape. Maximum porosity is achieved in a medium consisting of spherical grains. Angularity promotes the opportunity for interlocking grains, regardless of their size, and reduces porosity.

Grains oriented in one direction can increase rock permeability parallel to their long axes and reduce it normal to their long axes (Fig. 384). This is particularly true where small, flat shale grains provide no permeability normal to the fissility of the shale, but can allow the lateral movement of fluids along it during compaction.

Large, flat grains develop significantly greater lateral permeability than vertical permeability. Lateral and vertical permeability of large, flat grains is materially greater than for small, flat grains.

Vertical and lateral permeability for large, rounded grains is greater than for small, rounded grains and small, irregular grains. Lateral permeability in nonspherical, rounded grains is greater than vertical permeability because of compaction, however.

Grain Compaction. Spherical sand grains in the open

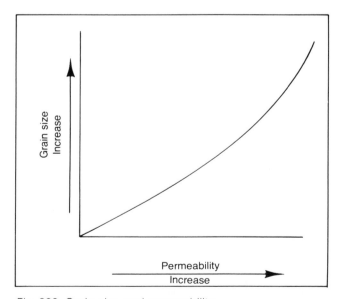

Fig. 383. Grain size and permeability

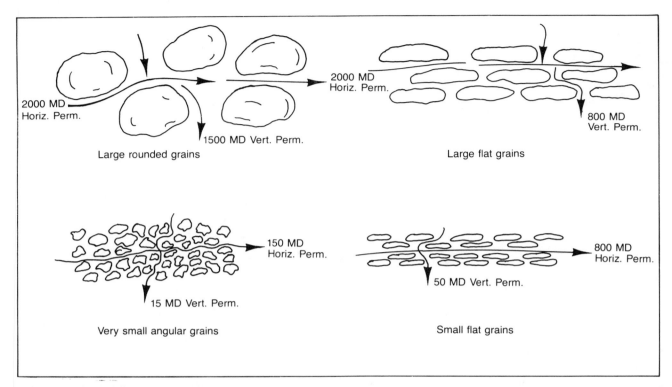

Fig. 384. Permeability and grain size and shape

packing configuration can produce nearly 48% porosity. The same grains, closely packed produce almost 26% porosity and compact no more, regardless of overburden.

Closely packed, well-sorted, spherical grains produce greater porosity than poorly sorted, angular grains but less than loosely packed spherical grains. Compaction of flat grains reduces vertical permeability, which is less than lateral permeability.

Matrix. Poorly sorted rock grains comprise coarse fragments with fine-grained matrix materials in the intergranular spaces (Fig. 385). Decrease in matrix grain size and an increase of matrix pervasiveness substantially reduces porosity and permeability as fine-grained materials effectively seal the pore spaces between the large grains. Matrix materials are often altered by diagenesis, subsurface water, or incipient metamorphism, and further reduce porosity and permeability by mineral and volume changes. Clays can be damaged by the introduction of extraneous fluids into wells and reduce well and reservoir potential.

Vugular Porosity. Carbonate rocks are susceptible to solution effects during diagenetic and post-diagenetic processes. Solution of carbonates may occur along frac-

tures, during cavern formation, or dissolving of granular or fossil components of the original rock. Dolomitization of limestone reduces the original rock volume and increases vugular porosity.

Improved reservoirs usually result where vugular porosity (Fig. 386) is formed, provided the new porosity remains. Some carbonate reservoirs develop good vugular porosity, which is subsequently eliminated by mineral infilling by precipitates from later solutions. Carbonate reservoirs in west Texas and eastern New Mexico have

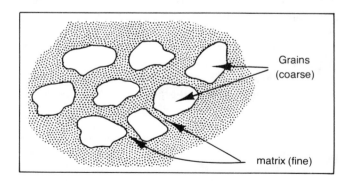

Fig. 385. Coarse grains and fine matrix

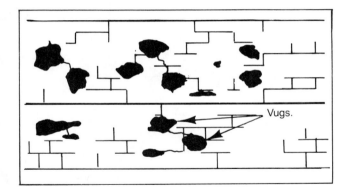

Fig. 386. Vugular porosity in carbonates

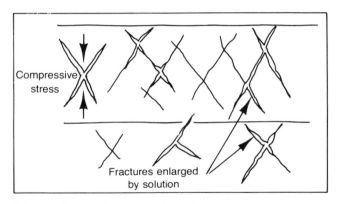

Fig. 387. Oriented fractures in rocks.

excellent vugular porosity in many areas. Much or all of the porosity is lost, however, where secondary anhydrite fills part or all of the vugs. Exploration and production programs in these reservoirs are designed to account for the amount of remaining vugular porosity.

Vugular porosity, leading to large cavern formation, offers drilling problems similar to those accompanying fracture-related caverns. Lost mud circulation and equipment damage are possible.

Fractures. All rock types are affected by fracturing, which can improve reservoir potential. Rock type and composition are important factors in determining how brittle the rock is and how much facturing will occur. Differentiation of natural tectonically related fractures from hydraulically induced fractures during drilling or well stimulation is important. Induced fractures are hydraulically held open to maintain reservoir conductivity and promote production.

Some very hard limestones and sandstones can have their intrinsically low reservoir potential greatly enhanced by fracturing. Igneous and metamorphic rocks, which only occasionally act as reservoirs, can act as reservoirs after they have been fractured.

Fracturing is related to regional tectonic stress and occurs in specific patterns (Fig. 387), which can be exploited in recovery programs. Fractures must be sufficiently developed to permit transmission of fluids through the rock, however.

Behavior of fracture sets is of importance to production operations. They can be open and permit fluid passage or closed and prevent fluid passage. Overburden-induced lithostatic stress usually causes natural fractures to be closed at depth. Under the circumstances, hydraulic fracture stimulation may be required to reopen and maintain the open-fracture condition to permit fluid production

through communication with the natural fracture system.

When naturally fractured rocks are encountered during drilling, partial or complete mud circulation loss may occur. Since carbonate rocks are very often fractured, solution can increase the net fluid transmissability of the rock. Lost mud circulation, resulting from cavern formation in fractured carbonate, is a significant drilling problem. In some areas, drilling into large caverns in carbonate rocks can result in dropping of the drill string and damage to equipment. Lost mud circulation can also be caused by drilling fluid hydrostatic pressure, which exceeds lithostatic pressure and creates an induced fracture system.

Inasmuch as fractures are usually oriented, they may create difficulties for waterflood recovery programs or they might illustrate preferred directions for fluid flow (Fig. 388). They can be held open while under pressure and reduce efficiency of the sweep by misdirecting the flow of the flood.

Cement. Intergranular cement can be composed of any mineral material capable of being distributed through a porous, uncemented, granular medium (Fig. 389). It has a direct effect upon the ultimate reservoir potential of the resulting rock. Cementation is a diagenetic process that occurs during the conversion of sediment to rock and usually accompanies compaction if effected at depth beneath a thickness of overburden. Cementation involves the precipitation of mineral material in the intergranular spaces of a sediment and locks the grains together. The amount of cement determines the amount of porosity and permeability reduction experienced by the compacted uncemented sediment.

Cementation can be partial or complete and as such is significant in the reservoir potential of any rock. If cementation is complete, the rock has no reservoir potential

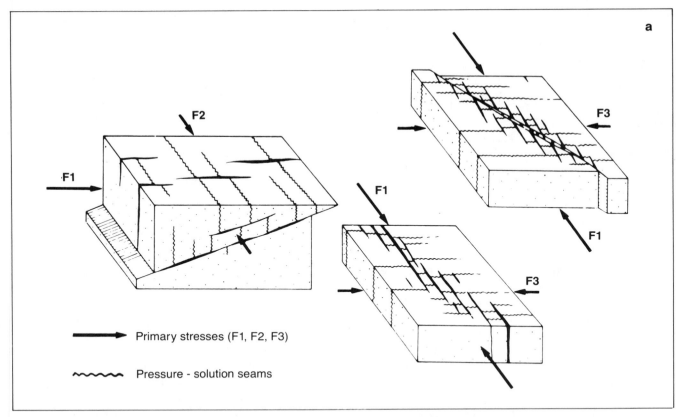

Primary stresses (F1, F2, F3)

~~~~~ Pressure - solution seams

Fig. 388. Model for development of extension fractures along which petroleum migration could occur. From du Rochet, 1981. Permission to publish by AAPG.

unless secondary solution of the cement or the rock grains takes place. Fracturing can also improve reservoir potential and may be a function of amount and type of cement.

Pore space remaining after cementation can be additionally reduced by secondary solutions disseminating

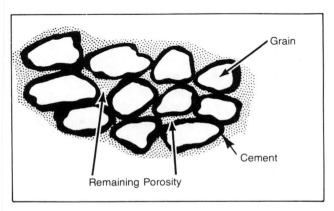

Fig. 389. Cemented clastic grains

mineral material into the available voids. The disseminated minerals can be compositionally different from the rock or its cement or can consist of slightly different minerals in which single ions are involved in chemical transfer and replacement.

Reservoir Type. Reservoirs are formed under many depositional conditions that relate directly to the parameters of rock type and grain size and shape. It is evident that stream channels are substantially different from offshore barrier islands, and within their very different depositional environments vary in dimension, shape, zonation, framework, and distribution. Since these factors are variable between the channel and barrier island, the internal characteristics of each are variable as well. Herein is the paramount reason for a complete understanding of the type of reservoir. A reservoir in an offshore barrier island cannot be produced in the same manner as a reservoir in a stream channel, which itself can not be exploited in the same way as a reef.

Reservoir Size. Petroleum recovery programs involve

well spacing and well perforation intervals, which are related to the size of the reservoir. Geologic features that occur as reservoirs in traps are a direct function of their depositional environments and the length of time those environments were extant relative to sediment source, transportation, etc. Recognition of geologic factors pertaining to framework of the reservoir body provides dimensional information that can be used to properly space wells and establish zones to be perforated while producing the maximum amounts of oil and gas.

Clastic Reservoirs

Clastic deposits vary significantly in origin and depositional environment. Their individual attributes can be very different though they have the common characteristic of consisting of transported and deposited fragments of pre-existing rock and organic materials. They are deposited by rivers, marine waves and currents, and by wind and gravity on land. Clastic depositional environments produce a wide variety of reservoir types (Fig. 390). Their geometries, framework, and internal characteristics must be properly considered in order to effectively produce oil and gas from them.

Most important clastic reservoirs consist of sandstone and conglomerate made of quartz, feldspar, and minor amounts of heavy mineral grains. Their grains are capable of being very mature, well-rounded, and well-sorted because they are resistant to mechanical and chemical breakdown. Other clastic reservoirs include those made of limestone grains or shell fragments that are not resistant to chemical and mechanical weathering and accumulate very nearly in place. Both types of clastic rocks can have good porosity and permeability and often perform as excellent reservoirs. Cement in the noncarbonate clastics can be silica, carbonate, a variety of oxides, and clay minerals (Fig. 391). Clastic carbonates are most often cemented by calcite or dolomite and can also be silicified at times. Clastic reservoirs are laid down under three primary depositional environments: continental, transitional, and marine (Figs. 392, 393, and 394).

Clastic sediments are made of fragments with size and shape that control the texture of the resulting clastic rock. These factors are important in determining the reservoir characteristics of the sediment. Compaction, cementation, and other diagenetic processes combine with the rock texture to ultimately determine its reservoir potential. These factors are also important in drilling technology as the formations are being penetrated. They affect drilling rates and how the formations react to drilling mud and circulation.

Drilling Rate. Tightly cemented, hard clastic formations often drill very slowly. A quartz sandstone containing silica cement will drill more slowly than a quartz sandstone with calcite cement. Drilling rates are a function of the competence of the formations encountered and the drilling engineering plan, which includes such items as hole diameter and coring point selection, bit weight, bit design, rpm, and properties of the drilling fluid to be used.

Some Gulf Coast sandstone formations are very poorly cemented and friable. They break up easily and drilling is accomplished by medium tooth jet bits that generate chips and provide jetting capacity sufficient to remove them from beneath the bit. Up to 6,000 feet of hole penetration per day have been made under favorable conditions. Hard, tightly cemented, siliceous quartz sandstones in the Rocky Mountains require button bits to drill 50 feet per day or less.

Most shale units drill more rapidly than most other rock types. Some hard, splintery shales effectively resist rapid drilling, and progress is made very slowly through them. Some very soft, sticky shales combine with water in the drilling fluid and clog the bit, preventing drilling progress. A bit change is warranted under the circumstances.

Shale Sloughing. Some soft shales will not maintain hole diameter because they are so poorly consolidated that they slough into the hole (Fig. 395) and create drilling problems. A formation of this type is often handled by altering the composition of the drilling mud to stabilize the shale.

Sand Washout. Unconsolidated sandstone formations are subject to caving and washing out in a manner similar to that affecting very soft shale (Fig. 395). If there is little intergranular cement, mud circulation and jet bit action may cause hole enlargement.

Circulation. Some clastic formations are very porous because they are well-sorted, very coarse-grained, poorly cemented or for other depositional and diagenetic reasons. Although most lost circulation results from fractures, loss of drilling fluid can occur in very porous formations, and in extreme cases the hole can be lost if improper mud is used. An important function of drilling mud is to build a filter cake on the sides of the wellbore to prevent or reduce loss of drilling fluid. Inasmuch as different formations have different characteristics, mud programs must be devised for specific drilling requirements.

Mud Composition. Soft shale and clay-rich sandstone can contribute clay constituents to drilling fluids during drilling operations. Changes in drilling mud composition,

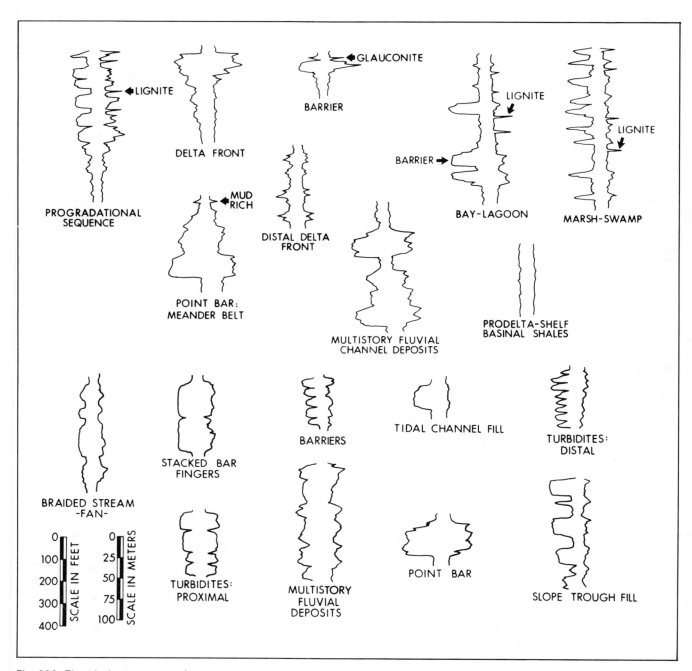

Fig. 390. Electric log patterns of some clastic depositional forms. From Garcia, 1981. Permission to publish by AAPG.

PRIMARY POROSITY MODELS

Initial Porosity

Probably close to 40%, and may have been short-lived. In outcrop, primary porosity is about 30% in the St. Peter and about 20% in the Mount Simon.

Early Cementation

Inferred from open packing and high proportion of cement. Early cements in the Mount Simon and St. Peter include quartz overgrowths and carbonate.

Stable Framework

Mechanical reduction of porosity by rotation and sliding of grains prior to later, possibly sutured contacts, which are very common where both sandstones are deeply buried.

SECONDARY POROSITY MODELS

Leached Cement

Cements dissolved and dissolution usually selective as to kind, place, and completeness. All cements in the Mount Simon and St. Peter Sandstones have been partially dissolved, especially where deeply buried.

Leached Framework

Framework grains dissolved and dissolution usually selective as to kind, place, and completeness. Quartz grains in the St. Peter are commonly etched, and in the Mount Simon feldspar and quartz grains are dissolved.

Framework Collapse

Dissolution has removed both cements and framework so that a large, irregular pore system exists with significant permeability. Both sandstones locally have framework collapse and oversize pores. Most advanced stage of secondary porosity.

Open network of etched calcite rhombohedra partially fill primary pores in the St. Peter Sandstone and generally occurs at shallow to moderate depths.

Dolomite cement fills primary pores and is preserved in the shallow St. Peter as a late, replacement cement in the Mount Simon. Dolomite has been dissolved, increasing porosity in both sandstones.

Anhydrite is present as a cement where the St. Peter is deeply buried, but is locally dissolved and enhances porosity. Maximum abundance of anhydrite cement is 4%.

Chert and chalcedony cement the St. Peter near outcrop in southern Wisconsin at depths less than 400 ft (121 m).

Quartz overgrowths are common almost everywhere in both the Mount Simon and St. Peter Sandstones and locally completely fill primary pore space.

Chlorite cement is present as a fracture filling in the St. Peter (right) and kaolinite and chlorite are pore fillings in the Mount Simon (left).

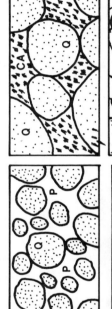
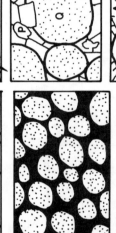

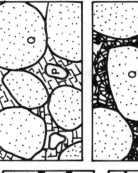
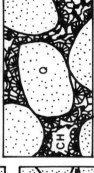

Fig. 391. Porosity and cement models for St. Peter and Mt. Simon sandstones of the Illinois Basin. From Moholick, et al., 1984. Permission to publish by AAPG.

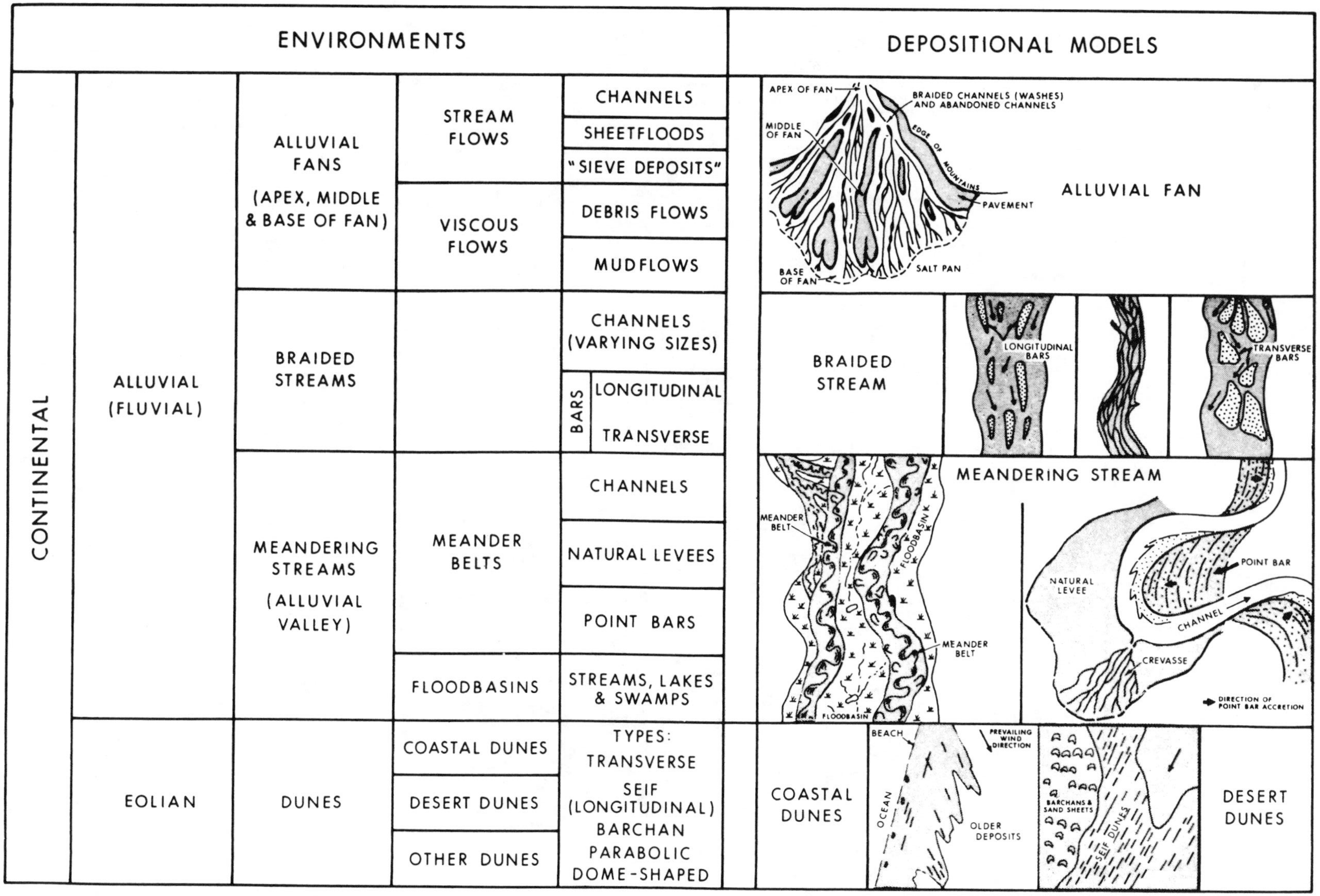

DEPOSITIONAL MODELS

Fig. 392. Alluvial and eolian clastic sedimentation models. From LeBlanc, 1973. Permission to publish by AAPG.

ENVIRONMENTS

CONTINENTAL			
ALLUVIAL (FLUVIAL)	ALLUVIAL FANS (APEX, MIDDLE & BASE OF FAN)	STREAM FLOWS	CHANNELS
			SHEETFLOODS
			"SIEVE DEPOSITS"
		VISCOUS FLOWS	DEBRIS FLOWS
			MUDFLOWS
	BRAIDED STREAMS		CHANNELS (VARYING SIZES)
		BARS	LONGITUDINAL
			TRANSVERSE
	MEANDERING STREAMS (ALLUVIAL VALLEY)	MEANDER BELTS	CHANNELS
			NATURAL LEVEES
			POINT BARS
		FLOODBASINS	STREAMS, LAKES & SWAMPS
EOLIAN	DUNES	COASTAL DUNES	
		DESERT DUNES	TYPES: TRANSVERSE
			SEIF (LONGITUDINAL)
		OTHER DUNES	BARCHAN
			PARABOLIC
			DOME-SHAPED

Depositional model diagrams include: ALLUVIAL FAN (EDGE OF MOUNTAINS, APEX OF FAN, MIDDLE OF FAN, BASE OF FAN, SALT PAN, PAVEMENT, BRAIDED CHANNELS (WASHES) AND ABANDONED CHANNELS); BRAIDED STREAM (TRANSVERSE BARS, LONGITUDINAL BARS); MEANDERING STREAM (POINT BAR, CHANNEL, CREVASSE, NATURAL LEVEE, DIRECTION OF POINT BAR ACCRETION, FLOODBASIN, MEANDER BELT); COASTAL DUNES (OCEAN, BEACH, PREVAILING WIND DIRECTION, OLDER DEPOSITS); DESERT DUNES (SEIF DUNES, BARCHANS & SAND SHEETS).

ENVIRONMENTS

DEPOSITIONAL MODELS

TYPES OF DELTAS

TRANSITIONAL

DELTAIC

UPPER DELTAIC PLAIN
- MEANDER BELTS — CHANNELS / NATURAL LEVEES / POINT BARS
- FLOODBASINS — STREAMS, LAKES & SWAMPS

LOWER DELTAIC PLAIN
- DISTRIBUTARY CHANNELS — CHANNELS / NATURAL LEVEES
- INTER-DISTRIBUTARY AREAS — MARSH, LAKES, TIDAL CHANNELS & TIDAL FLATS

DELTA FRONT
- FRINGE — INNER — RIVER-MOUTH BARS / BEACHES & BEACH RIDGES / TIDAL FLATS
- DISTAL — OUTER

BIRDFOOT-LOBATE DELTA

DELTAIC PLAIN ENVIRONMENTS
MEANDER BELT
OLDER COASTAL PLAIN
DISTRIBUTARY CHANNEL
RIVER MOUTH BARS
FLOOD BASIN
SWAMPS
LAKE
MARSH
ALLUVIAL PLAIN
UPPER DELTAIC PLAIN
LOWER DELTAIC PLAIN
INNER FRINGE
OUTER FRINGE
SUBAQUEOUS PORTION OF DELTA

CUSPATE-ARCUATE DELTA

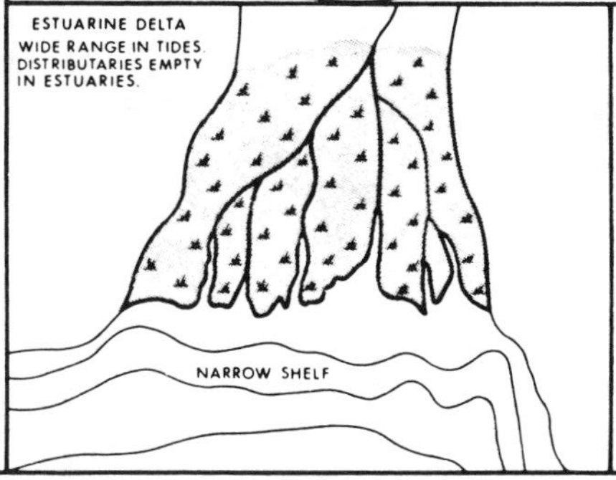

TIDAL CHANNELS
COASTAL SAND BARRIERS
MARINE CURRENTS
NARROW SHELF

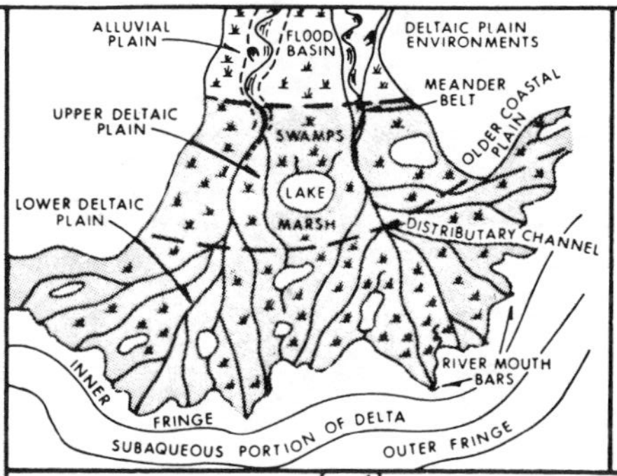

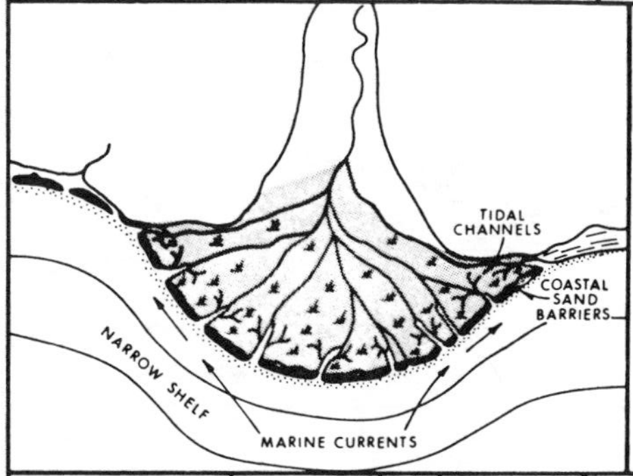

ESTUARINE DELTA

ESTUARINE DELTA WIDE RANGE IN TIDES DISTRIBUTARIES EMPTY IN ESTUARIES.
NARROW SHELF

Fig. 393. Deltaic clastic sedimentation models. From LeBlanc, 1973. Permission to publish by AAPG.

ENVIRONMENTS

DEPOSITIONAL MODELS

TRANSITIONAL

Coastal inter-deltaic:
- Coastal plain (subaerial): Barrier islands — BACK BAR, BARRIER, BEACH, BARRIER FACE, SPITS & FLATS, WASHOVER FANS; Chenier plains — BEACH & RIDGES; Tidal — TIDAL FLATS
- Subaqueous: Tidal — TIDAL FLATS, TIDAL DELTAS; LAGOONS — SHOALS & REEFS; TIDAL CHANNELS; SMALL ESTUARIES

MARINE

- Shallow marine: Shelf (neritic) — INNER, MIDDLE, OUTER; SHOALS & BANKS
- Deep marine: CANYONS; FANS (DELTAS); SLOPE & ABYSSAL; TRENCHES & TROUGHS

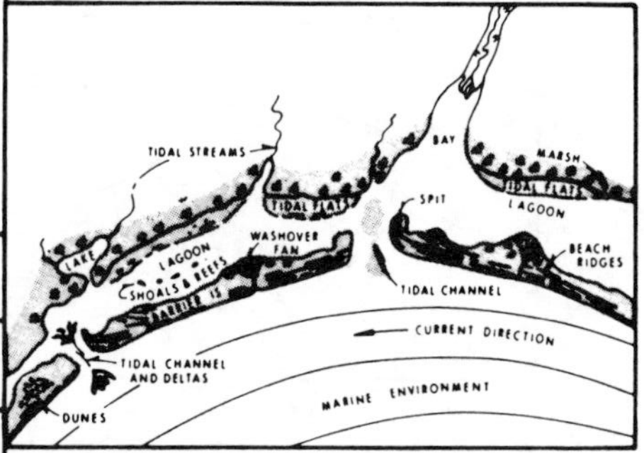

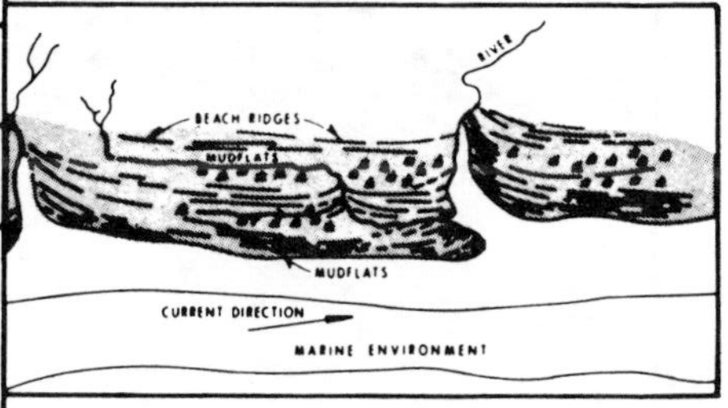

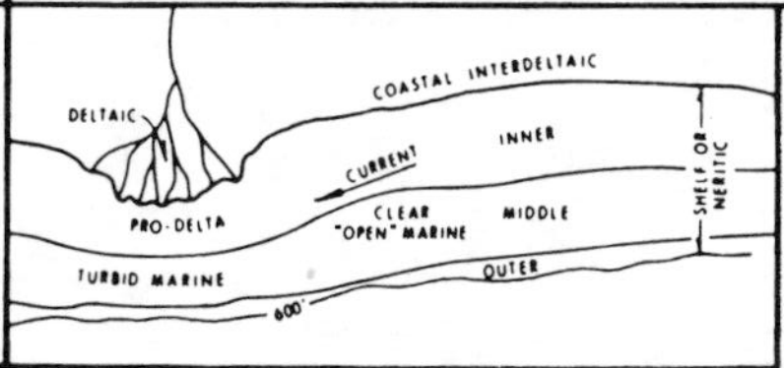

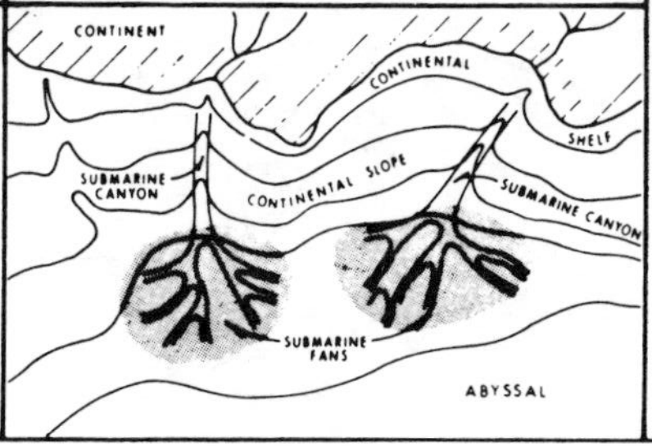

BARRIER IS. COMPLEX

CHENIER PLAIN

SHALLOW MARINE

DEEP MARINE

Fig. 394. Coastal-interdeltaic and marine sedimentation models. From LeBlanc, 1973. Permission to publish by AAPG.

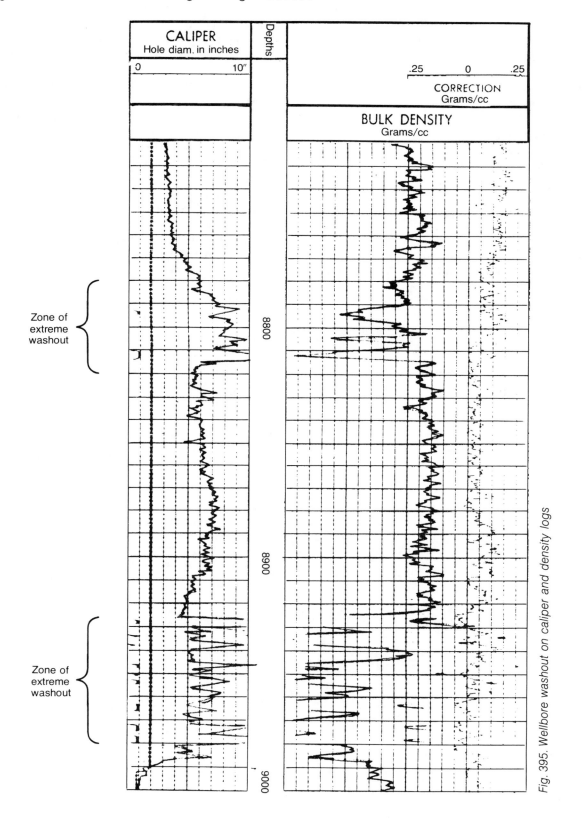

Fig. 395. Wellbore washout on caliper and density logs

particularly when high mud weights are reduced by formation contamination, can lead to drilling problems particularly where high pressures are encountered. Personnel trained in drilling fluid technology are often assigned to important drilling wells, particularly when formation characteristics are not well known.

Bit Effects. Hard, well-cemented clastic formations appreciably slow drilling rates. They also increase bit wear and require more frequent bit changes. Hard clastic formations are usually best penetrated by short-tooth or button bits, which create shattering impact in the rock instead of the cutting effect a long tooth bit has upon a soft formation.

Carbonate Reservoirs

Deposition of carbonate rocks can be clastic, biologic, or physico-chemical as a precipitate. Clastic carbonates comprise detrital fragments mechanically deposited as carbonate sand or gravel. Biologically deposited carbonates occur as sediment resulting from an association with the metabolic processes of plants or animals. Some overlap is possible in that sediments composed of cemented shell fragments, for example, comprise broken, biologically formed shells mechanically deposited in a clastic environment. Some carbonate rocks are precipitated directly from fresh or sea water because of temperature or solution concentration variations. Classification of carbonate rocks involves types of grains, cement, depositional texture and diagenesis to represent environments of growth, deposition and alteration. These parameters illustrate original depositional conditions, diagenetic alteration and present characteristics and processes that bear directly upon the reservoir potential of the rock (Figs. 396, 397, 398, 399, 400).

Grain/Micrite Ratio	% Grains	GRAIN TYPE					Organic Frame-Builders	No Organic Frame-Builders
		Detrital Grains	Skeletal Grains	Pellets	Lumps	Coated Grains		
9:1	~90%	Detrital Ls.	Skeletal Ls.	Pellet Ls.	Lump Ls.	Oolitic Ls. Pisolitic Ls. Algal Encr. Ls.	Coralline Ls. Algal Ls., Etc.	Caliche Travertine Tufa
1:1	~50%	Detrital-Micritic Ls.	Skeletal-Micritic Ls.	Pellet-Micritic Ls.	Lump-Micritic Ls.	Oolitic-(Pisolitic, Etc.) Micritic Ls.	Coralline-Micritic Ls. Algal-Micritic Ls., Etc.	
1:9	~10%	Micritic-Detrital Ls.	Micritic-Skeletal Ls.	Micritic-Pellet Ls.	Micritic-Lump Ls.	Micritic-Oolitic (Pisolitic, Etc.) Ls.	Micritic-Coralline Ls. Micritic-Algal Ls, Etc.	
		←		Micritic Limestone		→		

Fig. 396. *Limestone classification according to matrix, grains and origin. From Leighton and Pendexter, 1962. Permission to publish by AAPG.*

DEPOSITIONAL TEXTURE				
Original components not bound together during deposition				Original components were bound together during deposition . . . as shown by intergrown skeletal matter, lamination contrary to gravity, or sediment-floored cavities that are roofed over by organic or questionably organic matter and are too large to be interstices
Contains mud (particles of clay and fine silt size)		Grain supported	Lacks mud and is grain supported	
Mud supported				
Less than 10 percent grains	More than 10 percent grains			
Mudstone	Wackestone	Packstone	Grainstone	Boundstone

Fig. 397. *Limestone classification according to depositional texture. From Dunham, 1962. Permission to publish by AAPG.*

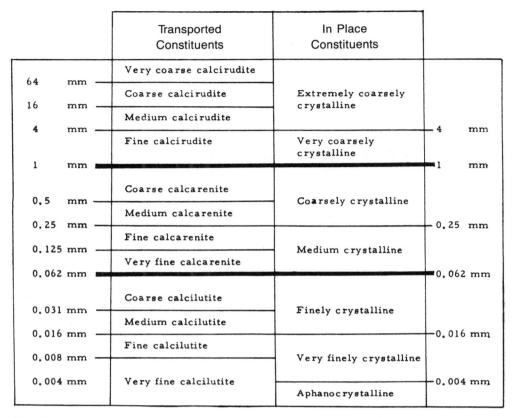

	Transported Constituents	In Place Constituents	
64 mm	Very coarse calcirudite	Extremely coarsely crystalline	
16 mm	Coarse calcirudite		
4 mm	Medium calcirudite		4 mm
	Fine calcirudite	Very coarsely crystalline	
1 mm			1 mm
0.5 mm	Coarse calcarenite	Coarsely crystalline	
0.25 mm	Medium calcarenite		0.25 mm
0.125 mm	Fine calcarenite	Medium crystalline	
0.062 mm	Very fine calcarenite		0.062 mm
0.031 mm	Coarse calcilutite	Finely crystalline	
0.016 mm	Medium calcilutite		0.016 mm
0.008 mm	Fine calcilutite	Very finely crystalline	
0.004 mm	Very fine calcilutite		0.004 mm
		Aphanocrystalline	

Fig. 398. Grain-size classification of carbonate rocks. From Folk, 1959. Permission to publish by AAPG.

All carbonate rock types display texture (Fig. 396) and some types of primary porosity and permeability, which are subject to secondary enhancement or reduction (Fig. 401). They can be described according to organic and/or inorganic parameters (Figs. 401 and 402). Carbonate porosity is initially a function of depositional environment. Subsequent diagenetic, biologic, or tectonic effects can significantly alter the ultimate type of porosity. Evolution of porosity and associated pore types can involve enlargement by solution (Fig. 403) or reduction by infill or a sequential combination of the two (Fig. 404).

Carbonate rocks are usually deposited as primary clastic, biologic, or precipitated limestone. Secondary dolomitization is important as porosity consideration in many carbonates. Dolomitization of porous limestone can reduce intrinsic porosity. Dense, low porosity lime mud can be favorably affected by dolomitization, which increases porosity.

Carbonate porosity is developed in three successive stages (Fig. 405). Predepositional primary porosity results when first-formed sedimentary material is finally deposited and buried. Depositional primary porosity develops during final sediment deposition at its ultimate burial site. After final deposition, secondary or post depositional porosity occurs. Burial stages accompanying deposition represent deep and shallow burial. Early, shallow burial is eogenetic, deeper burial is mesogenetic, and late-stage deep burial, subject to deformation and erosion, is telogenetic.

Each of the burial periods is subject to development of the type of porosity associated with the primary and secondary depositional and diagenetic processes. Any porosity created after final deposition is considered secondary porosity. Rock properties such as hardness, porosity, and permeability are related to primary and secondary factors. They are significant in drilling and production operations since they determine such factors as drilling rates and production characteristics.

Drilling Rate. Formation hardness affecting drilling rate relates to the type of carbonate rock being penetrated. A loosely cemented limestone composed of fossil fragments can be easily drilled. Dense, "lithographic" lime-

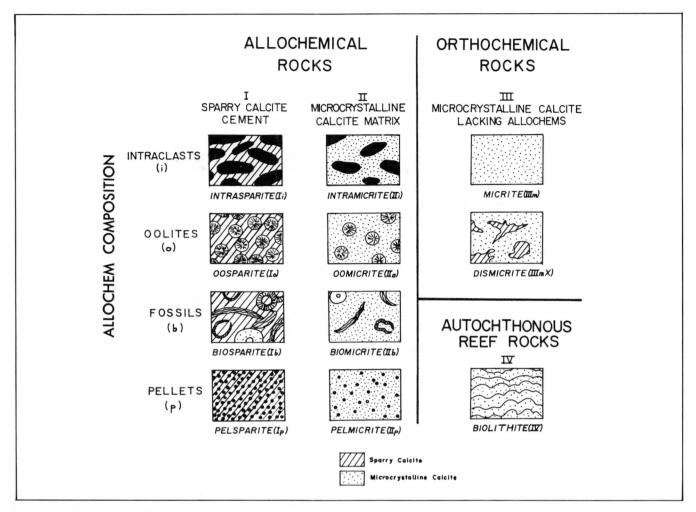

Fig. 399. Graphic classification of limestones. From Folk, 1959. Permission to publish by AAPG.

stone has low porosity, is very finely crystalline, can be very hard, and will drill very slowly.

Circulation. Carbonate formations usually produce a wellbore that remains in gauge with little or no washout. Cavernous porosity or fractures can disrupt or completely eliminate circulation as some or all of the drilling fluid flows into large voids in the formation. Special techniques are required to eliminate lost circulation problems.

Fractures. Fractures in carbonate rocks are subject to groundwater solution activity, which may produce a wide ranging network of geometrically oriented cavernous porosity. Carbonate rocks are brittle and shatter where stress is concentrated. Shattering carbonates can significantly improve reservoir performance (Fig. 406). Some shattered reservoirs have significant circulation problems, especially if solution is involved.

Bit Effects. Hard carbonates may react to drilling like hard metamorphic or igneous rocks or tightly cemented or silicified clastic sediments. A short-tooth or button bit may be necessary for best drilling rates and minimum bit wear. Soft, fragmental, and marly carbonate rocks are best drilled with longer tooth bits.

Production and Geologic Factors

Geological factors are of major importance in production operations, because reservoirs are so variable between depositional environments and within individual environments that manifest fundamental similarities. It is evident that a river channel deposit is geometrically, texturally, dimensionally, and depositionally different than a

CLASSIFICATION OF CARBONATE ROCKS

			Limestones, Partly Dolomitized Limestones, and Primary Dolomites (see Notes 1 to 6)					Replacement Dolomites[7] (V)	
			>10% Allochems Allochemical Rocks (I and II)		<10% Allochems Microcrystalline Rocks (III)		Undisturbed Bioherm Rocks (IV)	Allochem Ghosts	No Allochem Ghosts
			Sparry Calcite Cement > Microcrystalline Ooze Matrix	Microcrystalline Ooze Matrix > Sparry Calcite Cement	1–10% Allochems	<1% Allochems			
			Sparry Allochemical Rocks (I)	Microcrystalline Allochemical Rocks (II)					
Volumetric Allochem Composition — <25% Intraclasts / <25% Oölites / Volume Ratio of Fossils to Pellets		>25% Intraclasts (i)	Intrasparrudite (Ii:Lr) Intrasparite (Ii:La)	Intramicrudite* (IIi:Lr) Intramicrite* (IIi:La)	Intraclasts: Intraclast-bearing Micrite* (IIIi:Lr or La)			Finely Crystalline Intraclastic Dolomite (Vi:D3) etc.	Medium Crystalline Dolomite (V:D4)
		>25% Oölites (O)	Oösparrudite (Io:Lr) Oösparite (Io:La)	Oömicrudite* (IIo:Lr) Oömicrite* (IIo:La)	Oölites: Oölite-bearing Micrite* (IIIo:Lr or La)			Coarsely Crystalline Oölitic Dolomite (Vo:D5) etc.	Finely Crystalline Dolomite (V:D3)
		>3:1 (b)	Biosparrudite (Ib:Lr) Biosparite (Ib:La)	Biomicrudite (IIb:Lr) Biomicrite (IIb:La)	Fossils: Fossiliferous Micrite (IIIb: Lr, La, or Ll)			Aphanocrystalline Biogenic Dolomite (Vb:Dl) etc.	
		3:1–1:3 (bp)	Biopelsparite (Ibp:La)	Biopelmicrite (IIbp:La)					
		<1:3 (p)	Pelsparite (Ip:La)	Pelmicrite (IIp:La)	Pellets: Pelletiferous Micrite (IIIp:La)			Very Finely Crystalline Pellet Dolomite (Vp:D2) etc.	etc.

Most Abundant Allochem.

<1% Allochems column: Micrite (IIIm:L); if disturbed, Dismicrite (IIImX:L); if primary dolomite, Dolomicrite (IIIm:D)

Undisturbed Bioherm Rocks (IV): Biolithite (IV:L)

Evident Allochem / *Allochem Ghosts*

Fig. 400. *Classification of carbonate rocks. From Folk, 1959. Permission to publish by AAPG.*

barrier island deposit. Identification of the type of reservoir and its depositional environment permits development of a production program compatible with its known parameters. However, to consider all sandstone reservoirs, for example, as the same and to produce them in the same manner can result in reduced production and recovery.

Secondary factors affect reservoirs after deposition and must be considered part of the development program. Careful study of the reservoir and its primary and secondary features are important in any drilling, development, and production program.

Depositional Environment. Recognition of depositional environment permits the design of a production pattern based upon reservoir geometry and concentrates upon the most favorable portions of the reservoir. Beach or barrier island sandstones, for example, are coarsest, best-sorted, and most permeable in their upper portions. Channel and point bar sandstones are coarsest and most permeable in their lower portions. Depositional environment is responsible for the distribution of reservoir potential in each case.

Diagenetic History. Diagnesis controls the reservoir character of a sediment as it is converted to solid rock. Compaction, cementation, solution, and recrystallization (Table 32) are significant in the ultimate reservoir potential of any sediment as porosity and permeability are enhanced or diminshed. Reservoir potential of some carbonate rocks has been greatly enhanced by dolomitization porosity. The same rocks have been infilled by secondary anhydrite in some areas where they have no reservoir potential. Appreciation of the distribution of diagenetic effects can significantly affect exploration and development programs.

Tectonic–Structural History. Some rocks of low intrinsic reservoir potential become important reservoirs when structurally deformed. Regional tectonic stress is responsible for the development of fracture systems that provide porosity and permeability in rocks that otherwise have no reservoir potential. Igneous and metamorphic rocks not usually considered as primary reservoir rocks have been fractured in many places around the world where they produce oil and gas.

Local structural activity often induces fracture porosity

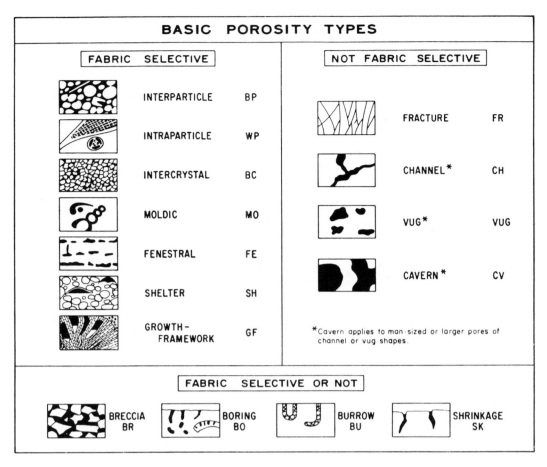

BASIC POROSITY TYPES

FABRIC SELECTIVE

	INTERPARTICLE	BP
INTRAPARTICLE	WP
INTERCRYSTAL	BC
MOLDIC	MO
FENESTRAL	FE
SHELTER	SH
GROWTH-FRAMEWORK	GF

NOT FABRIC SELECTIVE

	FRACTURE	FR
CHANNEL*	CH
VUG*	VUG
CAVERN*	CV

*Cavern applies to man-sized or larger pores of channel or vug shapes.

FABRIC SELECTIVE OR NOT

BRECCIA BR   BORING BO   BURROW BU   SHRINKAGE SK

MODIFYING TERMS

GENETIC MODIFIERS

PROCESS

SOLUTION	s
CEMENTATION	c
INTERNAL SEDIMENT	i

DIRECTION OR STAGE

ENLARGED	x
REDUCED	r
FILLED	f

TIME OF FORMATION

PRIMARY	P
pre-depositional	Pp
depositional	Pd
SECONDARY	S
eogenetic	Se
mesogenetic	Sm
telogenetic	St

Genetic modifiers are combined as follows:

PROCESS + DIRECTION + TIME

EXAMPLES:
solution-enlarged	sx
cement-reduced primary	crP
sediment-filled eogenetic	ifSe

SIZE* MODIFIERS

CLASSES			mm†
			— 256 —
MEGAPORE	mg	large	lmg
			— 32 —
		small	smg
			— 4 —
MESOPORE	ms	large	lms
			— 1/2 —
		small	sms
			— 1/16 —
MICROPORE	mc		

Use size prefixes with basic porosity types:
mesovug	msVUG
small mesomold	smsMO
microinterparticle	mcBP

* For regular-shaped pores smaller than cavern size.

† Measures refer to average pore diameter of a single pore or the range in size of a pore assemblage. For tubular pores use average cross-section. For platy pores use width and note shape.

ABUNDANCE MODIFIERS

percent porosity	(15%)
or	
ratio of porosity types	(1:2)
or	
ratio and percent	(1:2) (15%)

Fig. 401. Carbonate rock pore system classification. From Choquette and Pray, 1970. Permission to publish by AAPG.

Class		Crystal or Grain Size (mm.)	Usual Appearance (Luster)	Approximate Matrix Porosity % Not Visible (12×–18×) A	Visible Porosity (% of Cutting Surface) Size of Pore—mm.			Approximate Total Porosity Per Cent	
					−0.1 B	0.1–2.0 C	+2.0 D	A + B	A + C
I Compact	L*	0.4 or more	Resinous to Vitreous†	2	e.g. 10	e.g. 15	‡	12	17
	M	0.2							
	F	0.1							
	VF	0.05		5	e.g. 10	e.g. 15		15	20
II Chalky	VF	0.05 or less	Chalky†	15	e.g. 10	e.g. 15		25	30
III Sucrose	F	0.1	Finely sucrose	10	e.g. 10	e.g. 15		20	25
	M	0.2							
	L	0.4	Coarsely sucrose	5	e.g. 10	e.g. 15		15	20

*L = large (coarse); M = medium; F = fine; VF = very fine.
†Where cuttings are between vitreous and chalky in appearance, designate as I-II or II-I. Samples are considered in VF group when grain or crystal size is difficult to distinguish (12×–18×). Place in F group if grains are easily distinguished.
‡When pores are greater than about 2.0 mm. and therefore occur at edge of cuttings (e.g., sub-cavernous pores), amount of such porosity is indicated by % of cuttings in an interval showing evidence of large pores.
Symbols:
III F-B$_{10}$ = Finely sucrose (therefore, matrix porosity about 10%), visible porosity about 10%, total porosity about 20%.
(II-I) VF-A = Chalky to vitreous, very fine texture (therefore, matrix porosity about 8%), no visible porosity, total porosity about 8%.

Fig. 402. Limestone classification including texture and porosity. From Archie, 1952. Permission to publish by AAPG.

in rocks with otherwise low reservoir potential. It is important that proper location of tectonically induced reservoir potential be understood and delineated.

Weathering and Erosion. Chemical and mechanical weathering and erosion often enhance reservoir potential of rocks exposed at the ground surface. Carbonate rocks subject to surface solution develop good porosity. Fracture systems can be enhanced by the freezing of percolating groundwater. Solution of intergranular cement in quartz sandstone can improve porosity.

Intergranular Material. Material between grains in clastic rocks, whether cement or pore filling, directly affects porosity and permeability (Fig. 407). Some intergranular materials are sufficiently mobile to migrate into

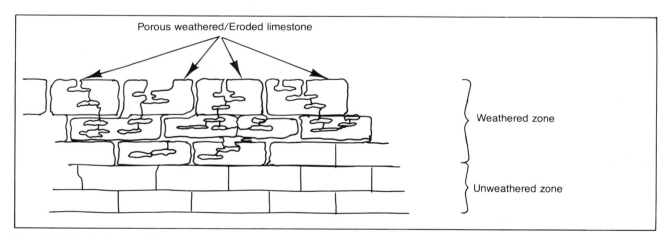

Fig. 403. Weathering and limestone porosity

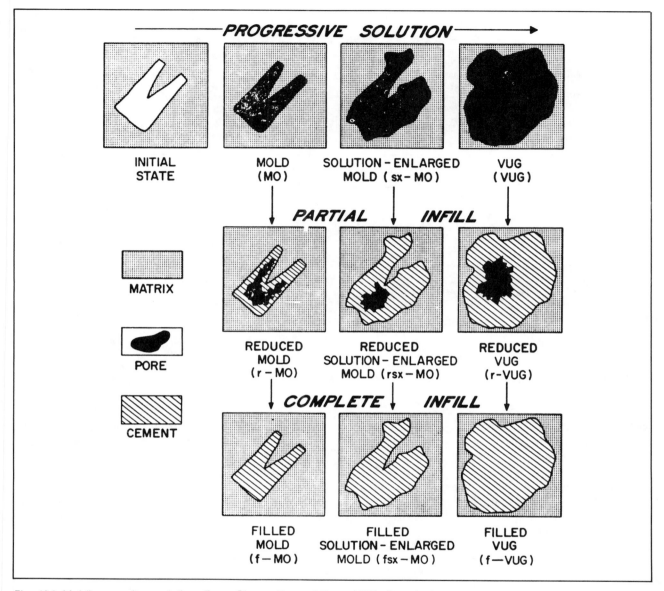

Fig. 404. Moldic porosity evolution. From Choquette and Pray, 1970. Permission to publish by AAPG.

pores where they obstruct fluid flow. This may be particularly true at fluid withdrawal points, such as casing perforations or production screens and gravel packs where material accumulates and restricts flow.

Some types of intergranular clay materials react adversely to drilling or completion fluids. They can absorb water, swell, and thoroughly clog a production system. Uncemented sand bodies may require sand control techniques to prevent movement of loose sand to the wellbore.

Hydrocarbon Traps

Oil and gas accumulate in porous, permeable reservoir rocks. Modes of accumulation vary with environments of deposition and deformational history. Accumulation styles can involve structural or stratigraphic parameters or a combination of them.

Structural Traps. Deformation of reservoir rocks often produces mechanically derived containers in which oil and gas accumulate (Fig. 408). The dome or doubly

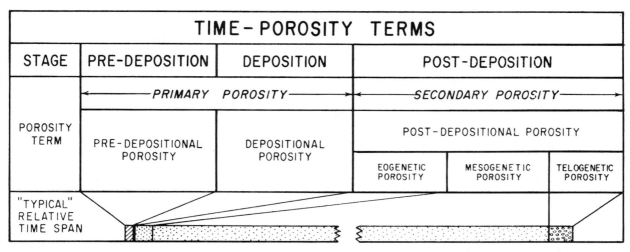

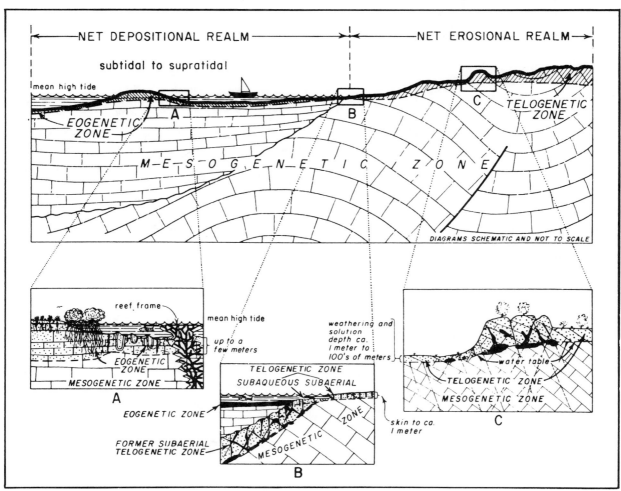

Fig. 405. Porosity creation and modification in carbonates. From Choquette and Pray, 1970. Permission to publish by AAPG.

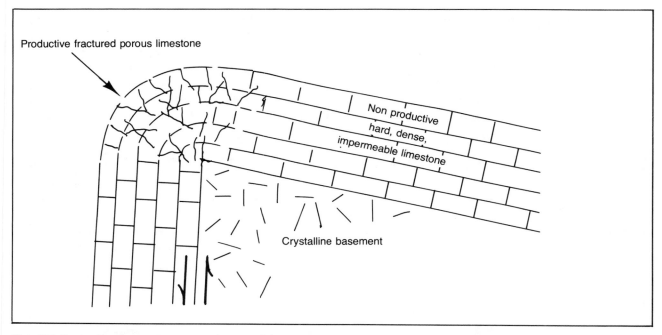

Fig. 406. Tectonic effect upon reservoir potential

plunging anticline is the simplest type of structural trap and has produced great amounts of oil and gas throughout the world. A dome or anticline can be gently and simply folded, severely and complexly folded, or faulted and fractured. Increased anticlinal complexity promotes interpretational complexity, which can complicate exploration and development procedures.

Faults can effectively trap oil and gas. Fault blocks can be simple, or complex as multiple faulted traps. Dipping reservoirs can be effectively sealed by simple or complex faults. Faulting often enhances the trapping capacity of anticlines or anticlinal structural noses.

Stratigraphic Traps. Traps are often developed by depositional processes that produce discrete sedimentary bodies with finite limits and hydrocarbon retention integrity (Fig. 409). A barrier island, beach, stream channel, or reef are limited by their modes of deposition, which control their geometries and reservoir parameters. Porosity and permeability are distributed within sedimentary bodies according to their types of deposition, which con-

Table 32
Diagenetic Effects and Porosity

	Porosity	
	Increase	*Decrease*
Compaction		x
Precipitation		
Cementation		x
Dissemination		x
Solution	x	
Recrystallization	x	x
Fracturing	x	

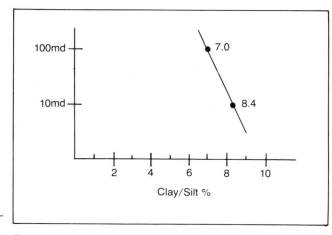

Fig. 407. Intergranular clay/slit and sandstone permeability

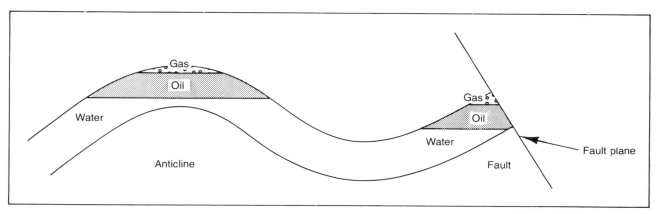

Fig. 408. Structural traps

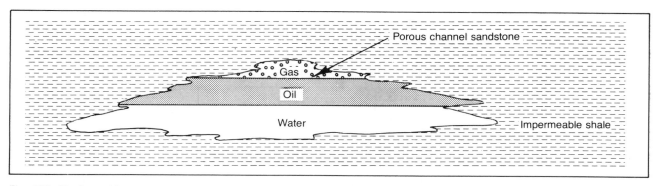

Fig. 409. Stratigraphic trap

trol their shapes, sizes, and hydrocarbon accumulation capabilities.

Inasmuch as depositional geometry can be variable, exploration and production programs should reflect the parameters of the individual reservoir under consideration. A river channel varies in length, width, thickness, and sinuosity. The dimensional and internal parameters of a reef vary similarly. Both potential reservoir bodies have characteristics that must be identified to achieve maximum oil and gas production and ultimate recovery.

Combination Traps. Traps that combine structure and stratigraphy are combination traps (Fig. 410). A faulted channel retains hydrocarbons by changes in porosity and permeability and by fault closure along the dimension not so restricted. Anticlinally deformed beaches or offshore bars combine structural closure with lateral permeability limitations. A faulted reef provides similar retention potential.

Unconformities. Oil and gas accumulations can associate with unconformities, which involve combinations of stratigraphy, deformation, and erosion. Angular unconformities (Fig. 411) involve tilting before erosion and subsequent deposition.

Deposition over an erosion surface in flat-lying strata produces a disconformity (Fig. 412). Oil and gas accumulate in the high portions of the old topography below the unconformity.

Sediments that unconformably overlie an eroded igneous or metamorphic surface produce a nonconformity (Fig. 413). If the crystalline rocks below the unconformity are fractured or weathered, they can act as a reservoir. The overlying sediments can provide the hydrocarbons to the crystalline rocks and seal them in as well.

Salt and Shale Structures. Salt and shale move in response to sediment-loading and deform the rocks associated with them. Traps form with the deformation and occur along the margins and crests of salt and shale domes and pillow structures (Fig. 414).

Reservoir Fractures. Rock porosity and permeability derive from intergranular, intercrystalline, or fracturing

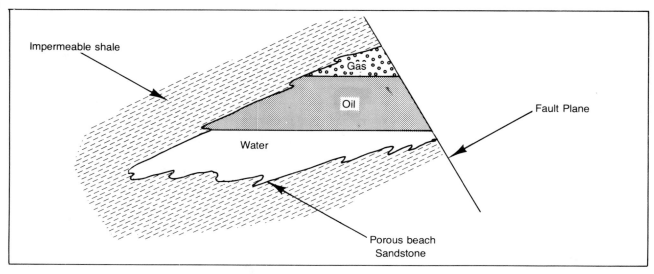

Fig. 410. Combination stratigraphic/structural trap

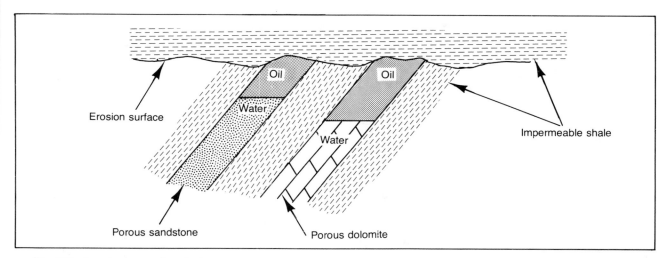

Fig. 411. Angular unconformity traps

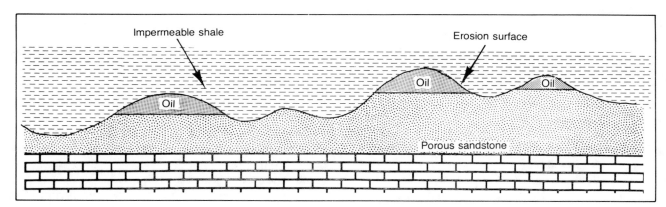

Fig. 412. Disconformity traps

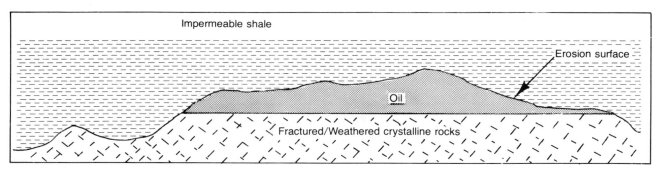

Fig. 413. Nonconformity trap

considerations. Fractures can affect rocks with high or low interparticle porosity and permeability and often improve reservoir potential. Dense, impermeable rocks benefit most from fracturing, which is oriented according to regional or local stress and provides avenues of fluid flow.

Open fractures enhance reservoirs. Closed or mineralized fractures do not. Drilling, hydraulic fracturing, waterflooding, and acidizing can force fracture systems to open and permit fluid flow.

Fractures cause drilling problems when they are sufficiently developed to permit loss of drilling fluids. Large

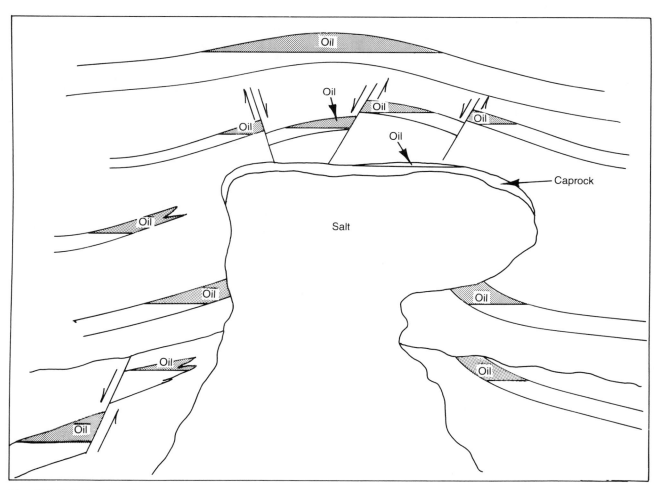

Fig. 414. Salt structure traps

fractures may cause complete circulation mud loss, particularly if they occur in rocks with cavern formation.

Surface manifestations of fractures are common, especially in hard, brittle cap rock. However, fractures occur in all rock types and often control surface drainage patterns that develop in softer rocks.

Fracture reservoir enhancement has improved recovery potential in the Elk Basin Field in Montana and Wyoming. Fracturing of the chalk reservoir in the North Sea Ekofisk Field resulted from salt intrusion from below (Fig. 415).

Seals

Production and drilling personnel are particularly concerned with trap seals in reservoir evaluation.

> . . . *Effective seals for hydrocarbon accumulations are typically thick, laterally continuous, ductile rocks with high capillary entry pressures. Seals need to be evaluated at two differing scales: a **micro** scale and a **mega** or prospect scale . . . Geologic work can be focused on the characteristic seal problems that plague classes of prospects. Anticlines have relatively little seal risk. Stratigraphic traps and faulted prospects have substantial seal risks* . . . (From Downey, 1984)

Micro Characteristics of Seals

> . . . *the quality of a seal . . . is determined by the minimum pressure required to displace connate water from pores or fractures in the seal, thereby allowing leakage.* (From Downey, 1984).

Macro Characteristics of Seals

Rock materials form seals of petroleum traps. Their properties determine how effectively they act as seals and comprise:

> Lithology
> Ductility
> Thickness
> Stability

Petroleum traps consist of structural stratigraphic and combination styles. Structural traps require deformation. Stratigraphic traps are depositional and combination traps include deformation and deposition. All three categories require stratigraphic integrity to isolate petroleum accumulations. No structural trap can exist without a strati-

graphic seal of some sort. Stratigraphic traps, which intrinsically have seals, often do not require deformation or petroleum accumulations.

Cut-and-fill structures including channels and tidal inlets can provide effective seals (Figs. 416 and 417).

Unless potential reservoir reefs are effectively sealed they will not accumulate petroleum (Figs. 418 and 419).

Variability of sealing and trapping mechanisms can involve deposition (Figs. 416–423) and deformation (Figs. 424–427).

Geologic Studies For Engineers

Production and reservoir engineers should approach each project with knowledge of the reservoirs with which they work. Inasmuch as reservoirs vary sedimentologically and are formed under a variety of depositional conditions, rock types, structures, geometries, and reservoir characteristics are different. Therefore, drilling and production procedures for a channel sandstone will be significantly different than those for a carbonate reef. Failure to recognize individual parameters can result in costly mistakes that may reduce or prevent oil and gas production from improperly treated reservoirs. It is important that geologic information be used in drilling and production to the extent that proper procedures are used. Appropriate data are often easily obtained from a geologist on location or a production geologist assigned to field development. If it is not already prepared, the engineer should develop sufficient data to permit him to understand his field, reservoir, and well.

Development of pertinent data involves examination of geologic reports, cores and samples, logs, and test data to provide information in the following parameters.

Stratigraphy

Rock sequences are important to the drilling of wells and how reservoirs will be produced. The engineer should be familiar with rock information with which to plan his program.

1. Rock type
2. Depositional environment
3. Depositional history
4. Postdepositional history
5. Bedding thickness
6. Grain composition
7. Grain size
8. Grain shape

9. Grain sorting
10. Grain maturity
11. Cement
12. Diagenetic parameters
13. Porosity
14. Permeability

Geometry

Reservoir geometry determines the dimensions and their orientations of the producing body.

1. Length, width, thickness, and shape
2. Lateral and vertical facies changes
3. Lateral and vertical porosity and permeability changes

Structure

Structural configuration of the reservoir affects its closure, volume of recoverable reserves, fluid content, and locations of wells.

1. Structural type
2. Structural history
3. Position of structural axis
4. Size of structure
5. Faulting
6. Fracturing
7. Closure
8. Strikes and dips of faults and reservoir beds

Fluid Content

Fluid content of a reservoir determines its segregation patterns, fluid contacts, and types of drive mechanisms.

1. Water composition
2. Oil type
3. Oil gravity
4. Oil pour point
5. Oil viscosity
6. Oil sulfur content
7. Gas type
8. Gas composition
9. Fluid phases
10. Fluid contacts

Pressure Data

Formation pressure and its maintenance can be significant in ultimate recovery of oil and gas. Pressures can sometimes be used to correlate tectonically separated formations.

Integration of geologic parameters by production and development engineers utilizes all of the above information applicable to a specific situation. However, study methods utilize similar sequences of development to permit the most effective data use in the production of a specific field (Table 33).

Table 33
Study Program for Field Development

Core and Sample Study

    Lithology
    Stratigraphy
    Facies Distribution
    Depositional Environment
    Reservoir Parameters

Log Study

    Stratigraphic Thickness
    Correlation
    Reservoir Geometry
    Structure

Map Study

    Lithology
    Facies Distribution
    Stratigraphic Thickness
    Structure

Drill Stem Test Study

    Fluid Content
    Fluid Composition
    Fluid Flow
    Formation Pressure
    Production Data

Engineering Study

    Reservoir Character
    Reservoir and Fluid Volume
    Production Data
    Recovery Data
    Recovery Potential

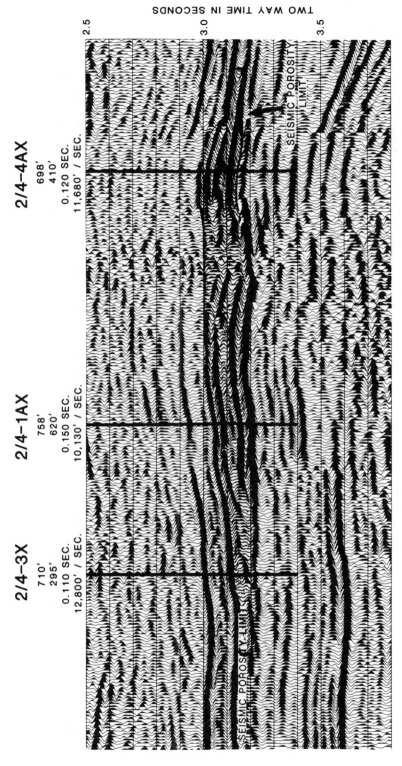

Fig. 415. Ekofish seismic section showing axial chalk fracturing. From Van den Burk, et al., 1980. Permission to publish by AAPG.

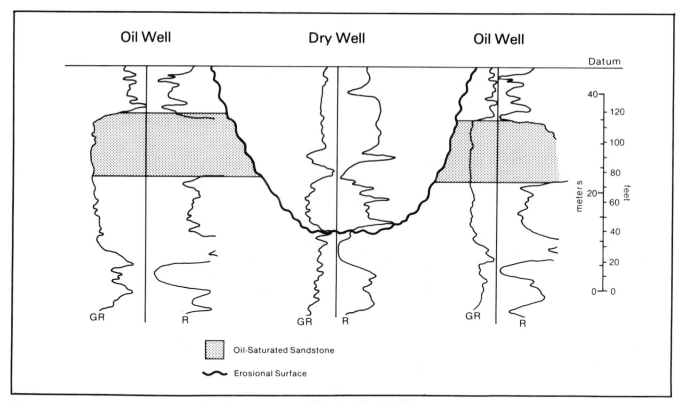

Fig. 416. Geophysical log cross section showing interruption in continuity of sandstone reservoir. Fine-grained sediments in dry well represent abandonment of either tidal inlet, which cut across beach trend, or later stage fluvial channel. From Tilley and Longstaffe, 1984. Permission to publish by AAPG.

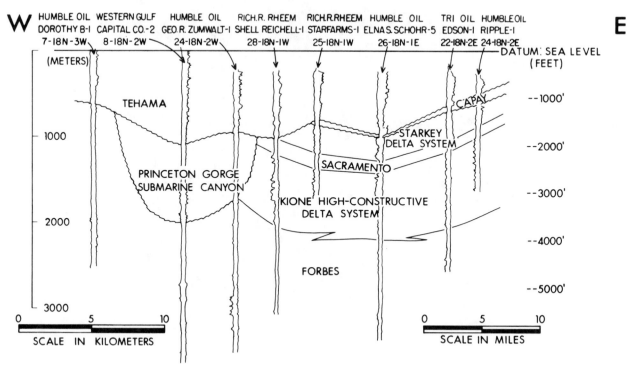

Fig. 417. Princeton Gorge fill seal for productive Kione delta system, Sacramento Valley, California. From Garcia, 1981. Permission to publish by AAPG.

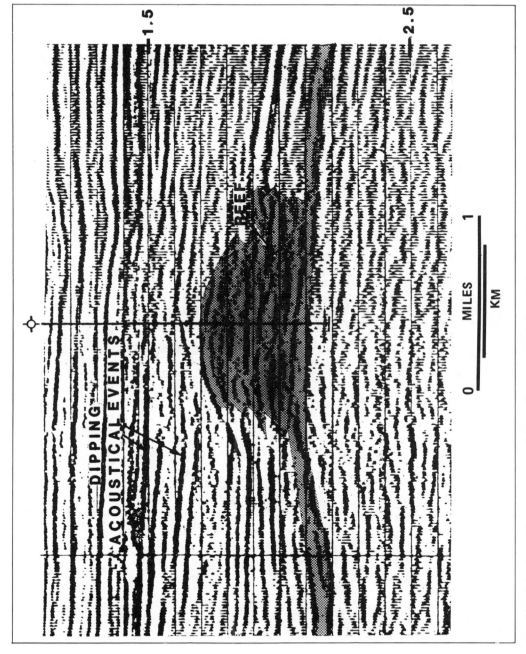

Fig. 418. Seismic section across a reef. From Downey, 1984. Permission to publish by AAPG.

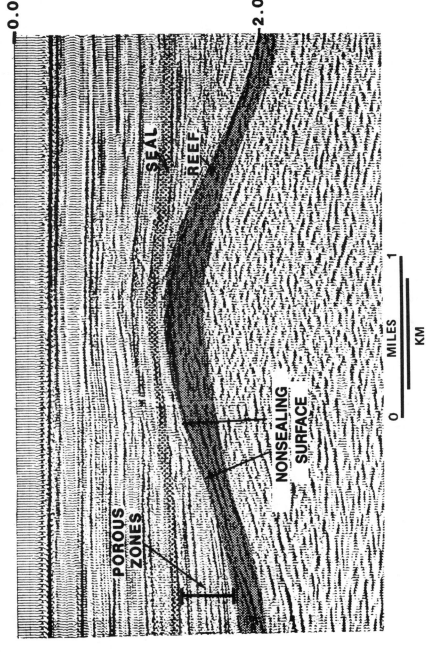

Fig. 419. Porous and permeable stratigraphic units onlapping reef indicate that flanks of reef probably lack a sealing surface. From Downey, 1984. Permission to publish by AAPG.

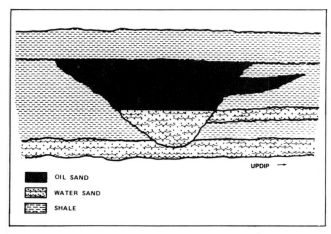

Fig. 420. Schematic cross section of channel sandstone incised into varying lithologic units; hydrocarbon column is limited by lateral seals. From Downey, 1984. Permission to publish by AAPG.

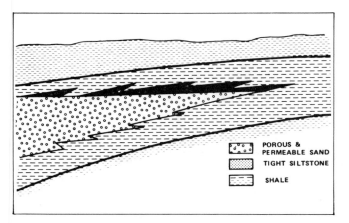

Fig. 421. If porous and permeable sands change abruptly into shales, hydrocarbon column will all be in reservoir quality rock. From Downey, 1984. Permission to publish by AAPG.

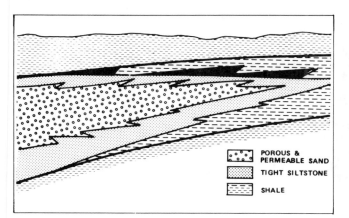

Fig. 422. If rate of change from reservoir to seal is gradual, a limited hydrocarbon column is restricted to nonreservoir "waste-zone" facies. From Downey, 1984. Permission to publish by AAPG.

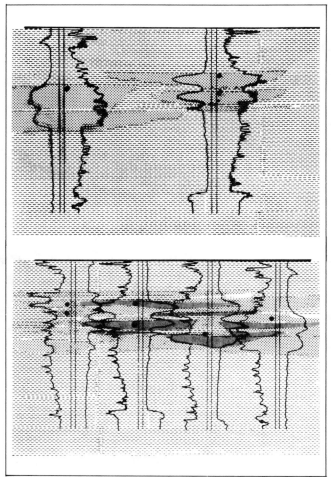

Fig. 423. Upper: Early, widely spaced drilling demonstrated oil staining in upper portions of cannel sandstones. Lower: In-fill drilling demonstrates such numerous cut-and-fill channel sequences that little lateral seal lithology could be demonstrated. From Downey, 1984. Permission to publish by AAPG.

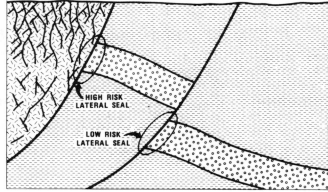

Fig. 424. Schematic cross section of potential fault closure traps; indicated trap against shallow basement block will have substantial lateral seal risk. From Downey, 1984. Permission to publish by AAPG.

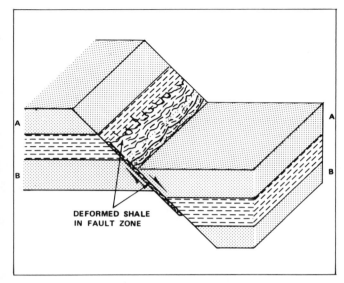

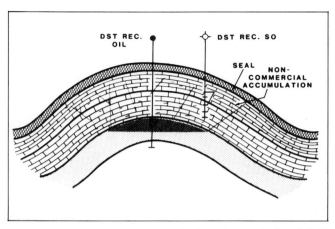

Fig. 427. Tight limestones in this fold are not seals for oil field, but are noncommercial parts of accumulation. From Downey, 1984. Permission to publish by AAPG.

Fig. 425. Dragging of undercompacted clays into fault plane can locally emplace sealing material along a fault. From Downey, 1984. Permission to publish by AAPG.

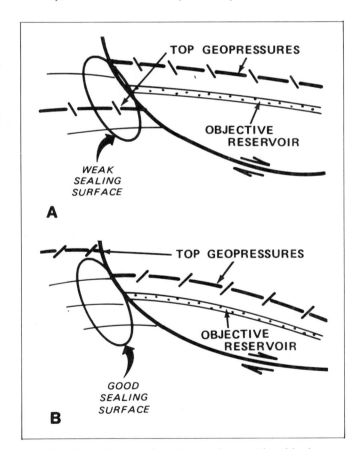

Fig. 426. Dip attitudes of sediments in trapping block provide information as to likelihood of a sealing surface. From Downey, 1984. Permission to publish by AAPG.

13

Rocks, Reservoirs, and Recovery Techniques

Rocks

We have discussed rock characteristics in various places in this text. This chapter is designed to summarize rock properties in the context of how they are obtained and how they are important to petroleum-related considerations. Emphasis on rock properties follows the primary theme of rocks as reservoirs and how reservoir characteristics are important to petroleum accumulations.

Rock Sampling

The rock-sampling techniques previously discussed offer examples to study in terms of their reservoir properties, how fluids can pass through and accumulate in the rocks, how fracture systems affect them, and ultimately how they can be best treated to maximize primary and subsequent production.

Some sampling methods are preferable to others but can cost more than the value of the information they provide. Regardless of cost, however, reservoir rock studies are best conducted on conventional cores and are least satisfactorily conducted on well cuttings. Sampling techniques in the order of favorability for rock studies are:

1. Conventional and diamond coring methods, which recover the maximum amount of rock material if the rock is hard and consolidated.

2. Rubber sleeve coring methods, which are best for the recovery of soft, unconsolidated rock. Rubber sleeve, conventional, and diamond-coring methods have best recoveries for soft and hard rocks.

3. Sidewall cores, which can provide good sample

recoveries. However, the cores are smaller than conventional cores.

4. Outcrop samples, which are easily obtained and when consisting of fresh rock can be useful in reservoir rock studies. There are decided limitations to outcrop samples because they may not truly represent the same formation in the subsurface many miles away. Outcrop samples are often weathered to an extent that they are not representative.

Outcrop samples can adequately represent the lithology of the formation they come from. Their use in representing absolute reservoir conditions in the subsurface is less than complete.

Porosity and Permeability

Reservoir porosity and permeability are controlled by many factors that relate to transportation and deposition of sediments, diagenesis, and deformation. The study of porosity and permeability, from the generic standpoint, is largely the function of the geologist as he examines the geologic processes involved in reservoir formation. Engineers, on the other hand, are not necessarily concerned with how the reservoirs were developed. They are, instead, interested in the present condition of the reservoir and how it can be made to perform at its maximum.

Geological considerations in reservoir analysis relate to how porosity and permeability are formed because geologic processes are involved. Engineers measure the porosity and permeability of the rock after it has formed and relate these parameters to recovery of hydrocarbons.

Total, effective, and non-effective porosity are the engineering considerations that bear directly upon how a reservoir can hold fluid. These terms apply to the physical condition of the reservoir and can be measured during core studies in terms of the rock mass.

The ability of the reservoir to transmit fluid determines its productive capabilities relative to its fluid content and ultimate recovery. Permeability is measurable as a reservoir characteristic as well.

Factors that control porosity and permeability have geologic rationales, which in their combination result in a certain type of reservoir. Many of these factors cannot be specifically measured because their actual history and the related passage of time are not easily and precisely determined.

Geologists know that geologic processes, constantly and variably in operation, produce certain types of rocks that have characteristics that occur within ranges. These processes are not easily measured because their actual occurrence has passed and only their results remain. In this regard the engineer is fortunate. His task is to observe and measure what is before him. The geologist, however, must deal with the past and make educated estimates of what he believes has happened.

Since reservoir characteristics are controlled by rock-related factors that are controlled by environmental processes it is evident that porosity and permeability are developed or eliminated by a continuum instead of a single event.

Grain size and shape are a function of mineral composition, transportation, maturity, deposition, and time, in approximately that order. Since these factors vary greatly, an absolute measurement of how a particular reservoir achieved certain degrees of these parameters is impossible.

Grain packing is a direct result of grain size, shape, overburden, and time. All of these are variable in the same context. These factors relate to porosity and permeability, which obviously can change as the factors change.

Particle movement in pore space can be physically or chemically controlled during and after diagenesis. These movements affect the reservoir and assist in the development of the reservoir into its condition when it is first drilled. Clogging or damaging of a reservoir by drilling activities are engineering considerations and take into account proper mud types and weights to be used.

Fluid Flow

Fluid flow within reservoirs is contingent upon the physical character of the rock and the types of fluids it contains. Both of these factors are geologically related because they are the result of geologic processes. However, fluid flow is measurable and the effects of multiple fluid phases are predictable. This applies to phase interference and the role of pore space and capillarity and is an engineering consideration. Note that all quantitative engineering considerations are controlled by geologic processes that are not quantitative. The processes are not quantitative; the results of the processes are.

Reservoirs

Reservoirs comprise different rock types, geometrics, and depositional processes. It is important to realize that reservoirs are indeed variable, that they are not of uniform thickness, porosity, and permeability, and that they are not of infinite lateral extent. Reservoirs change vertically and laterally and do so in response to the environments in which they were deposited.

Consider the differences between a beach and a channel. Their depositional environments are completely different, which means that sediment distribution and individual dimensions are different as well. The same condition applies to a barrier island and a reef. However, if the beach consists of quartz sand and the reef consists of limestone, significant compositional variations are added to the geometric and environmental differences. So, with all of the possible variations in reservoirs caused by depositional differences, the basic make-up of any reservoir is very specific and virtually unique.

Reservoir Properties

Geologically-related reservoir properties are caused by their transportational and depositional modes, which control all of the following factors:

1. Depositional and environmental regimes, which determine rock parameters, composition, and type of deposit.
2. Lateral and vertical extent of the depositional body.
3. Homogeneity or heterogeneity of the deposit.
4. Types of facies changes internal or external to the depositional body.
5. Types of sedimentary features and how they relate to the depositional geometry.
6. Diagenesis of the deposit and its affect upon the ability of the body to act as a reservoir.
7. Fluid content and saturation of the depositional body, which are related to rock character, grain size cement,

and compaction. Fluid content and saturation control fluid contacts and react according to fluid density, fluid phases, capillarity of the reservoir, and reservoir pressure.

8. Reservoir fluid characteristics are related to original groundwater composition, host rock composition, secondary water composition and temperature, pressure, and viscosity factors.

If reservoir heterogeneity and variability are recognized, the geologist and engineer who are evaluating and producing the reservoir can understand that specifically applicable procedures are necessary to achieve desired objectives. Otherwise, reservoir-treatment mistakes are certain to occur. Mistakes can cost an operator a great deal of time and money if, in fact, the reservoir is not damaged beyond reclamation.

Geologic Considerations

Reservoir characteristics are like geological parameters in that there are methods of study that can contribute to subsequent qualitative and quantitative measurements. Geologic reservoir analysis can lead into engineering reservoir analysis by providing fundamental data in a manner similar to exploration of a prospective area. Geologic study of a reservoir produces information upon which future engineering analysis can be based.

The geology of a reservoir involves its thickness, lithology, composition, size, rounding and sorting of grains, cement, depositional geometry, structure, diagenetic history, and depositional history. These parameters are fundamental to the understanding of the reservoir and how it came into being. They relate to the physical and chemical condition of the reservoir and promote an understanding of its engineering condition. These factors permit not only the definition of rock properties but can ultimately control well spacing and production programs.

Geologic reservoir analysis concerns the total physical evaluation of the reservoir and resembles an exploration program. It makes use of geologic methods to define the reservoir.

1. Well-sample logs, electric logs, core data, and test data can be correlated to define vertical and lateral reservoir characteristics (Fig. 428). Lithology, thickness, and distribution of the reservoir can be portrayed on cross-sections constructed from correlations.

2. Stratigraphic and structural cross-sections show what the composition and shape of the reservoir are (Fig. 429).

3. Structure, lithologic, facies, and isopach maps show pertinent variations in a reservoir group in a particular field (Fig. 430).

4. Fossil identification and zonation permits precise age determinations of reservoir rocks and is an aid in establishing the depositional and deformational histories.

Recovery Techniques

Once an exploration and wildcat drilling program confirms the presence of petroleum, the objective turns to maximum production of the petroleum in the reservoir. An initial recovery program is designed to produce the maximum petroleum at the lowest cost and is a function of the reservoir and its characteristics. Obviously, the primary recovery program will be influenced by the size of the reservoir, its thickness, type of drive, and how reservoir pressures can be conserved to promote maximum recovery. In the long term, the secondary and tertiary recovery programs must also be considered during the primary program.

Production methods vary with reservoirs, types of drive and reservoir pressures. High pressure reservoirs can be produced by allowing the petroleum to flow. Low or no pressure reservoirs are produced by the installation of a pumping system (Fig. 431).

Primary Recovery

Any development program may require drilling of a number of wells to fully appreciate the field characteristics. When these are determined, spacing, reservoir treatments, production rates, etc. can be established. First consideration in field development concerns primary recovery and consists of one or more of the following methods.

1. Reservoir production by depletion drive is accomplished by fluid movement resulting from solution of dissolved gas drive (Fig. 432). Well placement should be located to permit completion intervals in the lower part of the oil column. Since dissolved gas drives lose pressure rapidly and may form a gas cap upon production, low completions permit greater production from gas cap effects as well as later stage gravity drainage.

2. Low completions and regular well spacing produce the most oil from fields with external gas or gas cap drives (Figs. 433 and 434). Pressure depletion, expansion of the gas cap, and late-stage gravity drainage are characteristic.

3. High completions and regular spacing maximize production from water drive reservoirs (Fig. 435). This

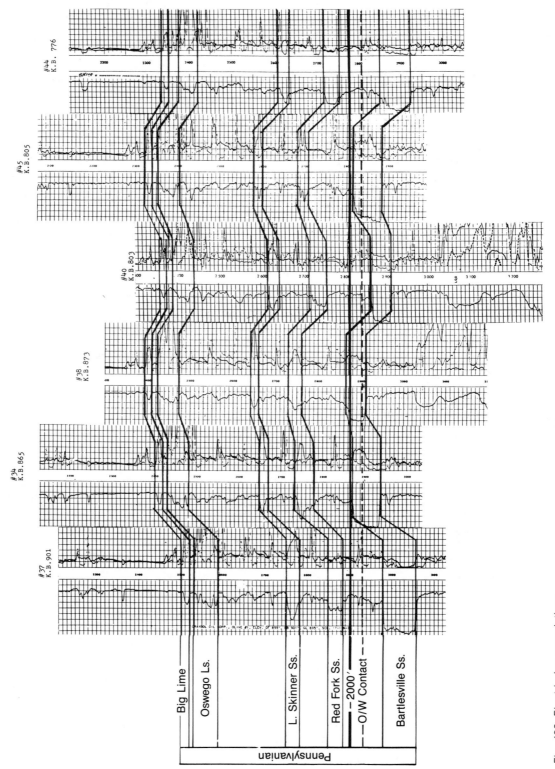

Fig. 428. Electric log correlations

means that oil is initially concentrated in the apex of the structure and will remain there as the water column displaces the oil upward.

4. Low completions peripherally located on the field produce most gravity drained oil. Inasmuch as there are no fluid dynamics to concentrate the oil, it drains away from the apex of the structure.

Secondary Recovery

Reservoirs produced under primary recovery practices often respond to secondary methods to produce additional remaining oil. Primary recovery methods sometimes produce limited amounts of oil in places that often respond to further recovery efforts.

1. Water injection into a depleted reservoir has the effect of moving oil to the wellbore in a manner similar to that involved in a normal water drive. Waterflood re-

covery programs can be very efficient in moving the oil remaining in the reservoir.

Injection of water into the reservoir is done according to carefully planned patterns that should reflect the geometry of the reservoir body (Fig. 436). Several types of injection well arrays include line drive, 5-spot, 7-spot, and 9-spot patterns. Note that several injection wells are often required to move oil to producing wells.

2. Gas injection operations move oil in the same way as waterfloods do. However, gas injection operations, like low-pressure gas cap drives, are usually less efficient than waterfloods or water drives.

3. Miscible flood operations inject a solvent into a depleted reservoir to place the oil in a solution in which it can move to the wellbore. Capillary forces of the oil are reduced by addition of the solvent, which allows passage through the reservoir pores (Figs. 437, 438, and 439). Miscible flood operations are often very effective

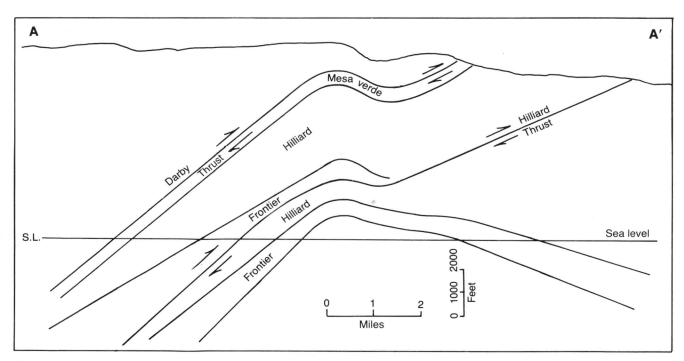

Fig. 429. Overthrust belt, Wyoming, cross-section. (See Figure 430 for location of section.)

at moving oil from the reservoir.

4. Thermal processes use heat to mobilize high viscosity oil that will not normally flow. Heat is provided by steam injection or in-situ combustion.

High temperatures cause oil to flow when it is otherwise immobile at normal temperatures. Production of oil by steam injection can range from three to twenty times the normal rate before treatment (Fig. 440).

Heat can be introduced to a reservoir by air injection, igniting the oil in the reservoir and sweeping heated oil

in advance of the moving, burning front. This system develops the burning front, a thermally-cracked hydrocarbon zone, and condensing steam that heats the cool sand in advance of the front and mobilizes the oil (Fig. 441).

Tertiary Recovery

Tertiary recovery methods require efficient oil-moving ability and advanced techniques. Miscible flood and thermal recovery methods are often used.

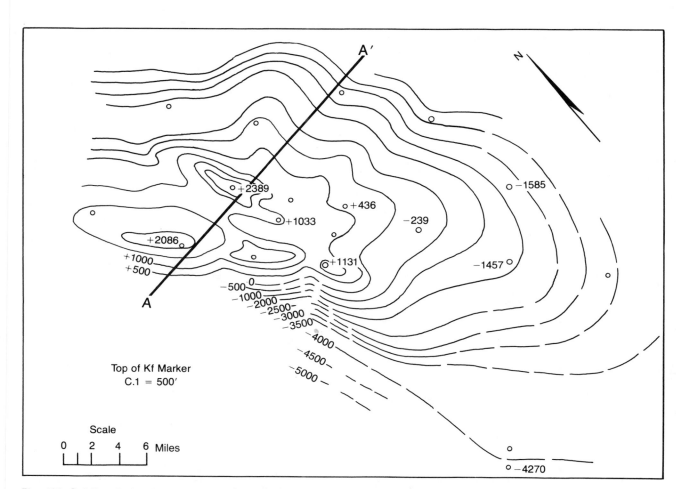

Fig. 430. Subthrust structure map, overthrust belt, Wyoming. (See Figure 429 for section).

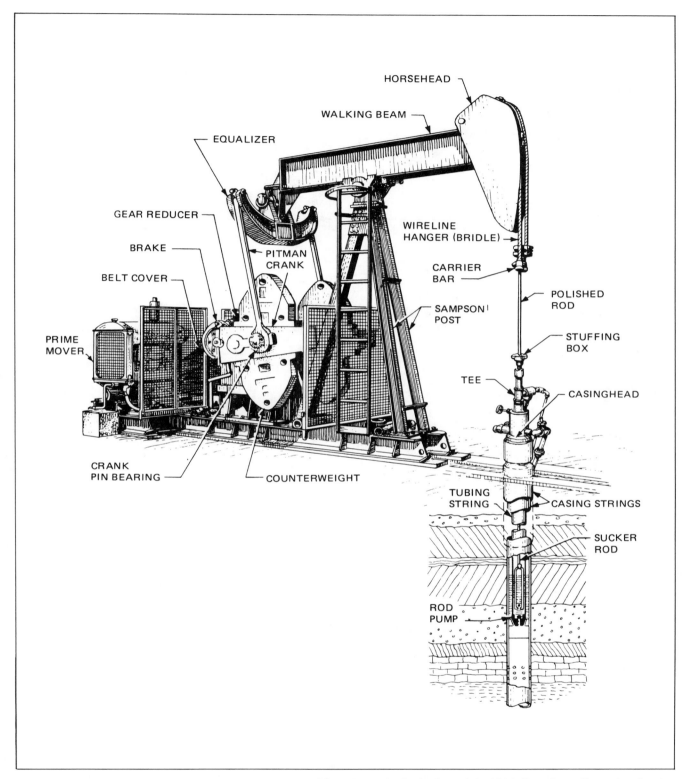

Fig. 431. Typical pumping system. From Fundamentals of Petroleum, 2nd ed., Copyright 1981, Petroleum Extension Service, The University of Texas at Austin (PETEX). Reprinted by permission.

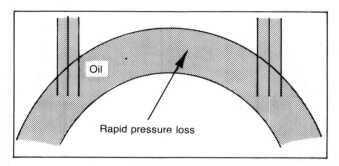

Fig. 432. Dissolved gas drive reservoir

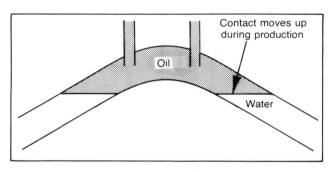

Fig. 435. Water drive reservoir

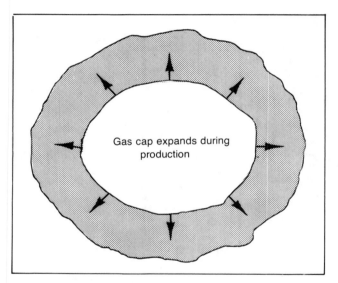

Fig. 433. Gas cap drive reservoir

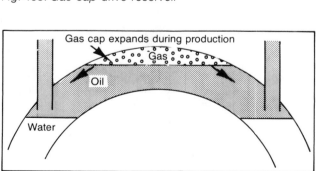

Fig. 434. Gas cap drive reservoir

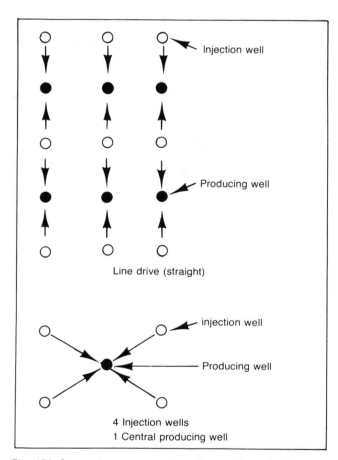

Fig. 436. Secondary recovery injection and production patterns

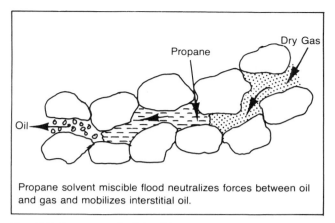

Propane solvent miscible flood neutralizes forces between oil and gas and mobilizes interstitial oil.

Fig. 437. Propane miscible drive

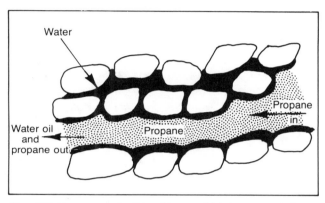

Propane miscible flood eliminates pressure between residual oil and tight porosity mobilizing the oil.

Fig. 438. Propane miscible drive and tight channels

Fig. 439. Propane miscible drive and high permeability

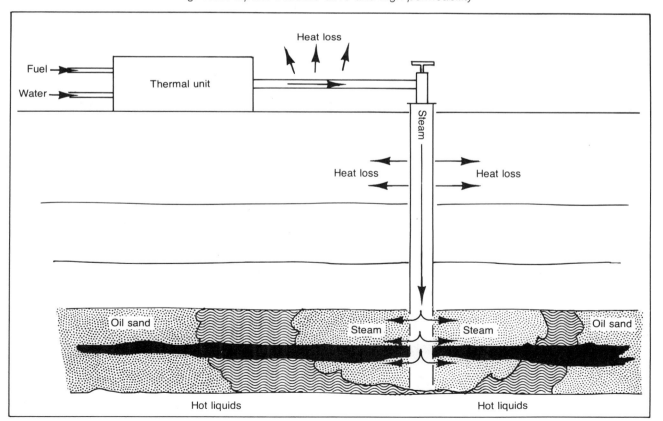

Fig. 440. Steam injection process

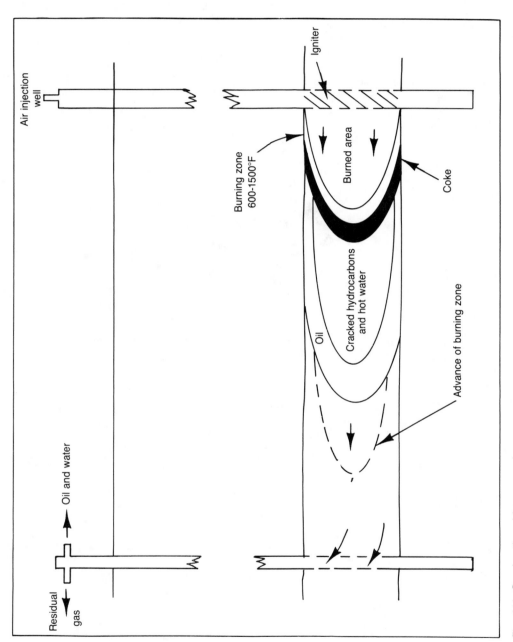

Fig. 441. Combustion drive secondary recovery

14

Exploration Techniques for Petroleum

Since the beginning of exploration for oil and gas, evolution of the art consisted of refinements of older methods and the addition of new techniques. Yet there has been no deletion of techniques through the years. Development of the art of looking for oil and gas involves the realization that there is no substitute for the imaginative approach and that there is no such thing as an oil finder or "black box."

Geology is a science that brings virtually every scientific discipline to bear in its application, whether it is the sorting out of geologic events and processes or the exploration for important raw materials. The proper orchestration of other sciences to the performance of geologic application requires substantial skill motivated by an analytic sense of imaginative investigation. Geology is not like mathematics, which produces results in absolute numbers. Geology uses mathematics and many other sciences, however, as adjuncts to the understanding of a large scenario of multiple components.

Exploration, as any other geologic investigation, uses all parameters. It is the duty of the explorationist to recognize the parameters as they are encountered. Without recognitive ability, the explorationist can not function properly. Petroleum exploration is not the type of endeavor where all definitive factors are obvious at all times. The explorationist must use personal experience as well as that of others to solve the problem. No information is discarded; the explorationist uses it all.

Surface Geology

Exploration techniques start with methods that involve the obvious: look for oil on the surface of the ground.

They continue with more esoteric procedures that produce sophisticated data suggestive of occurrences of oil and gas, which require analysis and imagination including geophysics, geochemistry, and aerial photography. The latter are substantially more indirect and circumstantial than the former. However, even the tactile and visual reality of oil on the surface of the ground does not insure its occurrence, in quantity, in the subsurface.

Direct Indications

Oil and gas evident upon the ground surface may promote some degree of certainty of their being in the subsurface. Surface indications are only that, however, because there is no definition of what will be found in the subsurface.

1. Oil and gas come to the surface along porous beds, fault planes, or in springs (Fig. 442). Mobile petroleum occurrences at the surface are called seeps. Seeps are abundant and well-documented worldwide (Figs. 443 and 444).

2. Porous rock outcrops can be impregnated with oil. They are suggestive of oil, present or past, in the subsurface. Oil-soaked rocks have provoked exploration for many years in portions of the Great Basin, Nevada, and Utah.

3. Oxidation of petroleum reduces the volatile constituents and causes it to become very viscous or immobile. The asphaltic or tarry residue at the surface seals off the unoxidized petroleum remaining in the reservoir from the atmosphere. Asphalt/tar deposits in Asphalt Ridge west of Vernal, Utah are indicative of subsurface oil in Tertiary rocks of the Uinta Basin.

4. Some plant forms have a particular affinity for petro-

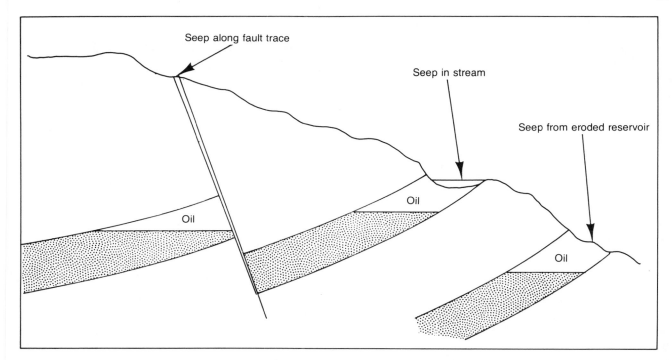

Fig. 442. Oil seeps

leum. They grow on the ground surface over subsurface petroleum accumulations. Identification of the particular plant varieties is essential. Vegetative growth over circular or nearly circular petroleum-associated salt domes can result in recognizable ground surface patterns. They may be detected by surface observation or on aerial photographs.

An adjunct to vegetative exploration involves exploration on "buffalo chip highs." Buffalo and cattle grazing the range of the mid-American plains were disposed to feeding upon vegetation with affinities for petroleum. Higher concentrations of buffalo or cow "chips" resulted from this geographic selection process and permitted successful exploration based upon the habits of grazing animals.

Surface Data

Exploration by surface geological activity is essential where rocks appear at the surface. In offshore areas, however, surface exploration is impossible.

The surface geologist must map all visible geological features to fully utilize the data toward the exploration program. The geologist's judgment can set the entire tone and control the direction of the exploratory effort. His job is to examine the rocks, evaluate their suitability as source and reservoir beds, and map their structure. Poor source beds, inadequate reservoir rocks, and improper structure do not combine to promote a favorable exploration program.

1. Geologic mapping involves plotting geologic data on a map (Fig. 445). The geologist measures and describes the rock sections and plots the different formations on a map which shows their distribution. He also plots surface structure on the geologic map or may construct a separate structure map. Other maps that are useful include thickness maps, which illustrate thickness variations of individual formations over the area of interest, lithologic maps, which show variations in rock types within a formation, and porosity maps, which illustrate how porosity in a particular rock unit may vary throughout the exploration area.

2. Without source and reservoir rocks, an area has no petroleum potential. Identification of source and reservoir rocks, their distribution, and their thicknesses are essential in an exploration program.

3. Exploration, particularly over large areas, requires correlation of geologic sections (Fig. 446). Correlations

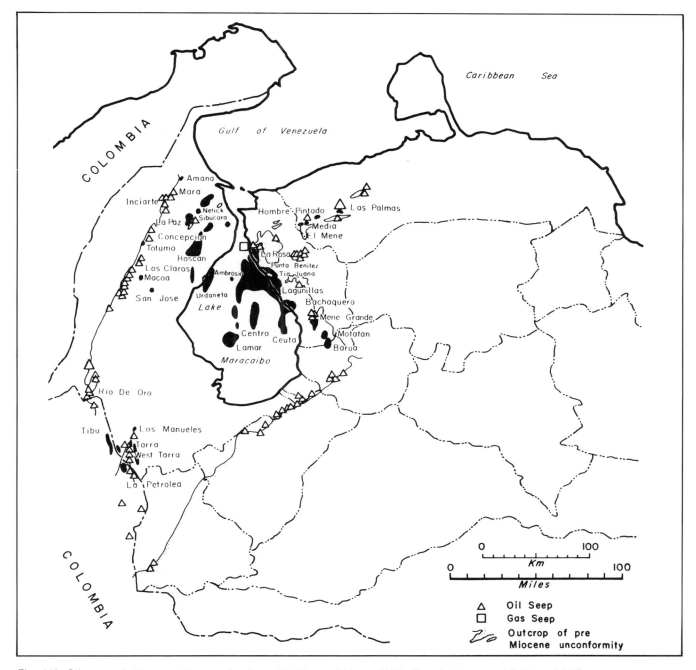

Fig. 443. Oil seeps in Western Venezuela. From Dickey and Hunt, 1972. Permission to publish by AAPG.

produce cross-sections that give visual information on structure, stratigraphy, and thickness.

4. Photogeology is particularly useful in conducting mapping operations. Aerial photographs allow accurate geographic plotting of contacts, structures, and topography (Fig. 447), which can later be transferred to a map.

Not all areas have aerial photographic coverage. However, photogeology is an important interpretive tool where there is coverage.

5. Cross-sections can incorporate a variety of geologic

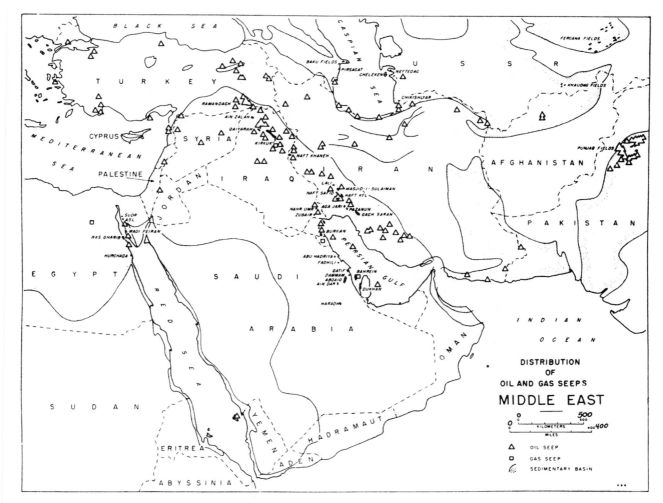

Fig. 444. Distribution of oil and gas seeps in Middle East. From Link, 1952. Permission to publish by AAPG.

information including stratigraphy, structure, porosity, lithology, and thickness of important formations or groups of formations (Fig. 448). They are often used as adjuncts to maps and enable three-dimensional visualization of the area covered.

Subsurface Geology

Large areas of the world have no outcrops or surface features or are located offshore. Surface geology is of little use in these places, which require subsurface analysis of whatever data are available.

Subsurface techniques use the same parameters as surface methods but do not have the benefit of continuity typical of surface exposures. Subsurface data quality improves as the density of data increases. However, increased amounts of data often require more specific interpretations. Interpretation of subsurface geology integrates data from a number of sources. Some sources provide direct information and others provide more indirect data.

Drilling Operations

Wells are drilled to explore for and produce petroleum. Drilling equipment has been developed by extensive experience throughout the long history of the petroleum business. It represents a complex and expensive technology that has evolved through operations all over the world in every conceivable environmental circumstance.

Onshore drilling equipment consists of a derrick floor above a substructure that elevates it from the ground. The diesel engines, drawworks, cable and block systems and

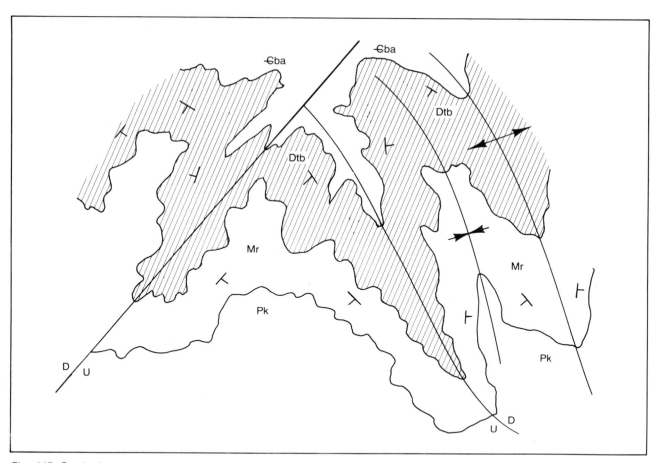

Fig. 445. Geologic map

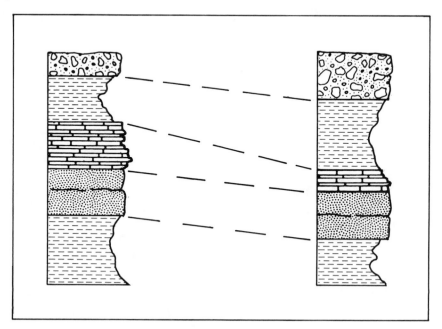

Fig. 446. Correlated sections. Note thickening and thinning of the stratigraphic units.

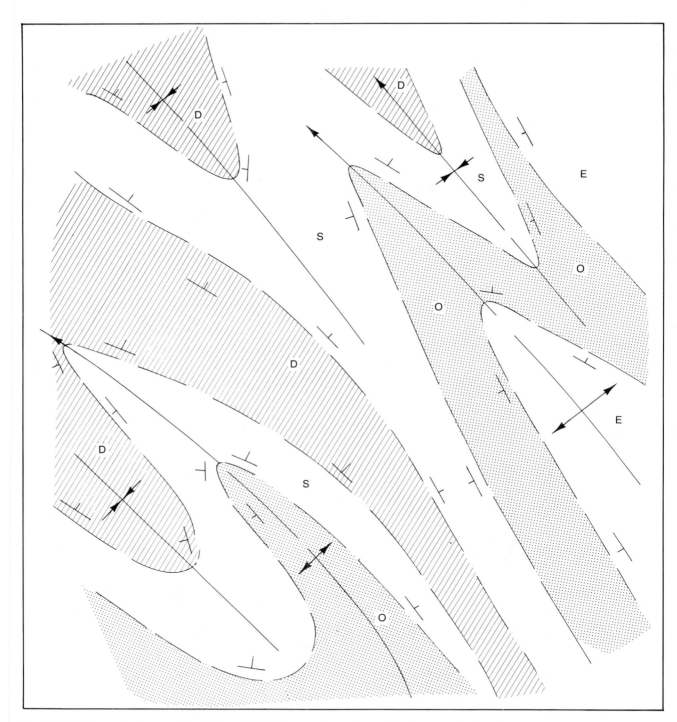

Fig. 447. Photogeologic map illustrating distribution and structure of various geologic units.

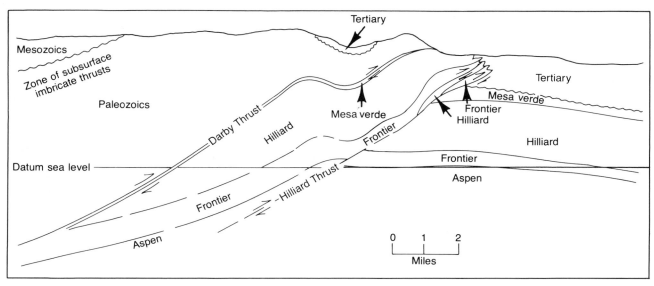

Fig. 448. Geologic cross-section

derrick are attached to the derrick floor (Figs. 449, 450). A blowout preventer is installed below the derrick floor to control high-pressure zones and to prevent equipment damage during drilling operations.

Offshore drilling equipment is the same as that used onshore except that it is situated on a platform (Fig. 451), which may rest on the sea bottom or be attached to a submerged or semi-submerged substructure. Many platform types permit drilling of multiple directional wells from a single location.

Drill pipe, drill collars and a drill bit make up the drill string that is lowered into the ground to drill the well (Fig. 449). Power for the drill string is provided by the diesel engines through a chain drive that turns the rotary table.

Drilling mud is pumped from the mud tanks, through the standpipe, rotary hose, and swivel into the kelly and the drillstring (Fig. 452). The mud passes through the drillstring and exits the jets in the bit. It cleans and conditions the well bore as it rises to the surface in the annulus between the drillstring and the walls of the well bore. As drilling progresses the drill string is extended by adding more lengths of drill pipe. The amount of mud required to maintain the well bore is increased as the depth increases.

Drilling mud is used as a control for high pressure formation at depth. If high pressure zones are anticipated or encountered the weight of the mud is increased to control the formation pressure. If low pressure or lost circulations zones are encountered mud weight is reduced.

Well Cuttings

Well samples are produced from drilling operations, by the drill bit penetrating the formation encountered in the subsurface. They are taken at regular intervals from the shale shaker of the mud circulation system and described to establish a lithologic record of the well (Fig. 453).

Cuttings are examined for lithologic characteristics and plotted on a strip sample log (Figs. 454 and 455). Oil and gas shows in the samples are noted on the strip log.

Cores

Cores are cut where specific lithologic and rock parameter data are required. They are cut by a hollow core barrel which goes down around the rock core as drilling proceeds (Fig. 456). When the core barrel is full and the length of the core occupies the entire interior of the core barrel, it is brought to the surface, and the core is removed and laid out in stratigraphic sequence. Cores are preferable to well cuttings because they produce coherent rock. They are significantly more expensive to obtain, however.

Sidewall cores are small samples of rock obtained by shooting small metal cylinders from a gun into the walls

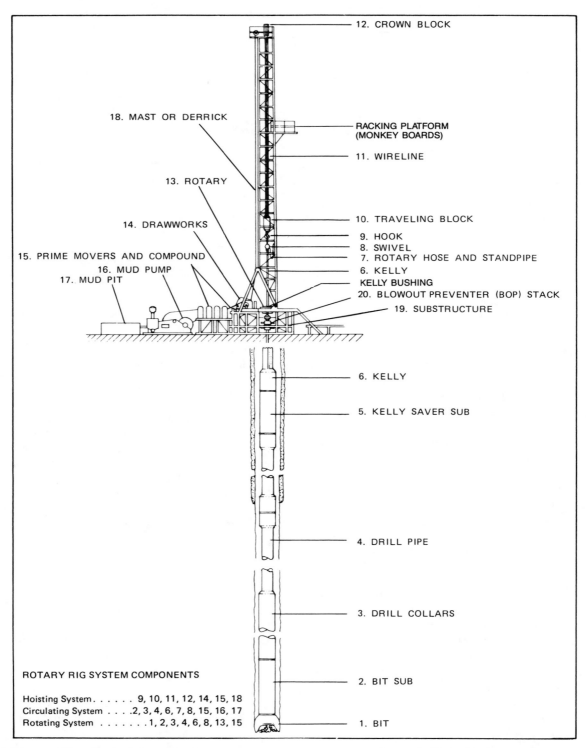

Fig. 449. Rotary drilling system. From Fundamentals of Petroleum, 2nd ed., Copyright 1981, Petroleum Extension Service, The University of Texas at Austin (PETEX). Reprinted by permission.

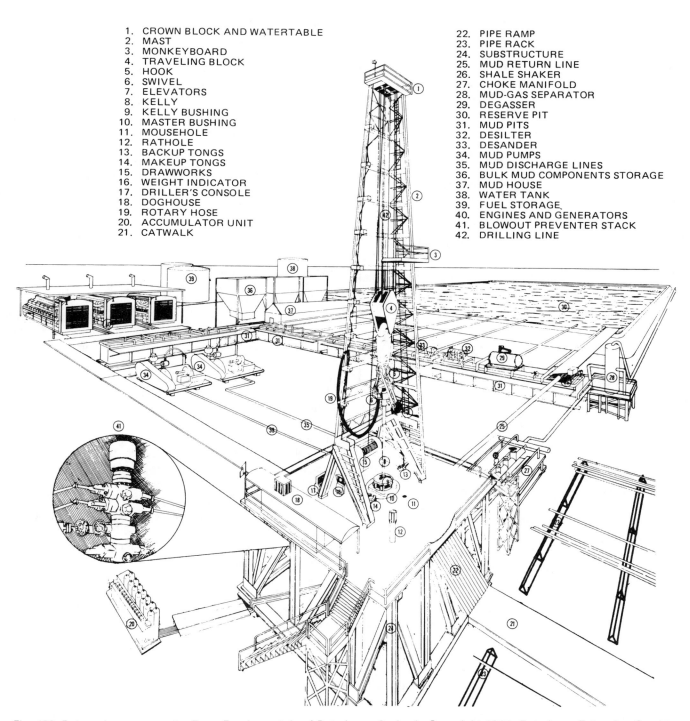

1. CROWN BLOCK AND WATERTABLE
2. MAST
3. MONKEYBOARD
4. TRAVELING BLOCK
5. HOOK
6. SWIVEL
7. ELEVATORS
8. KELLY
9. KELLY BUSHING
10. MASTER BUSHING
11. MOUSEHOLE
12. RATHOLE
13. BACKUP TONGS
14. MAKEUP TONGS
15. DRAWWORKS
16. WEIGHT INDICATOR
17. DRILLER'S CONSOLE
18. DOGHOUSE
19. ROTARY HOSE
20. ACCUMULATOR UNIT
21. CATWALK

22. PIPE RAMP
23. PIPE RACK
24. SUBSTRUCTURE
25. MUD RETURN LINE
26. SHALE SHAKER
27. CHOKE MANIFOLD
28. MUD-GAS SEPARATOR
29. DEGASSER
30. RESERVE PIT
31. MUD PITS
32. DESILTER
33. DESANDER
34. MUD PUMPS
35. MUD DISCHARGE LINES
36. BULK MUD COMPONENTS STORAGE
37. MUD HOUSE
38. WATER TANK
39. FUEL STORAGE
40. ENGINES AND GENERATORS
41. BLOWOUT PREVENTER STACK
42. DRILLING LINE

Fig. 450. Rotary rig components. From Fundamentals of Petroleum, 2nd ed., Copyright 1981, Petroleum Extension Service, The University of Texas at Austin (PETEX). Reprinted by permission.

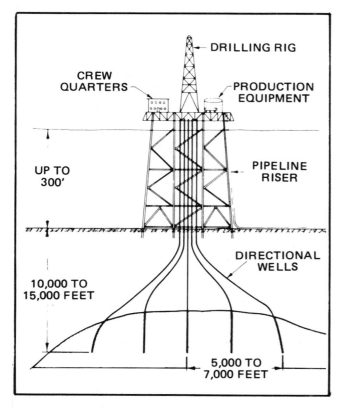

Fig. 451. Drilling platform schematic. From Fundamentals of Petroleum, 2nd ed., Copyright 1981, Petroleum Extension Service, The University of Texas at Austin (PETEX). Reprinted by permission.

of a drill hole. The tool for obtaining sidewall cores contains a number of guns oriented around its perimeter and arranged in several tiers (Fig. 457). Sidewall cores can be taken from several levels and at different locations using this versatile tool.

Electric, Radioactivity and Acoustic (Sonic) Logging

Subsurface geological information can be obtained by wireline well-logging techniques. Measurements are made of the electrical, radioactive and acoustic properties of rocks and their contained fluids encountered in the wellbore. Several types of measurements produce information on formation rock acoustic velocity, density, radioactivity, porosity, conductivity, resistivity, fluid saturation and permeability. Rock lithology, formation depth and thickness, and fluid type can also be determined. Caliper logs measure borehole diameter. Geologic maps and cross-sections are readily constructed from a variety of well-

log data and assist in understanding facies and geometric relationships and the locations of wildcat and development drilling sites.

Logs are obtained by lowering a sonde or tool (Fig. 458) attached to a cable or wire to the bottom of a wellbore filled with drilling mud. Electrical, nuclear, or acoustic energy is sent into the rock and return to the sonde or are obtained from the rock and measured as the sonde is continuously raised from the wellbore bottom at a specific rate. The well is logged when the sonde arrives at the top of the interval to be investigated. Formation water saturation, porosity, permeability radioactivity and resistivity are rock properties that affect logging and the types of logs to be obtained.

As a wellbore is drilled the rock formations and their contained fluids are penetrated by the bit and affected by the drilling process. Drilling mud invades the rock surrounding the wellbore, affects the logging of the hole and must be accounted for.

A permeable, porous formation, which has been penetrated and affected by drilling and invasion by drilling mud, develops parameters important to logging (Fig. 459).

Significant of these parameters from the center of the wellbore outward into the formation are hole diameter, drilling mud, mudcake, mud filtrate, flushed zone, invaded zone, and uninvaded zone (Fig. 459).

Hole Diameter (d_h)—The size of the borehole determined by the diameter of the drill bit.

Drilling Mud (R_m)—Fluid used to drill a borehole and which lubricates the bit, removes cuttings, maintains the walls of the borehole, and maintains borehole over formation pressure. Drilling mud consists of a variety of clay and other materials in a fresh or saline aqueous solution.

Mudcake (R_{mc})—The mineral residue formed by accumulation of drilling mud constituents on the wellbore walls as the mud fluids invade the formations penetrated by the borehole.

Mud Filtrate (R_{mf})—Mud fluid that penetrates the formation while leaving the mudcake on the walls of the borehole.

Flushed Zone (R_{xo})—The portion of the invaded zone immediately adjacent to the borehole in which mud filtrate has removed most or all of the formation water and/or petroleum.

Invaded Zone—That part of the formation between the borehole and unaltered formation rock penetrated by mud filtrate.

Annulus (R_i)—The portion of the invaded zone where mud filtrate mixes with formation water and/or petro-

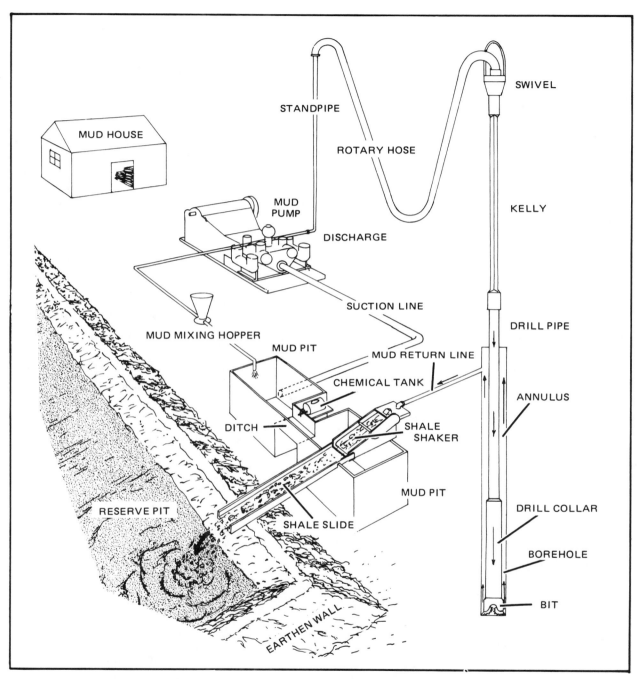

Fig. 452. Mud circulation system. From Fundamentals of Petroleum, 2nd ed., Copyright 1981, Petroleum Extension Service, The University of Texas at Austin (PETEX). Reprinted by permission.

DRILLING TIME (MINUTES PER FOOT) PENETRATION RATE INCREASES →	DEPTH	ACCESSORIES	LITHOLOGY	POROUS ZONES	OIL/GAS SHOWS	SAMPLE DESCRIPTION	REMARKS INCLUDING FORMATION TOPS TESTING DATA MUD RECORD BIT RECORD CORE DESCRIPTION

Fig. 453. Typical strip log

ROCK TYPES	ACCESSORIES		PRODUCTION		
SANDSTONE	F	Fossiliferous	HORIZON	INTERVAL	POTENTIAL
SILTSTONE	△	Chert			
SHALE	~	Glauconite			
LIMESTONE	&	Hematite,Etc.			
DOLOMITE	#	Pyrite			
ANHYDRITE	oo	Oolites			
GYPSUM	φφ	Oolicasts			
SALT	⊞	Salt Casts			
COAL	COLORS		REMARKS		
ARKOSE	G	Gray			
	B	Black			
	R	Red			
	O	Orange			
	Y	Yellow			
	Gn	Green			
	Bu	Blue			
	Bn	Brown			
	Bf	Buff			
	L	Light			
SAMPLES MISSING	D	Dark			

Fig. 455. Typical strip log sample designations

GEOLOGISTS LOG — Geological Services Of Tulsa

OPERATOR_____
WELL_____
LOCATION_____ SEC___ T___ R___
COUNTY_____ STATE_____

ELEVATIONS___ KB___ DF___ GL DATUM___
SAMPLES SAVED FROM___ TO___
SAMPLES LOGGED FROM___ TO___
CORED INTERVAL FROM___ TO___
GEOLOGICAL SUPERVISION FROM___ TO___
WELL-SITE GEOLOGIST_____
SAMPLES LOGGED BY_____

CONTRACTOR_____
SPUD___ DRLG. COMPL.___ WELL COMPL.___
RTD.___ LTD.___ PBTD.___
MUD UP AT___ TYPE MUD___
TYPE RIG_____
CASING RECORD_____

ELECTRICAL SURVEYS_____

Fig. 454. Typical strip log heading

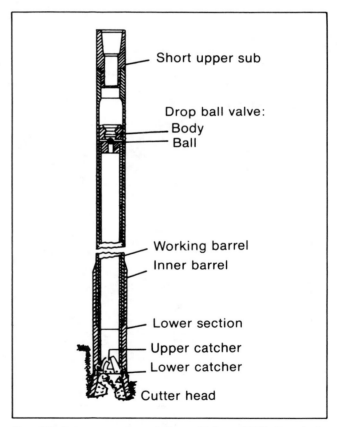

Short upper sub
Drop ball valve: Body / Ball
Working barrel
Inner barrel
Lower section
Upper catcher
Lower catcher
Cutter head

Fig. 456. Rotary core barrel. From Dickey, 1979. Permission to publish by PennWell Publishing Company.

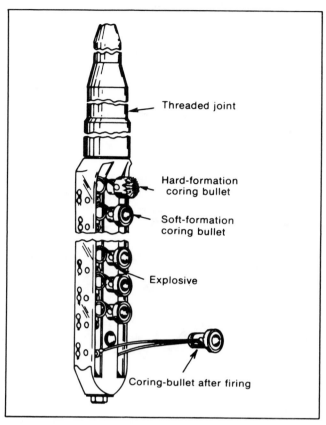

Fig. 457. Sidewall coring gun. From Dickey, 1979. Permission to publish by PennWell Publishing Company.

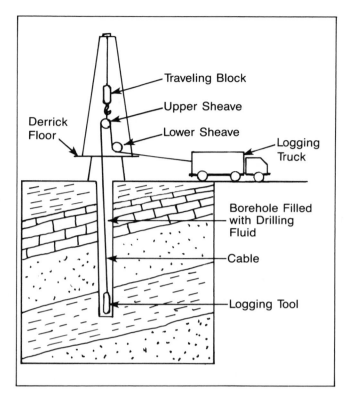

Fig. 458. Electric logging schematic.

leum. It is the portion of the invaded zone farthest from the wellbore.

Uninvaded Zone (R_i)—Formation rock materials away from and unaltered by mud filtrate and containing uncontaminated formation fluids.

Spontaneous (Self) Potential Logs (SP Logs)

SP logs (Figs. 460 and 461) are used to detect permeable formations and their upper and lower contacts, volume of shale, where present, in permeable formations, and to determine the resistivity of water in permeable formations. They are recorded in the first track on the left side of the log (Fig. 461).

SP log response is generated by salinity and resistivity differences between the mud filtrate and the formation water. Electrochemical processes create electrical current as drilling fluid invades the formations and causes the SP response (Fig. 460) to the left (negative) or right (positive) of the shale baseline (Figs. 460 and 461).

In SP and resistivity logs the shale baseline results from the consistent straight line response of shale formations and is located within a few measurement units of the control or interior portions of the respective logs (Fig. 461). It is from these lines that positive and negative deflections are evaluated.

It is evident that permeable formations produce greater negative SP response than impermeable ones that have little or no response (Fig. 460). Similarly, variations in resistivity between the mud filtrate and the formation water produce variation in response as well.

Resistivity Logs

Resistivity logs illustrate permeable formations, formation fluid (water versus petroleum) content, and the porosity characteristics of formation resistivity.

Resistivity, the inverse of conductivity, represents the tendency of rock materials and their contained fluids to resist the flow of electrical current. Salt water contains dissolved salt and, because it conducts electricity very

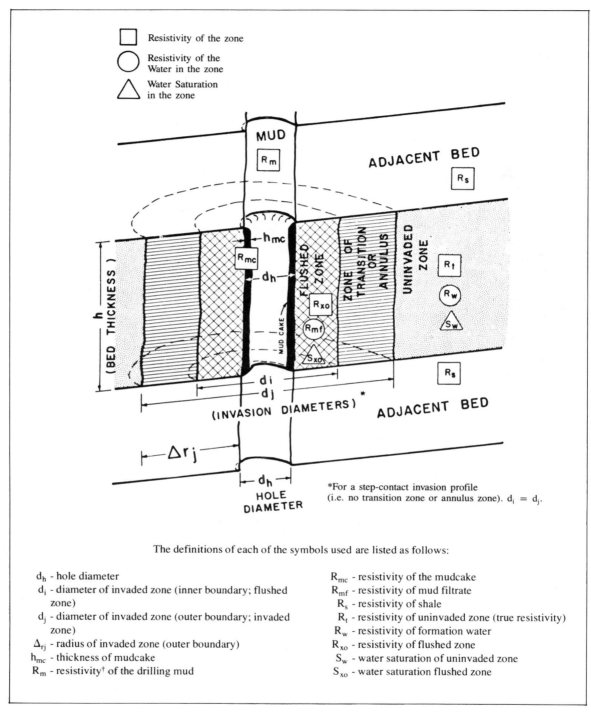

Fig. 459. The borehole environment and symbols used in log interpretation. This schematic diagram illustrates an idealized version of what happens when fluids from the borehole invade the surrounding rock. Dotted lines indicate the cylindrical nature of the invasion. From Asquith, 1982. Permission to publish by AAPG.

easily, has low resistivity. Fresh water contains no salt and demonstrates low conductivity and high resistivity. Rock materials that contain salty or fresh water offer differing degrees of resistivity and response on resistivity logs.

Formation resistivity is measured by induction and electrode logs. Increased resistivity is indicated by a positive deflection to the right from the shale baseline (Fig. 461).

The induction electric log and the dual induction-focused log represent the two types of induction logging surveys. Electrode logging includes normal, Laterolog, Microlaterolog, and Microlog surveys.

Induction Logs

Induction, short-normal, and SP curves (Fig. 461) comprise the induction log. The induction and short-normal curves represent the resistivity side of the log and deflect positively to the right of the shale baseline. Induction logs measured with a 40 in. transmitter-receiver spacing investigate the deep formation and provide true formation resistivity. Short-normal logs, taken with a 16 in. electrode spacing, investigate the shallow formation and measure the resistivity of the invaded zone that illustrates formation permeability.

Dual Induction-Focused Log

Three curves make up the Dual Induction-Focused Log (Fig. 462). Deep induction log resistivity (R_{ILO}) indicates the deep resistivity of the formation (R_t). Medium induction log resistivity (R_{ILM}) illustrates the medium investigation depth resistivity of the invaded zone (R_i). The Spherically Focused Log resistivity (R_{SFL}) indicates the shallow zone of investigation and the resistivity of the flushed zone (R_{XO}).

Electrode Logs

Several types of electrode logs can be implemented to illustrate specific parameters under specific mud and borehole conditions.

The short-normal log (Fig. 461) taken with a 16 in. electrode spacing, investigates the shallow formation, measures the resistivity of the invaded zone, and illustrates the presence of formation permeability.

True formation resistivity is measured by the Laterolog (Fig. 463). The depth to which measurements are made is controlled by the intensity of the focus of the surveying electrical current.

Resistivity of the flushed zone is measured by the Microlaterolog (Fig. 463).

Microlog measurements are made to illustrate buildup of mudcake that indicates invasion and suggests a permeable formation (Fig. 464).

Resistivity logs can be classified by several parameters (Fig. 465).

Radioactivity Logs

Commonly used radioactivity surveys result in gamma-ray, neutron and density logs, which are often obtained together.

Gamma-ray logs (Fig. 466) measure formation radioactivity and are useful in identification and correlation of formation rock types (Fig. 467). Shale normally contains radioactive materials and produces a positive deflection from left to right on the left hand log track. Gamma-ray logs are useful in estimating shale volume in potential or actual reservoir sandstone or carbonate.

Neutron logs (Fig. 466) illustrate formation porosity by measuring hydrogen ions. Water and/or oil-filled, shale-free, clean formations will be logged as liquid filled porosity. Zones of low porosity on the neutron log correspond to zones of higher radioactivity on the gamma-ray log and are reflective as approximate mirror images of each other (Fig. 467).

The density log evaluates formation porosity. It detects gas, evaluates hydrocarbon density and complex rock sequences, indentifies evaporite minerals and shale-bearing sandstone units. It is often taken in the same log suite as gamma-ray logs (Fig. 468).

Acoustic Logs

Like density and neutron logs, acoustic (sonic) logs illustrate formation porosity. The acoustic log measures the velocity of a sound wave through a rock medium (Fig. 469). Sound wave velocity is dependent upon lithology and porosity (Fig. 470). The sonic log (Fig. 471) illustrates both the sound wave transit time, which indicates rock velocity, and the related porosity of the rock.

Drill Stem Tests

Formation evaluation by obtaining samples of formation fluid and formation pressure data is made possible by drill stem testing procedures. The testing equipment is lowered into the wellbore on the drill pipe and put into place by seating a packer that seals off the formation from contamination by drilling mud (Fig. 472). The tool is opened and fluid samples and pressure data are obtained.

Drill stem tests are run in wells in which promising hydrocarbon shows (indications) are encountered in cores and samples. Segregation of the individual formations produces results from specific intervals.

Pressure data are evaluated to determine the productive potential of the formation being tested. These data and

fluid information can facilitate decisions on how the well is to be completed: as a producing well or as a dry hole to be plugged and abandoned.

Development of an individual reservoir can also be augmented by evaluation of pressure and fluid data (Fig. 473). Data similarities suggest the same reservoir. Dissimilar data are potentially indicative of separate reservoirs, permeability barriers, or contamination.

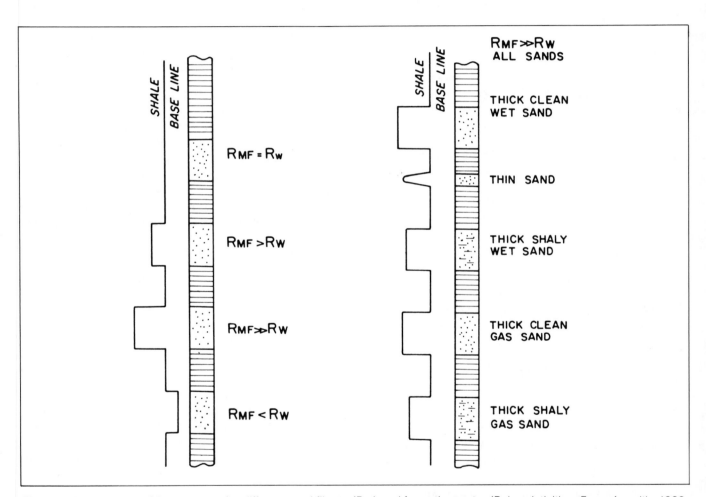

Fig. 460. Diagrammatic SP responses for different mud filtrate (R_{mf}) and formation water (R_w) resistivities. From Asquith, 1982. Permission to publish by AAPG.

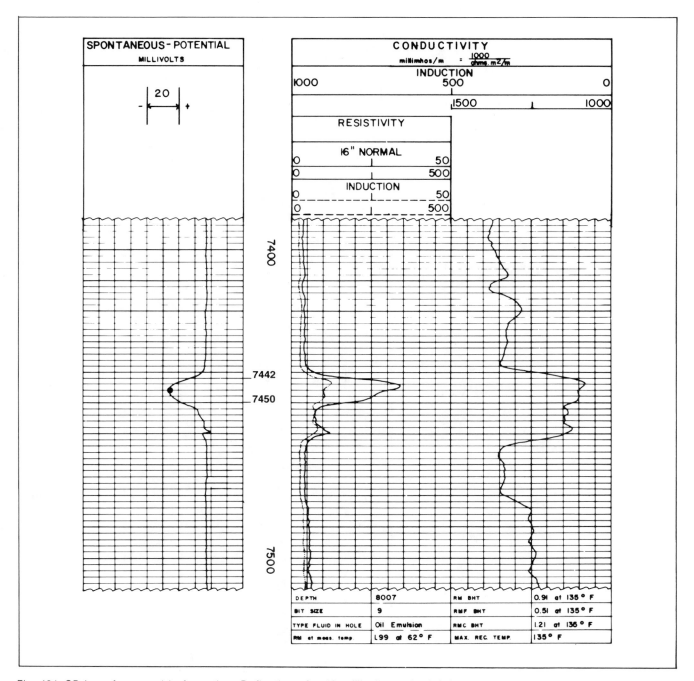

Fig. 461. SP log of permeable formation. Deflection of −40 millivolts to the left between 7442–7450. From Asquith, 1982. Permission to publish by AAPG.

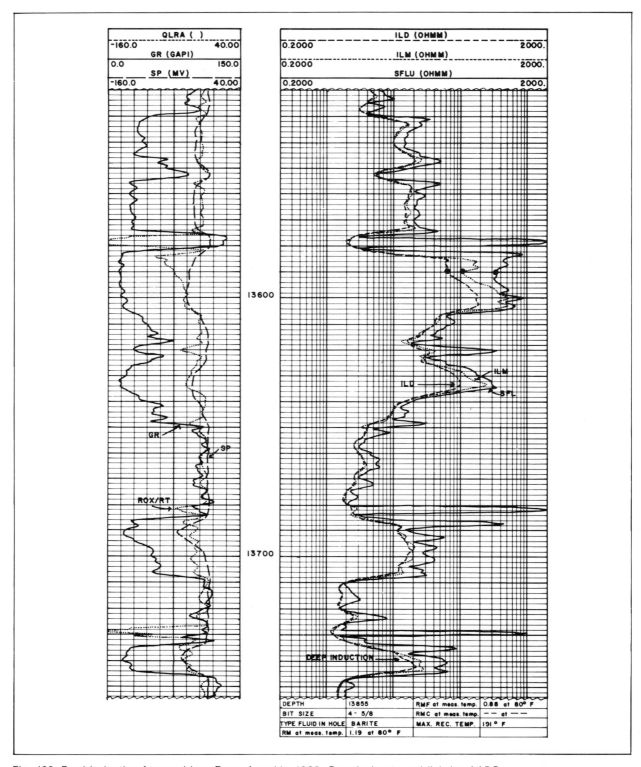

Fig. 462. Dual induction-focused log. From Asquith, 1982. Permission to publish by AAPG.

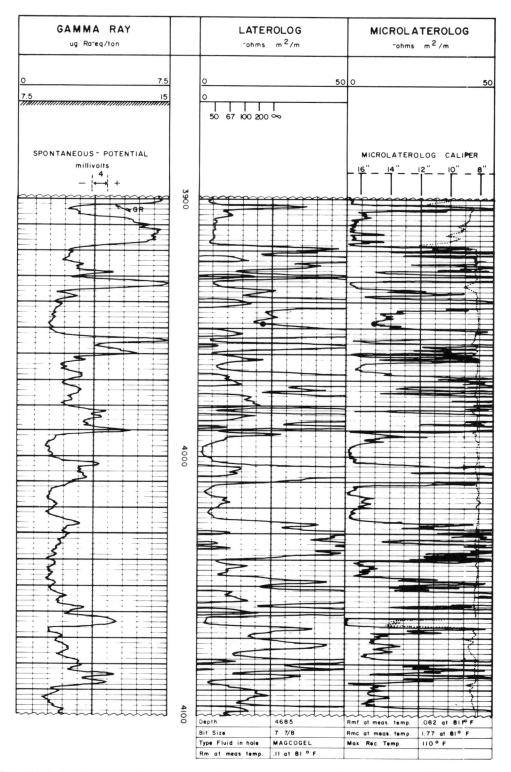

Fig. 463. Laterolog and microlaterolog with gamma-ray track. From Asquith, 1982. Permission to publish by AAPG.

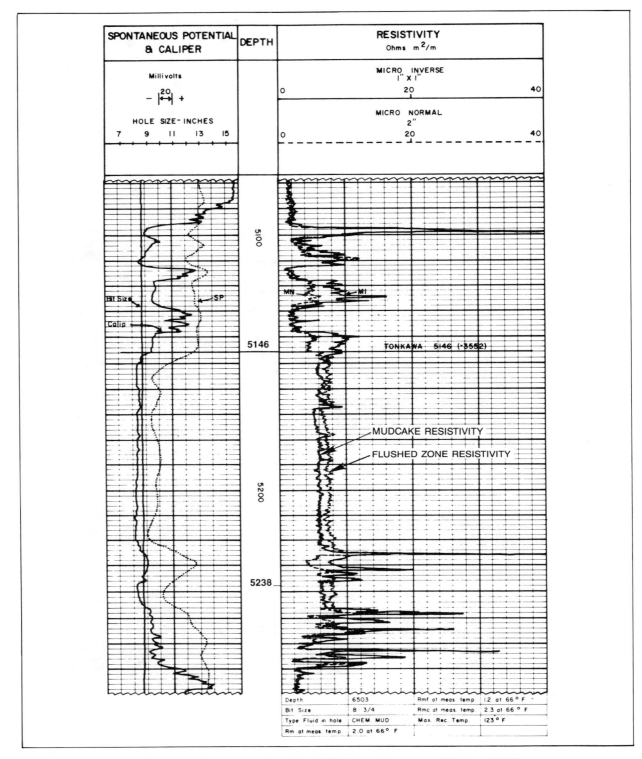

Fig. 464. Microlog with SP and caliper. From Asquith, 1982. Permission to publish by AAPG.

Classification of Resistivity Logs.

INDUCTION LOGS

ELECTRODE LOGS

A. Normal logs
B. Lateral Log
C. Laterologs
D. Spherically Focused Log (SFL)

E. Microlaterolog (MLL)
F. Microlog (ML)
G. Proximity Log (PL)
H. Microspherically Focused Log (MSFL)

DEPTH OF RESISTIVITY LOG INVESTIGATION

Flushed Zone (R_{xo})	*Invaded Zone (R_i)*	*Uninvaded Zone (R_t)*
Microlog	Short Normal	Long Normal
Microlaterolog	Laterolog	Lateral Log
Proximity Log	Spherically Focused Log	Deep Induction Log
Microspherically Focused Log	Medium Induction Log	Deep Laterolog
	Shallow Laterolog	Laterolog
		Laterolog
		Induction Log

Fig. 465. Classification of resistivity logs. From Asquith, 1982. Permission to publish by AAPG.

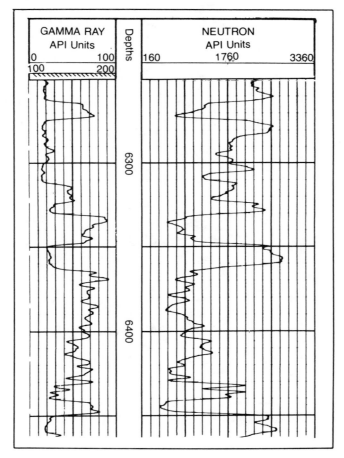

Fig. 466. Gamma-ray standard neutron log. From Schlumberger, Ltd., Published by permission.

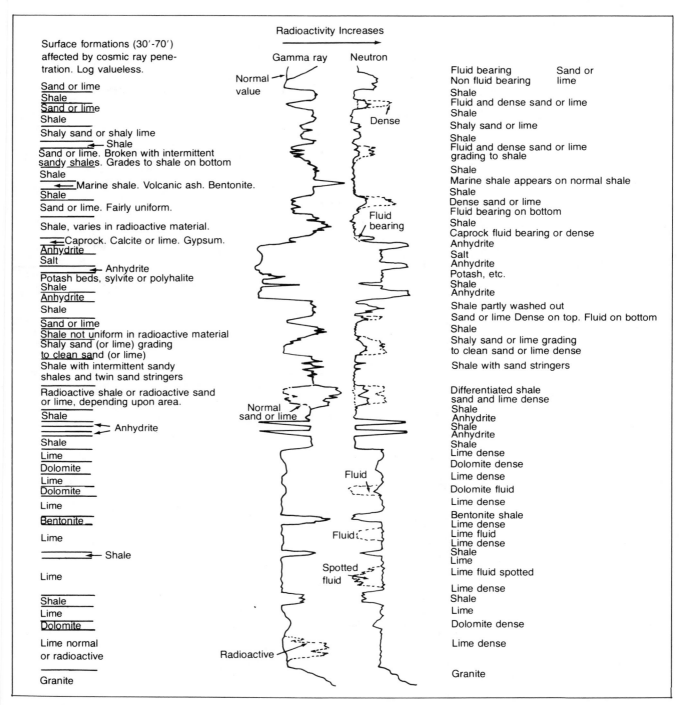

Fig. 467. Gamma-ray and neutron response curves to different types of formations. From Dresser Atlas, Dresser Industries. Published by permission.

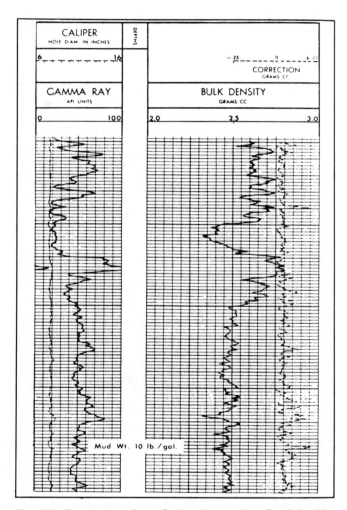

Fig. 468. Density log. From Schlumberger Ltd. Published by permission.

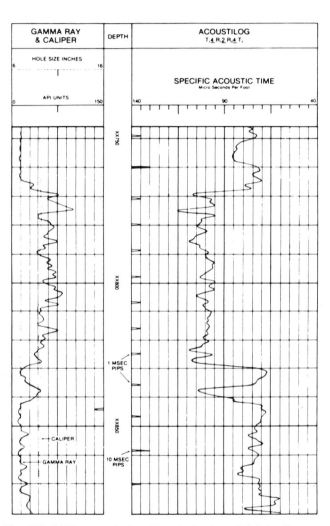

Fig. 469. BHC acoustic log. Permission to publish by Dresser Atlas, Dresser Industries, Inc.

Velocities of Various Substances

Material		v_c (ft/sec)
Non-porous solids		
Anhydrite		20,000
Calcite		20,100
Cement (cured)		12,000
Dolomite		23,000
Granite		19,700
Gypsum		19,000
Limestone		21,000
Quartz		18,900
Salt		15,000
Steel		20,000
Water-saturated porous rocks in situ		
	Porosity	
Dolomite	5–20%	20,000–15,000
Limestone	5–20%	18,500–13,000
Sandstone	5–20%	16,000–11,500
Sand (unconsolidated)	20–35%	11,500– 9,000
Shale		7,000–17,000
Liquids		
Water (pure)		4,800
Water (100,000 mg/l of NaCl)		5,200
Water (200,000 mg/l of NaCl)		5,500
Drilling mud		6,000
Petroleum		4,200
Gases		
Air (dry or moist)		1,100
Hydrogen		4,250
Methane		1,500

Fig. 470. Permission to publish by Hubert Guyod.

Geophysical Surveys

Several types of common geophysical surveys are in use in petroleum exploration. They provide some direct and some indirect structural data, and, in most instances, no direct indications of petroleum occurrences.

Differences in the density of crustal rocks are measured by the gravimeter. Low density rocks are represented as negative anomalies and high density rocks as positive anomalies (Fig. 474). Gravity surveys are useful in exploration of salt dome terrains. Salt domes have low density and appear as negative closed anomalies on gravity maps (Fig. 475). Gravity surveys provide indirect structural and lithologic data (Fig. 476).

Magnetometer exploration measures variations in magnetic intensity. Basement rocks usually contain more magnetically susceptible iron-bearing minerals. When basement rocks are deformed and raised as fault blocks, they are placed closer to the ground surface and produce stronger magnetic values (Fig. 477). They appear as positive magnetic anomalies and indirectly indicate possible basement-related structure (Fig. 478).

Unfortunately, variations in the magnetic susceptibility of the basement can be a function of compositional differences. Therefore, a level basement surface consisting of a variety of rock types can produce apparent positive anomalies, which can be confused with structural deformation. To overcome these problems, modeling of magnetic data to produce plausible structural patterns is essential to interpretation. Magnetic surveys provide indirect structural and lithologic data sections.

Seismic surveys are the best and most definitive geophysical means of structural representation currently in use. In very few special instances, seismic data can provide very nearly direct indications of gas or oil. However, these indications are very specialized and are intimately related to very local geologic conditions.

Two types of seismic surveys are available (Fig. 479). Refraction seismic surveys rely on specific knowledge of rock velocities and are not widely used. Reflection surveys are widely used and accurately represent subsurface structure.

Seismic exploration utilizes an acoustic source such as a dynamite explosion, travel time from the source to a receiver, and a recorder (Fig. 480). Seismic acoustic waves are reflected by discontinuities within the layered sedimentary section to be received by geophones on the ground surface and transmitted to the recorder. Variations in travel time of acoustic waves produce time-related records that indicate the positions of the reflections on their trace (Fig. 481). Correlations of the records produce a two-dimensional cross-section that illustrates the reflections and demonstrates the subsurface structure.

Refraction surveys differ from reflection surveys in that travel time through a rock unit is measured. A reflection survey measures time to the unit and from it to the receiver. A travel time map represents time-related depths to a particular reflecting surface and provides its structural configuration (Fig. 482).

Geochemical Surveys

Analysis of soil samples, river water, formation water, and oil can be useful in some exploration programs.

1. Soil analyses can indicate the presence of hydrocarbons beneath the surface. They can indicate little about the depth of the reservoir, however.

2. Analysis of surface water from rivers and streams might be indicative of the locations of oil seeps along the river banks. Sampling from several locations might be necessary to locate the seeps.

3. Chemical analysis of subsurface formation water is often useful in correlations within a reservoir or establishing the differentiation of reservoirs.

4. Oil composition analysis can provide age determinations of crude oil being produced from a well. In many cases, crude oil age data are required to establish source rocks.

Compositional variations of crude oil are also useful in discriminating between different reservoir horizons. Oil ages and migration histories can be determined by oil analyses, as well.

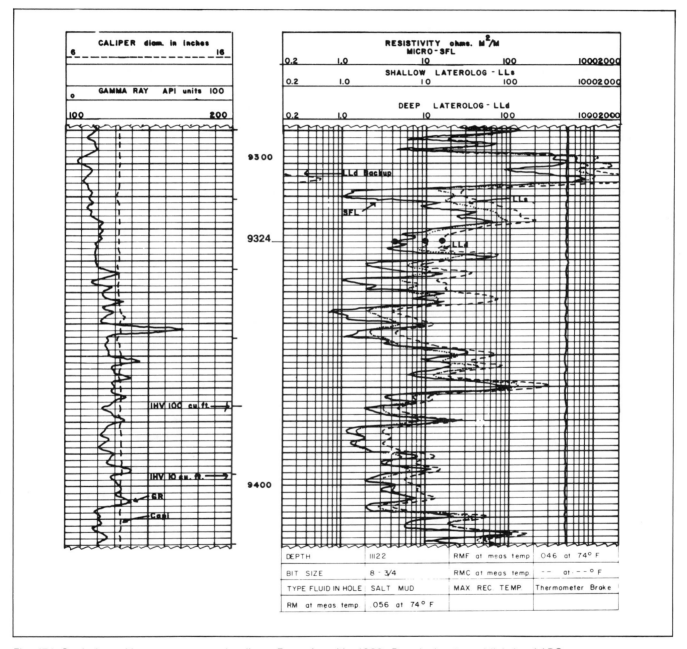

Fig. 471. Sonic log with gamma ray and caliper. From Asquith, 1982. Permission to publish by AAPG.

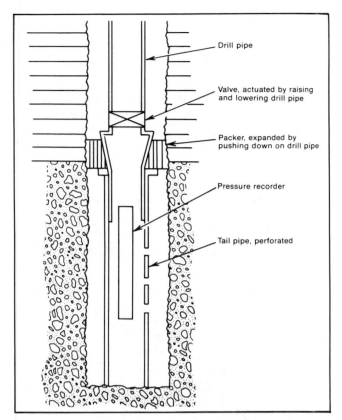

Drill pipe

Valve, actuated by raising and lowering drill pipe

Packer, expanded by pushing down on drill pipe

Pressure recorder

Tail pipe, perforated

Fig. 472. Drill-stem test tool. From Dickey, 1979. Permission to publish by PennWell Publishing Company.

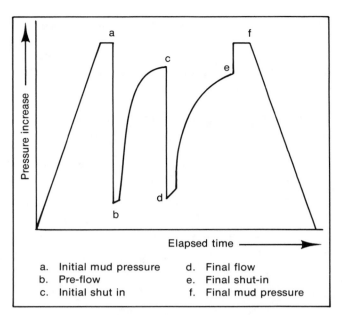

a. Initial mud pressure
b. Pre-flow
c. Initial shut in
d. Final flow
e. Final shut-in
f. Final mud pressure

Fig. 473. Drill-stem test pressure chart

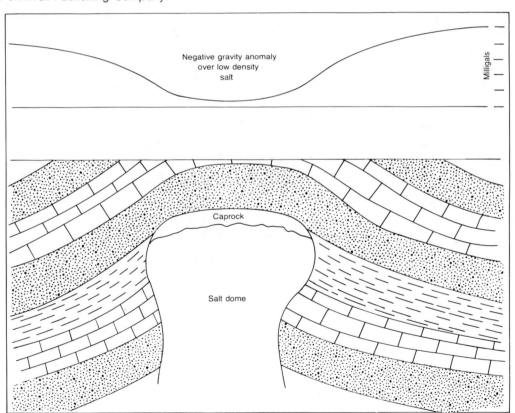

Negative gravity anomaly over low density salt

Caprock

Salt dome

Fig. 474. Salt dome gravity survey

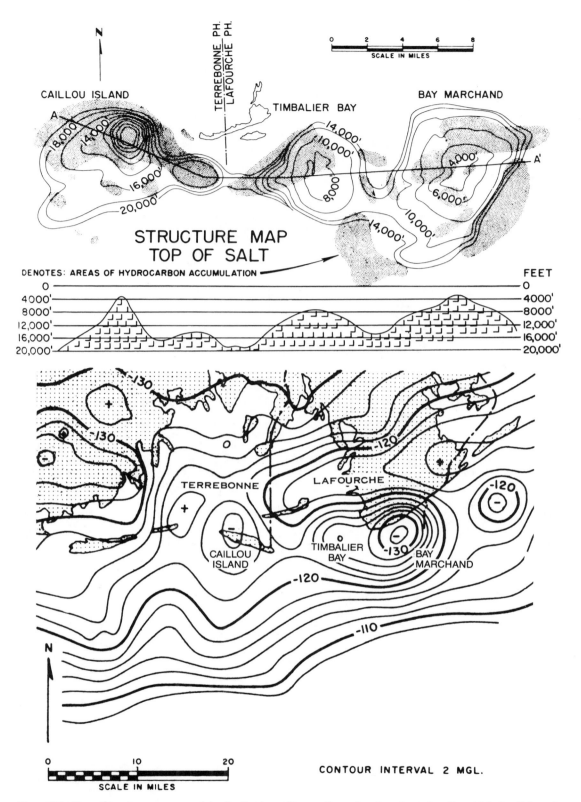

Fig. 475. Top: Structure map and logitudinal profile section showing top of salt, which is datum for structure contours. Cl = 2,000 ft. Stippling indicates hydrocarbon production as related to the salt massif. Bottom: Gravity map of Bay Marchand-Timbalier Bay-Cailou Island trend. Note strong negative anomalies caused by salt trend. From Frey and Grimes, 1970. Permission to publish by AAPG.

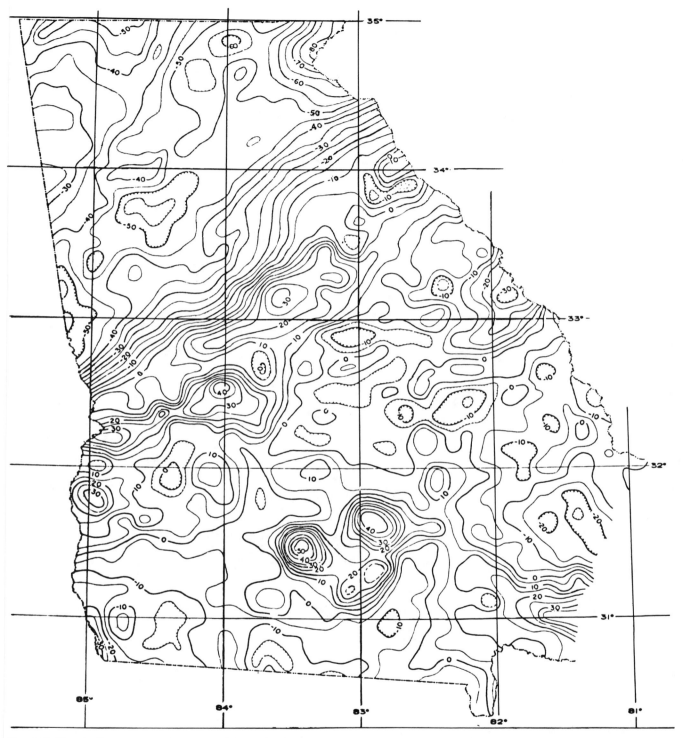

Fig. 476. Simple Bouguer gravity map of Georgia. From Long, et al., 1972. Permission to publish by Georgia Geological Survey.

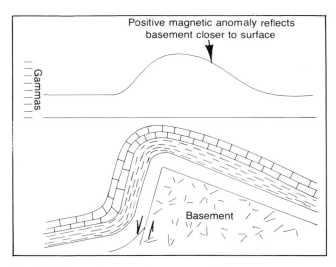

Fig. 477. *Magnetic survey over an anticline*

Maps and Cross-Sections

Practically any type of geologic data can be represented on a map. Some of the most useful maps are those that present clear pictures of the distribution of geologic parameters. A geologic map is an example of this because it shows the distribution of individual rock formations over the area of the map.

Contour maps can illustrate thickness, facies, percentages, topography, and structure (Figs. 483, 484, 485 and 486). They show variations that are useful in interpreting the complete geology of an area or individual characteristics within the entire data structure.

Contour Maps

Data upon which numerical values can be placed can be contoured, since contour lines connect points of equal value (Fig. 487). Contour maps are important interpretive aids and can represent anything from sequential geologic events to absolute values of individual parameters within single rock units.

1. Contour interval should adequately represent the data. Too large an interval overlooks some of the data and too small an interval clutters the map.

2. Contour lines should honor the data and be properly spaced relative to them.
3. Contour lines should be drawn smoothly and as parallel to each other as the data will allow.
4. Contour lines should never cross. Crossing contour lines are an impossibility.
5. Contour lines should be close together where gradients are steep and farther apart where gradients are shallow.
6. Contour lines should be labeled.

Geologic Maps

Geologic maps can include as much or as little data as desired (Fig. 488). Usually they include formations and their contacts and the most prominent faults. However, geologic maps can also include topographic and structural contours, as well as structural features.

Most geologic maps are colored. But data also can be represented by the use of symbols or different tones of gray. Geographic features are essential for purposes of location.

Cross-Sections

Structural, stratigraphic, and topographic information can be portrayed on cross-sections that reproduce horizontally represented map information in vertical section. Maps represent information in the plan view and provide a graphic view of distribution. Cross-sections present the same information in the vertical view and illustrate vertical relationships such as depth, thickness, superposition, and lateral and vertical changes of geologic features.

Raw data for cross-sections come from stratigraphic sections, structural data, well sample logs, cores, electric logs, and structural, stratigraphic, and topographic maps. Cross-sections are constructed as shown in Figures 489, 490 and 491. In this particular case a topographic section is shown. Datum for the section is sea level.

The diagramatic structure sections shown are used to illustrate development of thrust faults and folds (Fig. 492). They demonstrate the deformational sequence of a particular tectonic style.

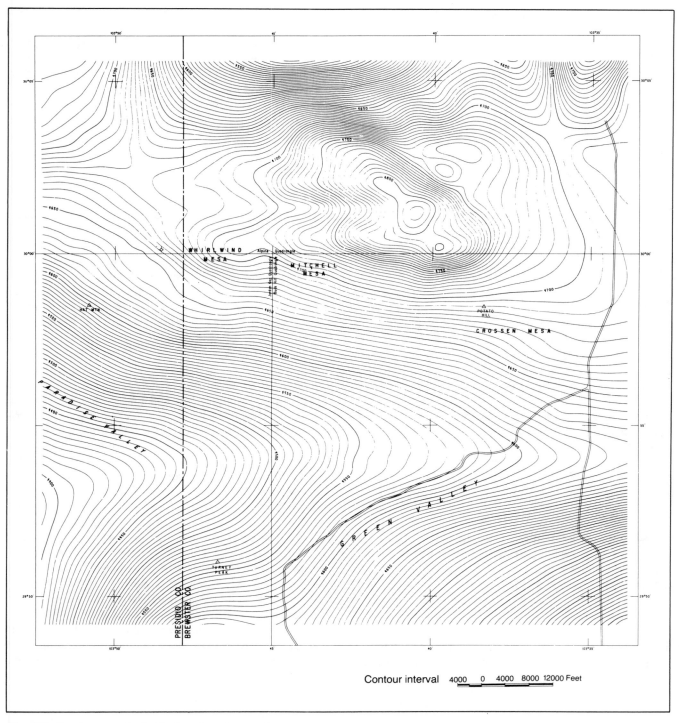

Fig. 478. Total magnetic field, Marathon Basin, Green Valley area. Permission to publish by Borehole Exploration Corp.

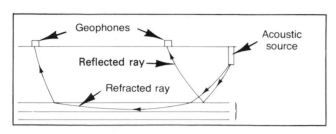

Fig. 479. Reflected and refracted seismic waves

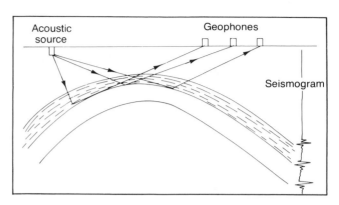

Fig. 480. Seismic exploration

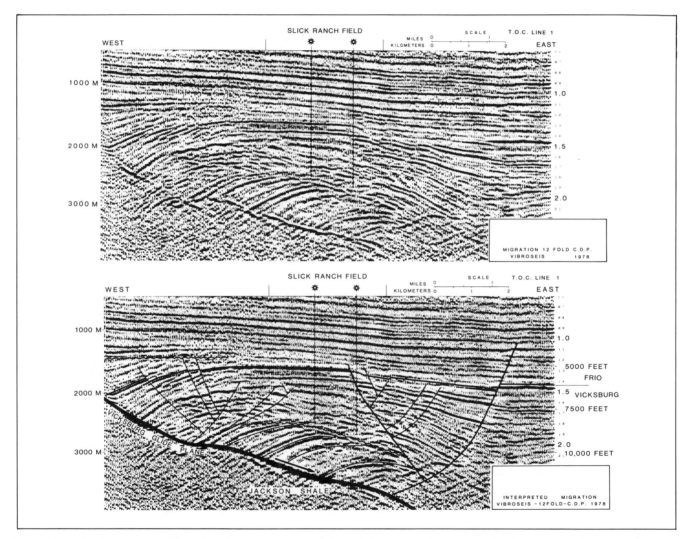

Fig. 481. Seismic line from Slick Ranch area, Starr County, south Texas, showing productive anticline in lower Oligocene sediments. From Erxleben and Carnahan, 1983. Permission to publish by AAPG.

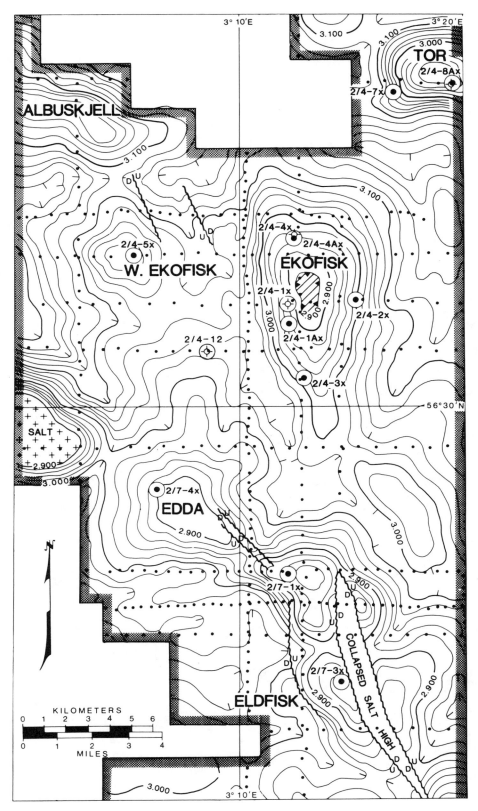

Fig. 482. Time interpretation within the Paleocene above the Ekofish Formation. Subsequent exploration wells have been plotted on the prospects. Contours in two-way time. From Van den Bark and Thomas, 1980. Permission to publish by AAPG.

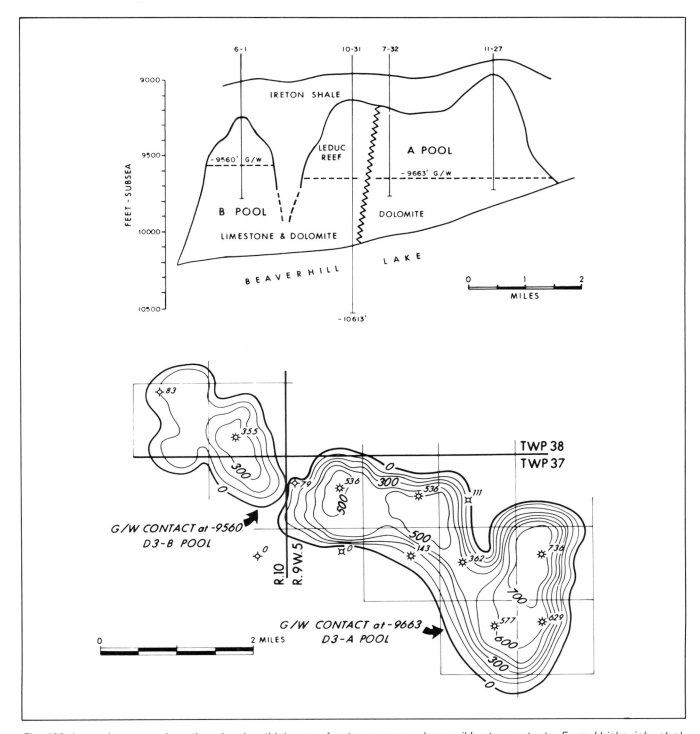

Fig. 483. Isopach map and section showing thickness of net pay zones above oil/water contacts. From Hriskevich, et al., 1980. Permission to publish by AAPG.

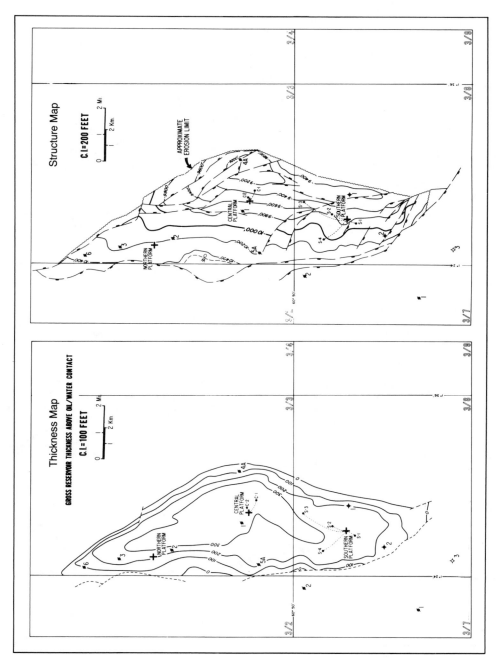

Fig. 484. Thickness and structure maps of Middle Jurassic Reservoir, Minian Field, North Sea. From Albright, et al., 1980. Permission to publish by AAPG.

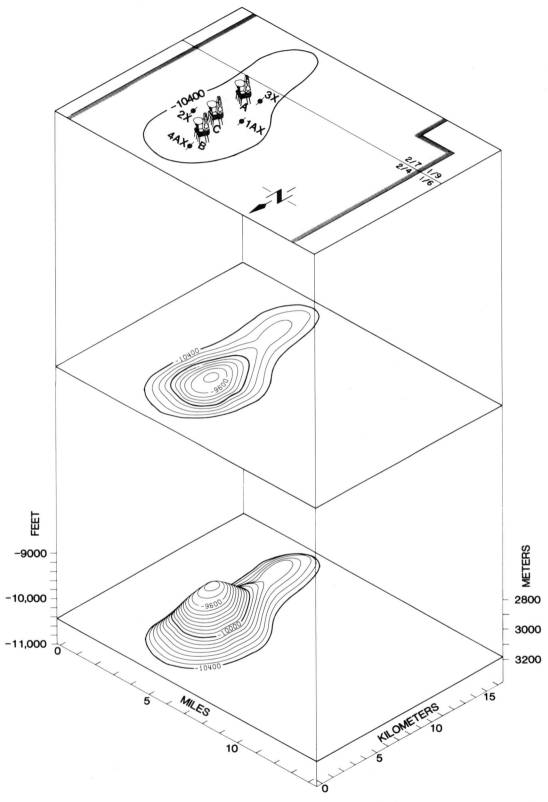

Fig. 485. Isometric view of structural configuration of the Ekofisk Field, North Sea. From Van den Bark and Thomas, 1980. Permission to publish by AAPG.

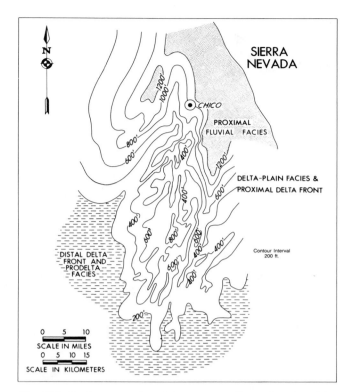

Fig. 486. Facies map of a delta system. From Garcia, 1981. Permission to publish by AAPG.

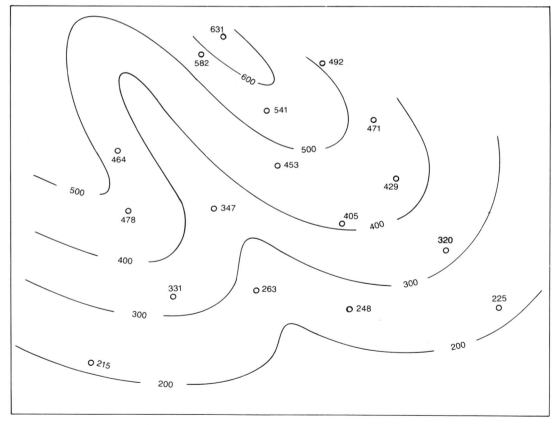

Fig. 487. Contour map

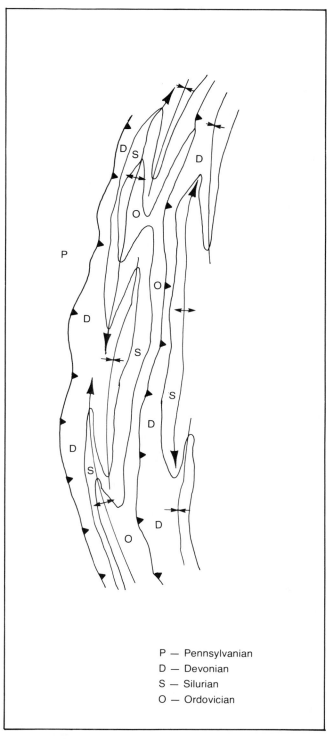

P — Pennsylvanian
D — Devonian
S — Silurian
O — Ordovician

Fig. 488. Geologic map

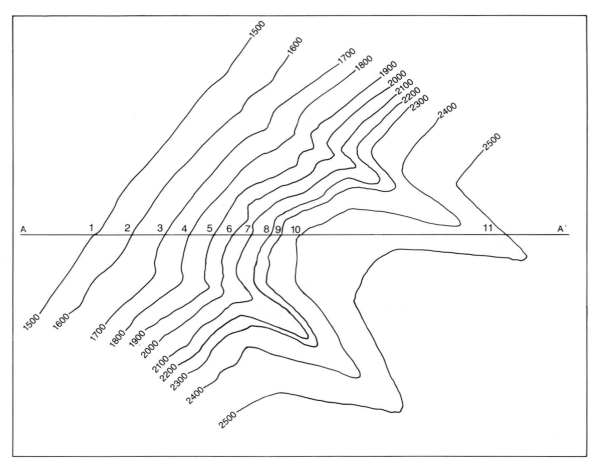

Fig. 489. Contour map and line of section

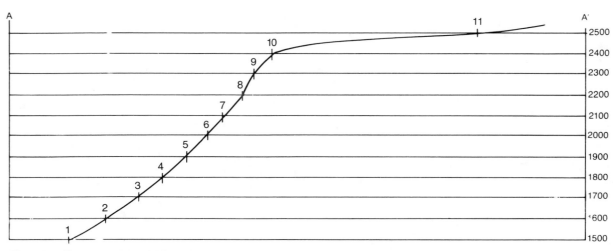

Fig. 490. Cross-section of Figure 489

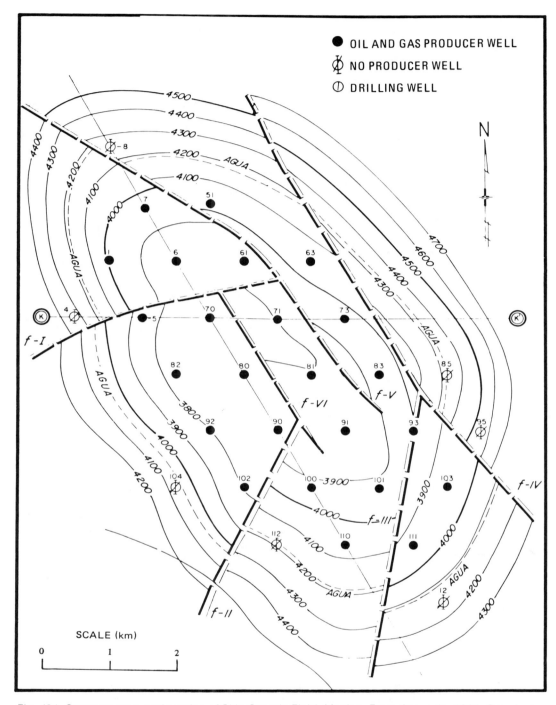

Fig. 491. Structure map and section of Sitio Grande Field, Mexico. From Acevedo, 1981. Permission to publish by AAPG.

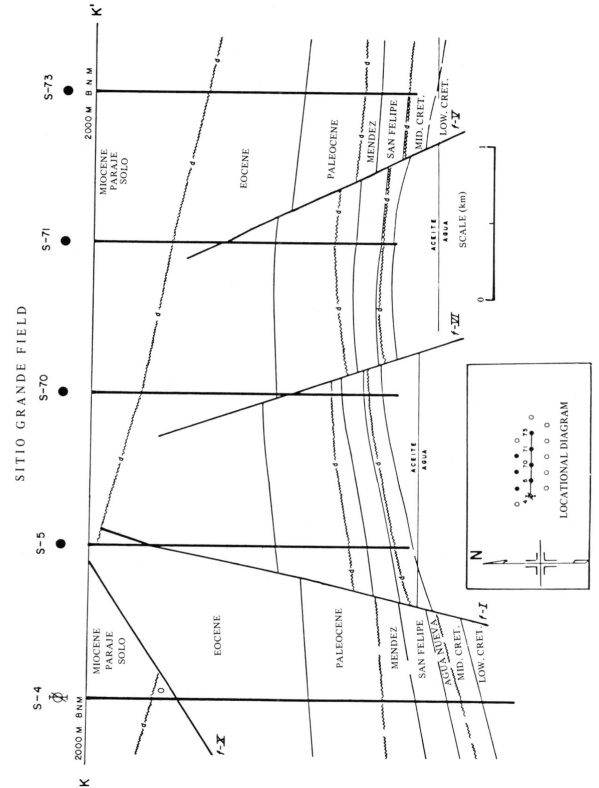

Fig. 492. Cross section across Sitio Grande Field, Mexico. From Acevedo, 1981. Permission to publish by AAPG.

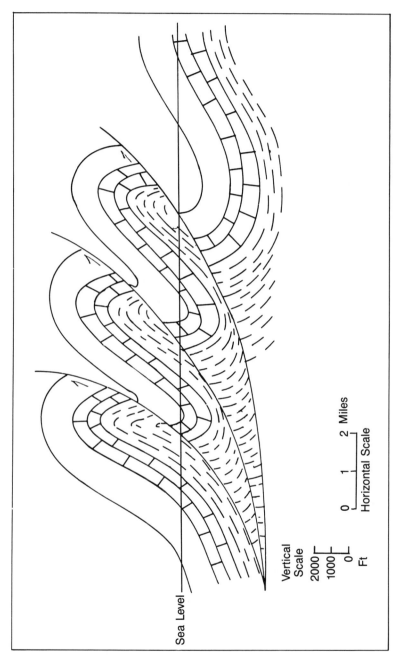

Fig. 493. Structural cross-section.

Bibliography

Acevedo, J. S., 1980, Giant fields of the southern zone—Mexico, *in* Giant oil and gas fields of the decade: 1968–1978: AAPG Mem. 30, p 339–389.

Albright, W. A., Turner, W. L., and Williamson, K. R., 1980, Ninian field, U.K. sector, North Sea, *in* Giant oil and gas fields of the decade: 1968–1978: AAPG Mem. 30, p. 173–193.

Allen, T. O., and A. P. Roberts, 1978, *Production Operations,* volumes 1 and 2, Third Edition, Tulsa: Oil & Gas Consultants International, Inc.

Amoco, 1976, Training Manual.

Archie, G. E., 1952, Classification of carbonate reservoir rock and petrophysical considerations: AAPG Bull., v 36, n 2, p 278–298.

Arthur, K. R., Cole, D. R., Henderson, G. G. L., and Kushnir, D. W., 1982, Geology of the Hibernia discovery, *in* The deliberate search for the stratigraphic trap: AAPG Mem. 32, p 181–195.

Asquith, G. B., 1982, *Basic Well Log Analysis For Geologists.* Tulsa, Oklahoma: AAPG.

Barker, C., 1979, Organic geochemistry in petroleum exploration: AAPG Continuing Education, Course Notes Series, n 10, 159 p.

Bates, R. L., and Jackson, J. A., 1980, Glossary of geology: American Geological Institute, Falls Church, Virginia, 751 p.

Bishop, R. A., 1982, Whitney Canyon-Carter Creek gas field, southwest Wyoming, *in* Geologic studies of the Cordilleran Thrust Belt: RMAG Symp., p 591–599.

Bolduchi, A., and Pommier, G., 1970, Cambrian oil field of Hassi Messaoud, Algerian, *in* Geology of giant petroleum fields: AAPG Mem. 14, p 477–488.

Bradley, W. H., "The Varves and Climate of the Green River Epoch," U.S. Geological Survey Professional Paper 158, p. 87–110.

Brown, L. F., Jr., 1979, Deltaic sandstone facies of the Mid-Continent, *in Pennsylvanian Sandstones of the Mid-Continent.* Tulsa Geological Society Special Publication no. 1, pp. 35–63.

Bruce, C. M., 1973, Pressured shale and related sediment deformation: Mechanism for development of regional contemporaneous faults: AAPG Bull., v 57, p 878–886.

Bubb, J. M., and Hatlelid, W. G., 1977, Seismic stratigraphy and global changes of sea level, part 10: Seismic recognition of carbonate buildups, *in* Seismic stratigraphy—applications to hydro-carbon exploration: AAPG Mem. 26, p 185–204.

Busch, D. A., 1974, Stratigraphic Traps in Sandstones—Exploration Techniques. AAPG Memoir 21, p. 174.

Buthod, P., 1962, Properties of Crude Oils and Liquid Condensate, *in Petroleum Production Handbook, Reservoir Engineering,* vol. 2. Society of Petroleum Engineers of AIME, pp. 18–23.

Butter, A. W., III, 1982, Case history of the North Caribou prospect, Bonneville County, Idaho, *in* Geologic studies of the Cordilleran Thrust Belt: RMAG Symp., p 649–655.

California Division of Oil and Gas, 1974, *California Oil and Gas Fields,* volume 2.

Cannon, J. L., 1966, Outcrop examination and interpretation of paleocurrent patterns of the Blackleaf Formation near Cretaceous stratigraphic traps, Sweet grass arch: Billings Geol. Soc. 17th Ann. Field Conf. Guidebook, p 71–111.

Casey, S. R., and Cantrell, R. B., 1941, Davis sand lens, Hardin field, Liberty County, Texas, *in* Stratigraphic type oil fields: AAPG, p 564–599.

Chang, S. S. L., and Pan, Y. S., 1983, Reflections near unconformity-southern Taiwan, *in* Seismic expression of structural styles: AAPG studies in Geology Series, n 15, v 1, p 1.2.2-3 -1.2.2-5.

Chapman, W. L., 1983, Erosional unconformity, river channel, *in* Seismic expression of structural styles: AAPG Studies in Geology Series, n 15, v 1, p 1.2.131 - 1.2.1-132.

Chapman, W. L., and Schafers, C. J., 1983, Shallow sand channel, coal exploration, Illinois Basin, *in* Seismic expression of structural styles: AAPG Studies in Geology Series, n 15, v 1, p 1.2.1-33 - 1.2.1-34.

Chatfield, J., 1972, Case history of Red Wash field, Uintah County, Utah, *in* Stratigraphic oil and gas fields—Classification, exploration methods, and case histories: AAPG Mem. 16, p 342–353.

Chen, S., and Wang, P., 1980, Geology of Gudao oilfield and surrounding areas, *in* Giant oil and gas fields of the decade: 1968–1978: AAPG Mem., n 30, p 471–486.

Chevron Oil Company, 1981. Personal Communication.

Choquette, P. W. and Pray, L. C., 1970, Geologic Nomenclature and Classification of Porosity in Sedimentary Carbonates. *AAPG Bulletin,* v. 54, no. 2, pp. 207–250.

Christensen, A. F., 1983, An example of a major syndepositional listric fault, *in* Giant oil and gas fields of the decade: 1968–1978: AAPG Studies in Geology Series, n 15, v 2, p 2.3.1-36 - 2.3.1-40.

Chuber, S., 1972, Milbur (Wilcox) field, Milam and Burleson counties, Texas, *in* Stratigraphic oil and gas fields—classification, exploration methods and case histories: AAPG Mem. 16, p 399–405.

Chung-Hsing P'An, 1982, Petroleum in basement rocks: AAPG Bull., v 66, n 10, p 1597–1643.

Clark, N. J., 1969, *Elements of Petroleum Reservoirs*. Society of Petroleum Engineers of AIME, H. L. Doherty series.

Cook, T. D., and Bally, A. W., eds., 1975, Stratigraphic atlas of North and Central America: Shell Oil Company (Exploration Department, Houston).

Crowe, J. C., and Buffler, R. T., 1983, Regional seismic reflection profiles across the Middle America Trench and convergent margin of Costa Rica, *in* Seismic expression of structural styles: AAPG Studies in Geology Series, n 15, v 3, p 3.4.2-147 - 3.4.2-162.

Curnelle, R., and Marco, R., 1983, Reflection profiles across the Aquitaine Basin (Sedimentary Sequences) *in* Seismic expression of structural styles: AAPG Studies in Geology, Series n 15, v 1, p 1.2.3-4 - 1.2.4-6.

Curry, W. H., and W. H. Curry III, 1972, South Glenrock Oilfield, Wyoming: Prediscovery Thinking and Post-Discovery Description, *in Stratigraphic oil and gas fields—Classification, Exploration Methods, and Case Histories*. AAPG Memoir 16, pp. 415–427.

DeChadenes, J. F., 1975, Frontier deltas of the western Green River Basin, Wyoming, *in Deep Drilling Frontiers in the Central Rocky Mountains*. RMAG Symposium, pp. 149–157.

de la Grandville, B., 1982, Appraisal and development of a structural and stratigraphic trap oil field with reservoirs in glacial to periglacial clastics, *in* The deliberate search for the stratigraphic trap: AAPG Mem. 32, p 267–286.

De Matharel, M., Lehman, P., and Oki, T., 1980, Geology of the Bekapai field, *in* Giant oil and gas fields of the decade: 1968–1978: AAPG Mem. 30, p 459–469.

DeSitter, L. U., 1956, *Structural Geology*. New York: McGraw Hill.

Dewey, J. F., and Bird, J. M., 1970, Mountain Belts and the New Global Tectonics. *Journal of Geophys. Res.*, v. 75, no. 14, pp. 2,625–2,647.

Dickey, P. A., 1979, *Petroleum Development Geology*. Tulsa: Petroleum Publishing Company.

Dickey, P. A., and Hunt, J. M., 1972, Geochemical and hydrogeologic methods of prospecting for stratigraphic traps, *in* Stratigraphic oil and gas fields—classification, exploration methods, and case histories: AAPG Mem. 16, p 136–167.

Dickinson, W. R., and Seely, D. R., 1979, Structure and stratigraphy of forearc regions. *AAPG Bulletin*, v. 63, no. 1, pp. 2–31.

Dobrin, M. B., 1977, Seismic exploration for stratigraphic traps, *in* Seismic stratigraphy—applications to hydrocarbon exploration: AAPG Mem. 26, p. 329–351.

Dolly, E. D., 1979, Geological techniques utilized in Trap Spring Field discovery, Railroad Valley, Nye County, Nevada, *in* Basin and range symposium and Great Basin field conference: Rocky Mountain Association of Geologists and Utah Geological Association, p 455–467.

Dow, W. G., 1977, Kerogen studies and geological interpretation: Geochem. Expl., v 7, p 79–99.

Downey, M. W., 1984, Evaluating seals for hydrocarbon accumulations: AAPG Bull., v 68, n 11, p 1752–1763.

Duey, H. D., 1979, Trap Spring Oil Field, *in* Basin and range symposium and Great Basin field conference: Rocky Mountain Association of Geologists and Utah Geological Association, p 469–476.

Duffy, G. E., 1975, Lay Creek Field, Moffat County, Colorado, *in Deep Drilling Frontiers in the Central Rocky Mountains*. RMAG Symposium, pp. 281–289.

Dunham, R. J., 1962, Classification of carbonate rocks according to depositional texture, *in* Classification of carbonate rocks: AAPG Mem. 1, p 108–121.

du Rochet, J., 1981, Stress fields, a key to oil migration: AAPG Bull., v 65, p. 74–85.

Dutton, S. P., 1982, Pennsylvanian fan-delta and carbonate deposition Mobeetie field, Texas Panhandle: AAPG Bull., v 66, p 389–407.

Dutton, S. P., and Land, L. S., 1985, Meteoric burial diagenesis of Pennsylvania arkosic sandstones, southwestern Anadarko basin, Texas: AAPG Bull., v 69, p 22–38.

Eardley, A. J., *Structural Geology of North America*, New York: Harper & Row, 1951.

Easton, W. H., 1960, Invertebrate paleontology: New York: Harper & Bros., 701 p.

Erxleben, A. W., and Carnahan, G., 1983, Slick Ranch area, Starr County, Texas *in* Seismic expression of structural styles: AAPG Studies in Geology, Series n 15.

Flint, R. F., and Skinner, F. J., 1974, *Physical Geology*. New York: John Wiley and Sons.

Folk, R. L., 1959, Practical petrographic classification of limestones: AAPG Bull., v 43, n 1, p 1–38.

Foster, R. J., 1979, *Physical Geology,* 3rd edition. Columbus, Ohio: C. E. Merrill.

Frazier, D. E., 1967, Recent deltaic deposits of the Mississippi River: their development and Chronology. *GCAGS Transactions*, v. 17, pp. 287–311.

Frey, M. G., and Grimes, W. H., 1970, Bay Marchand-Timbalier Bay-Caillou Island salt complex, Louisiana, *in* Geology of giant petroleum fields: AAPG Mem. 14, p 277–291.

Friedman, G. M., and Sanders, J. E., 1978, *Principles of Sedimentology*. New York: John Wiley and Sons.

Garcia, R., 1981, Depositional systems and their relation to gas accumulation in Sacramento Valley, California: AAPG Bull., v 65, p 653–673.

Gatewood, L. E., 1970, Oklahoma City Field—Anatomy of a giant, *in* Geology of Giant Petroleum Fields. AAPG Memoir 14, pp. 223–254.

Gignoux, M., 1955, *Stratigraphic geology*: San Francisco: W. H. Freeman & Co., 682 p.

Hagar, P. I., and Goodman, M. W., 1985, The Pennsylvanian Tyler Sandstone play, Central Montana Platform: A seismic-stratigraphic approach: Seismic exploration of the Rocky Mountain region: RMAG and Denver Geophysical Society, p 125–136.

Halbouty, M. T., 1979, *Salt Domes, Gulf Region, United States and Mexico,* 2nd edition. Houston: Gulf Publishing Company.

Halbouty, M. T., and Barber, T. D., 1972, Port Acres and Port Arthur Fields, Jefferson County, Texas: stratigraphic and structural traps in a Middle Tertiary delta, *in Stratigraphic Oil and Gas Fields—Classification, Exploration Methods, and CaseHistories*. AAPG Memoir 16, pp. 329–341.

Hamblin, W. K., 1978, *The Earth's Dynamic Systems,* 2nd edition, Minneapolis: Burgess Publishing Company.

Hamblin, W. K. and Howard, J. D., 1975, *Exercises in Physical Geology,* 4th edition. Minneapolis: Burgess Publishing Company.

Harding, T. P., Gregory, R. F., and Stephens, L. H., 1983, Convergent wrench fault and positive flower structure, Ardmore basin,

Oklahoma, *in* Seismic expression of structural styles: AAPG Studies in Geology, Series n 15, p 4.2-13 - 4.2-17.

Harding, T. P., and Lowell J. D., 1979, Structural styles, their plate tectonic habitats and hydrocarbon traps in petroleum provinces: AAPG Bull., v 63, p 1016–1058.

Harms, J. C., Tackenberg, P., Pickles, E., and Pollock, R. E., 1981, The Brae oilfield, *in* Illing, L. V., and Hobson, G. D., eds., Petroleum geology of the continental shelf of north west Europe: Institute of Petroleum, London, p 352–357.

Harwell, J. C., and Rector, W. E., 1972, North Knox City Field, Knox County, Texas, *in Stratigraphic Oil and Gas Fields—Classification, Exploration Methods, and Case Histories*. AAPG Memoir 16, pp. 453–459.

Hasan, M., Kamal, and Langitan, F. B., 1977, Discovery and development of Minas field: Ascope Conference, Jakarta, Indonesia.

Haun, J. D. and LeRoy, L. W., eds. 1958. *Subsurface Geology in Petroleum Exploration*. Golden, Colorado: Colorado School of Mines.

Helander, D. P., 1983, *Fundamentals of Formation Evaluation*. Tulsa, Oklahoma: Oil & Gas Consultants International, Inc.

Hemphill, C. R., Smith, R. I., and Szabo, F., 1970, Geology of Beaverhill Lake reefs, Swan Hills area, Alberta, *in* Geology of giant petroleum fields: AAPG Mem. 14, p 50–90.

Heritier, F. E., Horsel, P., and Watlane, E., 1980, Frigg Field large submarine fan trap in lower Eocene rocks of the Viking Graben, North Sea, *in* Giant oil and gas fields of the decade: 1968–1978: AAPG Mem. 30, p 59–79.

Hinz, K., 1983, Line BGR 76-11 from central east Greenland margin: Seismic expression of structural styles: AAPG Studies in Geology, Series n 15, v 2, p 2.2.3-45 - 2.2.3-46.

Hriskevich, M. E., Taber, J. M., and Langton, J. R., 1980, Strachan and Ricinus West gas fields, Alberta, Canada, *in* Giant oil and gas fields of the decade: 1968–1978: AAPG Mem. 30, p 315–327.

Hubbert, M. K., 1953, The entrapment of petroleum under hydrodynamic Conditions. *AAPG Bulletin*, v. 37, pp. 1,954–2,026.

Hughes, D. J., 1968, Salt tectonics as related to several smackover fields along the northeast rim of the Gulf of Mexico Basin. *GCAGS Transactions*, v. 18, pp. 320–330.

Hunt, J. M., Origin of the migration of hydrocarbons. AAPG Continuing Education Lecture Series.

Hunt, J. M., 1979, *Petroleum geochemistry and geology*. San Francisco: W. H. Freeman and Co., 617 p.

Jamison, H. C., Brockett, L. D., and McIntosh, R. A., 1980, Prudhoe Bay—A 10 year perspective, *in* Giant oil and gas fields of the decade: 1968–1978: AAPG Mem. 30, p 289–314.

Jones, R. W., 1978, Some mass balance and geological restraints on migration mechanisms, *in* Physical and chemical constraints on petroleum migration: AAPG Continuing Education Course Notes Series, n 8, v 2, p Ai-A43.

Kay, M. and Colbert, E. H., 1965, *Stratigraphy and Life History*. New York: John Wiley and Sons, 736 p.

King, P. B., 1969, Tectonic Map of North America. USGS Map G67154.

Kingston, D. R., Dishroon, C. P., and Williams, P. A., 1983, Global basin classification system: AAPG Bull., v 67, n 12, p 2175–2198.

Kirk, R. H., 1980, Statfjord field—North Sea, *in* Giant oil and gas fields of the decade: 1968–1978: AAPG Mem. 30, p 95–116.

Krumbein, W. C., and Sloss, L. L., 1953, *Stratigraphy and sedimentation*: San Francisco: W. H. Freeman and Co., 497 p.

Lamb, C. F., 1980, Painter Reservoir Field—Giant in Wyoming Thrust Belt. *AAPG Bulletin*, v. 64, no. 5, pp. 638–644.

LaRocque, Aurele, "Molluscan faunas of the Flagstaff Formation of Central Utah, Memoir 78, Geological Society of America, 1960.

Larson, V. F., 1975, Deep water coring for scientific purposes. *Journal of Petr. Tech.*, August, pp. 925–934.

LeBlanc, R. J., 1973, Geometry of sandstone reservoir bodies, *in Sandstone Reservoirs and Stratigraphic Concepts I*. AAPG Reprint Series no. 7, pp. 155–211.

Leighton, W. M., and Pendexter, C., 1962, Carbonate rock types, *in* Classification of carbonate rocks: AAPG Mem. 1, 279 p.

Lelek, J. J., 1982, Anschutz Ranch East field, northeast Utah and southwest Wyoming, *in* Geologic studies of the Cordilleran Thrust Belt: RMAG Symp., p 619–631.

LeMay, W. J., 1972, Empire Abo Field, southeast New Mexico, *in* Stratigraphic oil and gas fields—classification, exploration methods, and case histories: AAPG Mem. 16, p 472–480.

Leroy, L. W., 1950, *Subsurface Geologic Methods*, 2nd edition. Golden, Colorado: Colorado School of Mines.

Levorsen, A. I., 1967, *Geology of Petroleum*, 2nd edition. San Francisco: W. H. Freeman.

Link, P. K., 1982, *Basic Petroleum Geology*. Tulsa, Oklahoma: Oil & Gas Consultants International, Inc.

Link, W. K., 1952, Significance of oil and gas seeps in world oil exploration: AAPG Bull., v 36, p 1505–1540.

Long, L. T., Bridges, S. R., and Dorman, L., 1972, Simple Bouguer anomaly map of Georgia: Atlanta, Georgia, Georgia Geol. Surv.

Lowell, J. D., 1985, *Structural Styles in Petroleum Exploration*: Tulsa: Oil & Gas Consultants International.

Lu, B., Qian, S., and Chen, Z., 1982, Exploration in China: A look at geophysical activity in China: Oil and Gas Journ., Dec. 13, 1982, p 78–86.

Lukic, P., Maschek, W., and Bachmann, G. H., 1983, Salt plug Siegelsum in the Northwest-German basin, *in* Seismic expression of structural styles: AAPG Studies in Geology Series, n 15, v 2, p 2.3.2-3 - 2.3.2-6.

Mack, G. M., and Rasmussen, K. A., 1984, Alluvial fan sedimentation of the Cutler Formation (Permo-Pennsylvanian) near Gateway, Colorado: GSA Bull., v 95, p 109–116.

MacKenzie, D. B., 1972, Primary stratigraphic traps in sandstones, *in* Stratigraphic oil and gas fields—classification, exploration methods, and case histories: AAPG Mem. 16, p 47–63.

Mattavelli, L., Ricchiuto, T., Grignani, D., and Schoell, M., 1983, Geochemistry and habitat of natural gases in Po basin, northern Italy: AAPG Bull., v 67, p 2239–2254.

McClellan, B. D., and Storrusten, J. A., 1983, Utah-Wyoming overthrust line, *in* Seismic expression of structural styles: AAPG Studies in Geology Series, n 15, v 3, p 3.4.1-39 - 4.4.1-44.

McQuillen, R., Bacon, M., and Barclay, W., 1979, *An Introduction to Seismic Interpretation*. Houston: Gulf Publishing Company.

Millar, A. V., and Shiels, K. G., 1939, *Descriptive Geometry*. New York: D. C. Heath and Company.

Moholick, J. D., Metarko, T., and Potter, P. E., 1984, Regional variations of porosity and cement: St. Peter and Mount Simon sandstones in Illinois basin: AAPG Bull., v 68, n 6, p 753–764.

Momper, J. A., 1978, Oil migration limitations suggested by geological and geochemical considerations, *in* Physical and chemical constraints in petroleum migration: AAPG Continuing Education Course Notes Series, n 8, v 1, p B1-B60.

Nelson, P. H. H., 1980. Role of reflection seismic in development of Nembe Creek field, Nigeria, *in* Giant oil and gas fields of the decade: 1968–1978: AAPG Mem. 30, p 565–576.

Normark, W. R., 1978, Fan valleys, channels, and depositional lobes on modern submarine fans: characters for recognition of sandy turbidite environments: AAPG Bull., v 62, n 6, p 912–931.

Palmer, J. J., and Scott, A. J., 1984, Stacked shoreline sandstone of La Ventana Tongue (Companian), northwestern New Mexico: AAPG Bull., v 68, n 1, p 74–91.

Petroleum Information, 1978, *The Overthrust Belt,* p. 140.

Pettijohn, F. J., 1975, *Sedimentary Rocks.* New York: Harper & Row.

Picard, M. Dane, "Subsurface stratigraphy and lithology of Green River Formation in Uinta Basin, Utah," *AAPG Bulletin,* v. 39, pp. 75–102.

Ragan, D. L., 1968, *Structural Geology.* New York: John Wiley and Sons.

Ragan, D. L., 1973, *Structural Geology,* 2nd edition. New York: John Wiley and Sons.

Ragan, D. L., 1985, *Structural Geology,* 3rd Edition. New York: John Wiley and Sons.

Ray, R. R., and Berg, C. R., 1985, Seismic interpretation of the Casper Arch Thrust, Tepee Flats Field, Wyoming: Seismic exploration of the Rocky Mountain Region: RMAG and Denver Geophysical Society, p 51–58.

Rees, F. B., 1972, Methods of mapping and illustrating stratigraphic traps, *in* Stratigraphic oil and gas fields—classification, exploration methods and case histories: AAPG Mem. 16, p 168–221.

Reese, R. G., 1943, El Segundo Field: Calif. Dept. Nat. Res. Bull., 118, p 295–296.

Rice, D. D., and Claypool, G. E., 1981, Generation, accumulation and resource potential of biogenic gas: AAPG Bull., v 65, p 5–25.

Rocco, T., and D'Agostino, O., 1972, Sergnano Gas Field, Po Basin. Italy—A Typical Stratigraphic Trap, *in Stratigraphic Oil and Gas Fields—Classification, Exploration Methods, and Case Histories.* AAPG Memoir 16, pp. 271–285.

Rothrock, H. E., 1972, Davis-Gardner oil pool, Coleman County, Texas, *in* Stratigraphic oil and gas fields—classification, exploration methods, and case histories: AAPG Mem. 10, p 406–414.

Sabins, F. F., 1963, Anatomy of stratigraphic trap, Bisti Field, New Mexico, *AAPG Bulletin,* v. 47, no. 2, pp. 193–228.

Sabins, F. F., 1972, Comparison of Bisti and Horseshoe Canyon stratigraphic traps, San Juan Basin, New Mexico, *in Stratigraphic Oil and Gas Fields—Classification, Exploration Methods, and Case Histories.* AAPG Memoir 16, pp.610–622.

Saeland, G. T., and Simpson, G. W., 1982, Interpretation of 3-D data in delineating a subunconformity trap in Block 34/10, Norwegian North Sea, *in* The deliberate search for the stratigraphic trap: AAPG Mem. 32, p 217–235.

Schlumberger, Ltd., 1974, *Log Interpretation Principles,* vol. I. New York.

Shanmugam, G., 1985, Significance of coniferous rain forests and related organic matter in generating commercial quantities of oil, Gippsland Basin, Australia: AAPG Bull., v 69, n 8, p 1241–1254.

Shrock, R. E., 1948, Sequence in layered rocks: New York: McGraw-Hill, 507 p.

Smith, J. E., 1956, Basement reservoir of La Paz-Mara oil fields: AAPG Bull., v 40, p 380–385.

Spieker, Edmund M., "The transition between the Colorado Plateaus and the Great Basin in Central Utah," *Guidebook to the Geology of Utah,* no. 4, Utah Geological Society, 1949.

Stauble, A. J., and Milius, G., 1970, Geology of the Groningen Gas Field, Netherlands, *in Geology of Giant Petroleum Fields.* AAPG Memoir 14, pp. 359–369.

Stone, D. S., 1975, A Dynamic analysis of subsurface structure in Northwestern Colorado, *in Deep Drilling Frontiers in the Central Rocky Mountains.* RMAG Symposium, pp. 33–40.

Stone, D. S., 1983, Seismic profile: South Elk Basin *in* Seismic expression of structural styles: AAPG Studies in Geology Series, n 15, v 3, p 3.2.2-20 - 3.2.2-24.

Tian, Z., Chang, C., Huang, D., and Wu, C., 1983, Sedimentary facies, oil generation in Meso-Cenozoic continental basins in China: Oil and Gas Jour., May 16, 1983, p 120–126.

Tilley, B. J., and Longstaffe, F. J., 1984, Controls on hydrocarbon accumulations in glauconitic sandstone, Saffield heavy oil sands, southern Alberta: AAPG Bull., v 68, n 8, p 1004–1023.

Tissot, B. P., 1984, Recent advances in petroleum geochemistry applied to hydrocarbon exploration: AAPG Bull., v 68, n 5, p 545–563.

Todd, R. G., and Mitchum, R. M., Jr., 1977, Seismic stratigraphy and global changes of sea level, Part 8: Identification of Upper Triassic, Jurassic, and Lower Cretaceous seismic sequences in Gulf of Mexico and offshore West Africa *in* Seismic stratigraphy applications to hydrocarbon exploration: AAPG Mem. 26, p 145–163.

University of Texas, 1981, Fundamentals of petroleum: Petroleum Extension Service, University of Texas at Austin, 2nd Ed., 280 p.

Van den Bark, E., and Thomas, O. D., 1980, Ekofisk: first of the giant oil fields in western Europe, *in* Giant oil and gas fields of the decade: 1968–1978: AAPG Mem. 30, p 195–224.

Verdier, A. C., Oki, T., and Suardy, A., 1980, Geology of the Handil field (East Kalimantan-Indonesia) *in* Giant oil and gas fields of the decade: 1968–1978: AAPG Mem. 30, p 399–421.

Vest, E. L., Jr., 1970, Oil fields of Pennsylvanian-Permian Horseshoe Atoll, West Texas, *in* Geology of giant petroleum fields: AAPG Mem. 14, p 185–203.

Vincelette, R. R., and Soeparjadi, R. A., 1976, Oil-bearing reefs in Salwati Basin of Irian Jaya, Indonesia. *AAPG Bulletin,* v. 60, no. 9, pp. 1448–1462.

Viniegra, F., and Castillo-Tejero, C., 1970, Golden Lane fields, Veracruz, Mexico, *in* Geology of giant petroleum fields: AAPG Mem. 14, p 309–325.

Vreeland, J. H., and Berrong, B. H., 1979, Seismic exploration in Railroad Valley, Nevada, *in Basin and Range Symposium.* RMAG, pp. 557–569.

Walker, R. G., 1978, Deep water sandstone facies and ancient submarine fans: Models for exploration and stratigraphic traps: AAPG Bull., v 62, p 932–966.

Walters, R. F., Gutru, R. J., and A. James, III., 1979, Channel sandstone oil reservoirs of Pennsylvanian Age in Northwestern Hess County, Kansas, *in Pennsylvanian Sandstones of the Mid-Continent.* Tulsa Geology Society Special Publication no. 1, pp. 313–326.

Wegener, A., 1915, *Origins of Continents and Oceans.* New York: Dover.

Weimer, R. J., 1966, Time-stratigraphic analysis and petroleum accumulations, Patrick Draw Field, Sweetwater County, Wyoming, *in Sandstone Reservoirs and Stratigraphic Concepts I;* AAPG Reprint Series No. 7, pp. 4–29.

West, J., and Lewis, H., 1982, Structure and palinspastic reconstruction of the Absaroka thrust, Anschutz Ranch area, *in* Geologic Studies of the Cordilleran Thrust Belt: RMAG, pp 633–639.

Wilde, P., Normark, W. R., and Chase, T. E., 1978, Channel sands and petroleum potential of Monterey deep sea fan, California: AAPG Bull., v 62, p 967–983.

Williams, J. J., 1972, Augila Field, Libya: Depositional environment and diagenesis of sedimentary reservoir and description of igneous reservoir, *in Stratigraphic Oil and Gas Fields—Classification, Exploration Methods, and Case Histories*. AAPG Memoir 16, pp. 623–632.

Williams, W. D., and Dixon, J. S., 1985, Seismic interpretation of the Wyoming overthrust belt: Seismic exploration of the Rocky Mountain Region: RMAG and Denver Geophysical Society, p 13–22.

Ziegler, P. A., 1983, Inverted basins in the Alpine foreland *in* Seismic expression of structural styles: AAPG Studies in Geology, Series n 15, p 3.3-3 - 3.3-11.

Glossary

Abnormally high pressure—Subsurface pressure higher than rock over-burden or lithostatic pressure.

Abrasion—Mechanical wearing, grinding or scraping, by impact and friction, of rock surfaces or grains by gravity, water, ice or wind.

Absolute time—Specific geologic time in years.

Abyssal ooze—Fine-grained sediment of the abyssal marine zone often containing calcareous or siliceous pelagic skeletal remains.

Abyssal zone—The ocean, below 2000 meters deep.

Accretionary basin—A basin, which develops along a subduction zone in response to sediment accumulation and deformation to shift the subduction zone in the direction of the overridden plate.

Accumulation—The process of petroleum collection in a reservoir, or the collected body of petroleum itself.

Acidic igneous rock—Igneous rock containing a high proportion of quartz.

Acidization—A production enhancement procedure which uses acid to increase petroleum flow from a reservoir by partially dissolving it.

Acoustic wave—A sound or sonic wave. An earthquake wave.

Advection (meteorological)—Horizontal movement of air resulting in transfer of atmospheric properties.

Aerial photograph—Photograph of earth's surface taken from the air: may be vertically or obliquely taken.

Aerobic—Pertaining to the presence of oxygen. Aerobic bacteria require oxygenated conditions.

Age—A portion of geologic time when rocks of a specific stratigraphic unit were deposited, or designated a particular segment of the geologic time scale.

Algae—Primarily aquatic, diverse photosynthetic plants.

Calcareous algae can form carbonate banks and reefs.

Alluvium—Comparatively geologically recent, unconsolidated, poorly sorted, detrital gravel, sand, silt and clay deposited by often ephemeral, rapidly moving water under flood or flash-flood conditions: stream, flood-plain, delta and alluvial fan deposits.

Amorphous texture—Glassy.

Amphibole—Dark-colored, ferromagnesian silicate often found in the more acidic igneous rocks: hornblende.

Amphitheater (glacial)—Concave, bowl-shaped landform developed at the source of a mountain glacier where its ice and snow accumulate.

Anaerobic—Pertaining to the absence of oxygen. Anaerobic bacteria do not require oxygenated conditions.

Andesite—Medium, dark-colored, aphanitic, intermediate composition, extrusive igneous rock. Contains sodic plagioclase, biotite, hornblende, pyroxene, and some quartz.

Angular (grain)—A grain form with sharp edges, irregular shape, and no rounding.

Angular unconformity—An unconformity the erosion of surface of which lies between nonparallel beds. Flat, impermeable younger beds in erosional contact with tilted, permeable, petroleum-bearing older beds form an angular unconformity trap.

Anhydrite—An evaporite mineral of calcium sulfate.

Anoxic—Pertaining to an oxygen-deficient environment.

Anticline—Un upwardly convex fold with limbs that dip away from the axial position.

Antithetic fault—Conjugate, usually secondary faults, at approximately 60 degrees to a primary or synthetic fault.

Aphanitic (igneous)—Extrusive igneous rock texture too fine for identification of mineral constituents without magnification.

API gravity—Specific gravity of crude oil determined by a formula developed by the American Petroleum Institute comparing the specific gravity of oil to the specific gravity of water at the same temperature.

Apparent dip—Any dip measured at other than perpendicular to the strike of an inclined surface.

Aquifer—Water-bearing permeable rock.

Arch—A broad, regionally positive, anticlinal feature.

Arenite—Consolidated, clastic rock of sand sized particles: arkose, sandstone, etc.

Arete—Serrated ridge resulting from headward glacial erosion on both sides of a divide.

Argillaceous—Shaly, or containing clayey constituents.

Argillite—Compact mudstone or shale of greater coherence due to increased induration.

Arkose—Coarse-grained, feldspathic, variably sorted sandstone containing angular grains, representing rapid deposition and limited grain transport.

Artesian well—A well in which the water from the aquifer either flows (flowing artesian well) or rises in the pipe above the aquifer but below ground level (non-flowing artesian well).

Ash (volcanic)—Fine-grained, pyroclastic, volcanic material, often transported by wind.

Asphalt—Brown to black bituminous petroleum of high viscosity and low API gravity.

Asphaltic oil—Crude oil with asphaltic characteristics.

Associated gas—Natural gas occurring, as free gas or in solution, with liquid petroleum in a reservoir.

Asthenosphere—Upper mantle earth layer below the lithosphere.

Asymmetric fold—A fold with an inclined axial surface or nonidentical limbs.

Atmosphere—The gas envelope surrounding the earth.

Atoll—A circular, elliptical or U-shaped coral reef complex enclosing a central lagoon and often growing on sinking volcanic islands.

Avalanche—A large, often destructive mass of ice, snow, rock, soil or trees, flowing rapidly downhill in mountainous terrain.

Axial plane—The planar axial surface of a fold.

Axial surface—A not necessarily planar surface connecting hinge lines of folded beds.

Axis (fold)—A line which generates the form of a fold when moved parallel to itself in space.

Azimuth—Direction of a horizontal line measured clockwise from north on the 360 degree compass circle.

Backarc basin—Structural/depositional basin in the convex area behind the volcanic zone of an arch-trench subduction complex.

Backbar—Lagoon side of an offshore bar.

Backbeach—Backshore area of a beach between the upper shore zone limit and high-water line of mean spring tides.

Backshore—Beach or shore area between upper-most shore zone and high-water line of mean spring tides.

Backswamp—Poorly drained fluvial flood plain feature behind a natural levee.

Bar—An elongate detrital ridge, mound or bank deposited by marine waves and currents, or rivers and streams.

Barchan—Horseshoe-shaped, isolated sand dune.

Barrier bar—A detrital bar at or immediately below water level at a bay mouth (baymouth bar) or parallel to a shore.

Barrier island—Linear, narrow, detrital island, parallel to a coast and separated from it by a lagoon. Characterized by a beach, dunes, marshy areas and some vegetation.

Barrier reef—Linear reef trend, parallel to a coast and separated from it by a lagoon.

Basalt—Dark, basic, aphanitic, extrusive igneous rock of pyroxene, calcium feldspar and olivine.

Base level—The theoretical level below which streams will not erode: sea level.

Basement—Crustal rocks, usually crystalline (igneous, metamorphic), below the sedimentary section of any area.

Basic igneous rock—Igneous rock containing no quartz: Mafic.

Basin—A low area with no exterior drainage. Often an area of sedimentary deposition: lake basin; marine basin.

Batholith—Extensive, discordant intrusive igneous mass greater than 40 square miles in area and with no observed bottom.

Bathyal zone—Ocean depth zone between 200 and 2000 meters deep.

Bay—A coastal indentation between two headlands along the shore of a lake or ocean; smaller than a gulf and larger than a cove.

Baymouth bar—A sand or gravel bar deposited by waves and longshore current across the opening of a bay.

Beach—Unconsolidated detrital shoreline material deposited by waves and longshore currents between the vegetation line and the low water line: usually gently sloping towards the sea.

Bearing—Direction of a line or points connected by a

line and measured horizontally less than 90 degrees east or west of a reference meridian (north-south) from either end of the meridian.

Bed—A stratum or layer of rock.

Bedding—Layers of stratified rock.

Bedding plane—A surface separating stratified rocks.

Bed load—The sediment load carried along the channel bottom of a stream by dragging (traction) or jumping (saltation).

Bed rock—Solid rock beneath soil or unconsolidated surficial material.

Biogenic gas—Low temperature gas generated by an-aerobic bacterial decomposition of organic material.

Birdfoot delta—A delta consisting of a number of sea-ward projecting distributaries creating a pattern resembling an extended bird's foot.

Bit—A cutting device at the end of a drill string designed to break rock or soft material as a well bore is drilled.

Bitumen—Migratable hydrocarbon formed from thermal alteration of kerogen.

Black shale—Shale containing sufficient organic material to have a black color and sometimes have petroleum source rock potential.

Block diagram (structural)—A three-dimensional perspective diagram representing structural features.

Block fault—A type of normal fault which separates the crust into fault-bounded blocks of different elevations.

Block trellis—A rectilinear stream drainage pattern controlled by parallel hard ridges and intervening soft valleys.

Blowout (well)—Caused by drilling into a high-pressure formation which forces well-bore fluids to the ground surface. If the formation contains high-pressure gas or oil, ignition may occur at the surface and destroy the drilling equipment.

Blowout (wind)—A geomorphologic term for a small, wind-hollowed depression, usually in arid or semi-arid areas.

Blowout preventer—Equipment installed below a drilling derrick floor to prevent a well blowout.

Bottomset bed—Horizontal or gently inclined sediment layer deposited at the toe of a delta or in a stream channel. Can be deposited as a turbidite at the delta front.

Boulder—A clast larger than 256 mm in diameter.

Bounding bluff—The bluff or cliff on either side or both sides of, and establishing the limit of, a floodplain.

Boundstone—Sedimentary carbonate rock the original components of which were bound together in place during deposition: most algal bank and reef deposits.

Brachiopod—A bilaterally symmetrical marine inverte-brate characterized by two valves connected along a hinge.

Braided stream—A multiple channel stream divided because the alluvial material to be carried exceeds the capacity of the water to carry it.

Breaker—A marine or lacustrine translation wave.

Breccia—Angular, unworn rock fragments in mineral cement or fine matrix. A multi-genetic clastic rock formed by sedimentation and/or deformation.

Bubble point—The pressure at which gas dissolved in crude oil comes out of solution.

Butane—A paraffinic hydrocarbon gas of composition C_4H_{10}.

Button bit—A tri-cone rotary bit studded with steel buttons for drilling of hard formations.

Calcarenite—A clastic limestone comprising over 50 percent sand-size calcium carbonate particles cemented as a calcareous sandstone.

Calcareous—Rock or other material containing up to 50 percent calcium carbonate.

Calcilutite—A clastic limestone comprising over 50 percent silt/clay-size calcium carbonate particles cemented as calcareous mud.

Calcirudite—A clastic limestone comprising over 50 percent larger than sand-size calcium carbonate particles cemented as conglomerate.

Calcite—Calcium carbonate mineral.

Caldera—Volcanic depression formed by collapse and/or explosion of a volcano.

Capillary pressure—A variable pressure caused by the surface tension of a given liquid in capillaries of different diameters. It is higher in narrow capillaries and lower in wider capillaries.

Cap rock (salt dome)—An anhydrite, gypsum, calcite and sulfur body over the top of a salt dome.

Carbonate—Rock-forming minerals containing the carbonate ion which include calcite and dolomite.

Carbonate mud—Usually precipitated lime mud: calcium carbonate mud without reservoir or source potential.

Carbonate platform—A substantial limestone or dolomite substrate upon which a reef might be built.

Carbonate rock—Rock of carbonate minerals.

Carbonic acid—A calcium carbonate solvent comprising a chemical solution of carbon dioxide and water.

Carbon ratio (coal)—Fixed carbon percent in a coal.

Casing—The metal pipe lining placed in a well to protect the well bore from caving and fluid contamination.

Casing string—Joints of connected casing placed in a well to protect it.

Catagenesis—Middle level (between diagenesis and metagenesis) sediment consolidation and alteration when most oil and gas are generated.

Cave—An underground space most commonly associated with limestone solution.

Cavern—A large cave.

Cavern system—A network of connected caves and caverns.

Cement—Mineral material precipitated into intergranular or intercrystalline pore space.

Cementation—Precipitation of mineral material into intergranular or intercrystalline pore space.

Cenozoic Era—Geologic time from the end of the Mesozoic Era to the present.

Chalk—Fine-textured marine limestone formed by shallow water accumulation of calcareous remains of floating micro-organisms and algae.

Channel—A place through which a current (marine) can flow such as between two sand bars.

Channel (river)—A river bed.

Channel mouth bar—A detrital bar or mound formed where a stream enters an ocean or lake.

Chemical weathering—Weathering by chemical change of mineral constituents in rocks.

Chert—Sedimentary silicon dioxide.

Chronostatigraphic unit—A unit of layered rock related to a specific increment of geologic time.

Cinder—A small, volcanic, pyroclastic fragment.

Circulation (mud)—The process of pumping drilling mud from mud tanks, through a drill string, through the annulus between the drill string and the borehole wall, to the derrick floor and back to the tanks during well drilling operations.

Cirque—The large bowl feature at the head of a glacial valley.

Cirque lake—Lake in the floor of a cirque.

Clast—A grain or fragment.

Clastic (rock)—A rock composed of clasts.

Clay—Clastic rock mineral constituent less 1/256 mm in diameter.

Cleavage—Mineral parting along consistent zones of weakness in its molecular structure. A diagnostic mineral physical property.

Closure—The vertical distance between the highest point on an anticline or dome and its lowest closed structural contour.

Coal rank—The degree of coalification achieved by coal from low-rank peat to high-rank anthracite.

Coastal marsh—A marsh resulting from the in-filling of a lagoon behind a barrier island or in an estuary.

Cobble—A clastic particle between 64 and 256 mm in diameter.

Collar (drilling)—A weighted length of pipe attached to a drill string to increase its weight and maintain stability and hole size.

Colloid—Fine-grained material of diameter less than 0.00024 mm in aqueous suspension.

Color—A diagnostic physical property for mineral identification: its hue.

Combination drive—Crude oil reservoir drive mechanism which combines any of the individual drives: water drive plus gas in solution and gas cap drives.

Combination trap—A trap combining structural and stratigraphic parameters: a faulted channel sandstone.

Combustion drive—A secondary recovery technique in which crude oil is ignited in a reservoir in order to heat it and reduce its surface tension to mobilize it toward a producing well.

Compaction—Sediment volume decrease by increase in overburden pressure.

Completion program—A series of steps taken to prepare a producing well for long term production.

Complex fault trap—A fault trap involving multiple faulting and or fault block tilting.

Compressive stress—Pressure that pushes material together: operates in opposition toward a common point or plane.

Concentric fold—A parallel fold with constant orthogonal thickness of its folded layers.

Concretion—A hard spherical to oblate mineral mass formed by aqueous precipitation around a center or nucleus of often different composition.

Condensate—High API gravity (60°+ API) liquid hydrocarbon produced with associated wet gas.

Conductivity—Ability of a substance to transmit electricity.

Conglomerate—Coarse-grained clastic sedimentary rock with fragments larger than 2 mm diameter.

Conjugate angle—Angle between two related synthetic and antithetic shears bisected by the axis of maximum principal compressive stress: usually between 60° and 90°.

Conjugate fault—Either of two shears, the angle between which is bisected by the axis of maximum principal compressive stress.

Connate water—Original formation water trapped at time of deposition.

Contact—A plane or surface between rocks and/or fluids.

Contact metamorphism—Rock alteration caused by

immediate contact proximity to a hot, igneous body.

Continental crust—Crust of silicon and aluminum minerals which underlies continental masses: SIAL.

Continental deposition—Deposition of sediments under nonmarine and often subaerial conditions.

Continental drift—Movement of continental and oceanic crustal plates resulting in deformation and sedimentary deposition.

Continental glacier—An ice sheet that covers large, nonmountainous continental areas.

Continental margin—The edge of a continental crustal plate associated with convergent, divergent or transform activity.

Continental shelf—The area between the shore and the top of the continental slope: includes the neritic zone.

Continental slope—The inclined area between the continental shelf and the deep ocean: it averages about 6° from the horizontal.

Continental terrace—Sediment and rock material which underlies the continental slope, continental shelf, and coastal plain.

Contour interval—Difference in value between adjacent contour lines. Usually, but not always, constant over a single map. Selected to best illustrate features or values being contoured.

Contour line—A line of specific value which connects points of the same numerical value.

Contour map—A map of contour lines, spaced by a contour interval to illustrate topography, structure, density, percentages, ratios, etc.

Convection (atmospheric)—The temperature-motivated lateral and vertical movement of air: results in latitudinal winds, thunderstorms, etc.

Convection (mantle)—Heat exchange by heating and cooling circulation patterns in the earth's mantle.

Conventional core barrel—A hollow steel barrel with teeth on the circumference of its cutting end, attached to a drill string to cut and obtain a rock core.

Convergent margin (plate tectonics)—The leading edge of an advancing continental margin. Usually associated with subduction, deformation and volcanism.

Coral—Warm water, sessile, commonly calcareous, bottom-dwelling marine invertebrate animal. Colonial corals build coral reefs and banks.

Coralline limestone—Limestone composed of coral constituents.

Coral reef—A reef composed of coral.

Core—The solid and liquid center of the earth. It is about 7000 km in diameter, composed mostly of iron and nickel, and has an average specific gravity of approximately 10.5.

Core barrel—A tubular cutting device attached to a drill string to obtain a column of rock for detailed examination.

Core recovery—The amount of core obtained by cutting a core.

Correlate (stratigraphic)—To compare and demonstrate similarities of rocks and stratigraphic units from separate areas.

Correlation—The process of comparison of rocks and stratigraphic units. The results of comparison.

Cove—A coastal feature comprising a small bay, inlet or indentation.

Craton—A large, tectonically stable, interior continental area which has been little deformed for a prolonged period.

Creep—Downslope soil movement.

Crevasse splay—A break in the natural levee of a distributary channel through which water flows to deposit sediment and change the course of the channel.

Critical point—A point of temperature and pressure above which certain materials will not exist in two phases of gas or liquid.

Cross-bed—A bed inclined to primary, normally horizontal strata. Can be stream, wind or delta related.

Cross-cutting relations—Depositional, intrusive or erosional disharmony representing sequences of geological events.

Cross-section—A diagram along a specific, usually vertical plane or surface, which shows the distribution of structural and/or stratigraphic features.

Crown block—The complex of cable sheaves at the top of a derrick instrumental to the operation of the draw works.

Crude oil—Liquid natural petroleum.

Crust (earth)—The outer shell of the earth comprising oceanic and continental components with a specific gravity of 2.6–2.65.

Cryptocrystalline—Crystallinity too fine to be distinguished under an ordinary optical microscope.

Crystal—A plane face figure representing any element or compound of consistent composition.

Crystalline (igneous/metamorphic)—A term describing crystallized igneous or metamorphic rocks: often implicit of basement.

Crystallization—Physical and/or chemical conversion of gaseous or liquid material to solid crystal.

Crystal system—One of six groups of crystalline symmetry.

Current—The continuous directed movement of a fluid.

Cycle of erosion—Progressive development of a landscape or seashore, from youthful through mature to old age by continental and marine weathering and erosion. The cycles are never complete because of crustal activity and interruption.

Cyclic stratigraphy—Layered sediments representing repetition of recognizable depositional patterns.

Cycloparaffin—A crude oil with the formula C_6H_{12}.

Darcy—A permeability unit based upon the passage of one cubic centimeter of fluid of one centipoise viscosity for a distance of one centimeter through an area of one square centimeter in one second under a pressure differential of one atmosphere.

Daughter product—A nuclide formed by radioactive decay of a parent mineral. May be an isotope of the parent or a different product.

Dead oil—Oil with no volatiles or which will not produce a fluorescent cut in a solvent.

Debris flow—A mobile, soil, rock fragment and mud mass of more than 50 percent larger-than-sand size particles.

Deflation (wind)—Degradation of land surface by wind.

Deformation (strain)—Changing shape by stress application: folding, shearing.

Delta—The flat, commonly triangular, alluvial deposit occurring at the mouth of a river at its entry to a quiet body of water, i.e., lake or ocean.

Delta fringe—The deeper water leading edge of a delta.

Delta front—The slope deposit; foreset part of an advancing delta.

Delta lobe—A locus of active distributary deposition in the delta front position.

Delta plain—The level top-set, mostly subaerial, upper part of the delta.

Dendritic pattern—A branching stream pattern resembling that of the veins in a leaf.

Density—Mass per unit volume: grams per cubic centimeter; pounds per cubic foot.

Density current—Gravity motivated current resulting from density variations caused by suspended material, and/or salinity and/or temperature.

Deposition—The laying down or emplacement of material, especially sedimentary, as stratified or unstratified accumulations.

Depositional basin—A place where deposition occurs, usually under subaqueous conditions.

Depositional environment—The conditions under which deposition occurs: fluvial, marine, glacial, deltaic, etc.

Depositional sequence—A series of rock layers representative of a depositional event or group of events.

Descriptive geometry—The accurate representation of objects in three dimensions and the graphical solving of related problems.

Desert—A geographic area with little or no vegetation and less than 10 inches of annual rain fall.

Desert pavement—A roughly interlocking mosaic of pebbles left on a desert floor by wind or water removal of surrounding fine-grained materials.

Derrick—The vertical mast or block and cable support of an integrated oil well drilling system.

Derrick floor—The platform on which the derrick rests.

Detrital—Pertaining to detritus.

Detritus—An accumulation of mechanically derived rock and mineral fragments including gravel, sand and silt.

Diagenesis—The process of converting sediment to rock.

Diamond core barrel—A hollow pipe studded with diamonds at its cutting end and attached to a drilling string to obtain rock cores.

Diapir—An anticlinal fold ridge or dome formed by the squeezing of shale, salt or other mobile material into the core of the feature.

Dike—A cross-cutting, tabular igneous or sedimentary intrusion.

Diorite—A plutonic, intermediate composition igneous rock of some quartz, sodium feldspar, pyroxene and hornblende, and limited mica.

Dip—The departure normal to the strike in degrees of an inclined plane from the horizontal.

Dip slip—Fault movement parallel to the dip direction of the fault plane.

Disconformity—An unconformity with sediments parallel to each other above and below the erosion surface which is not necessarily planar.

Disconformity trap—A petroleum trap associated with a disconformity.

Disharmonic fold—A nonparallel fold in which beds of varying competence demonstrate variable deformation intensity.

Disintegration—Chemical and mechanical breakdown of rock and mineral material: weathering.

Displacement—Relative movement of two sides of a fault.

Dissemination—Precipitation of solution mineral into rock pore space remaining after original cementation.

Dissolved gas—Gas in a crude oil solution.

Dissolved gas drive—Reservoir drive associated with pressure released by gas coming out of petroleum solution.

Dissolved load—Solution load in a stream.

Distal—Remote or far away.

Distributary channel—A stream channel that diverges from the main channel not to return, as on a delta.

Divergent plate margin—The margin of a crustal plate that moves away from a spreading center: the trailing edge of a plate.

Dolomite—Calcium magnesium carbonate: $CaMg(CO_3)_2$.

Dolomitization—A volume-reducing recrystallization process which adds the magnesium ion to calcium carbonate to form dolomite: can occur contemporaneously with deposition or diagenetically.

Dome—An anticlinal structure with all dips away from the apex.

Downslope movement—Gravity induced movement of rock, debris, mud, etc.

Downthrown block—The fault block which is displaced downward relative to the upthrown block.

"Down to the coast" fault—A fault type common in areas of rapid deposition where fault blocks are displaced downward in the direction of depositional transport.

Drainage—The discharge or passage of water from an area, such as the stream drainage of rainfall on the ground surface.

Drainage pattern—The configuration of drainage channels of a particular area.

Drape fold—A fold in layered rocks produced by movement of underlying brittle rocks at high angles to the layers: a forced fold.

Draw works—The system of power equipment on a drilling installation which controls the raising and lowering of the drill string.

Drift—All glacially originated, transported and deposited material.

Drilling mud—Variably formulated, heavy, liquid suspension used in rotary drilling.

Drilling program—An integrated schedule of drilling parameters to most effectively drill an oil well.

Drilling rate—The time required for rotary drilling to penetrate a rock formation.

Drilling rig—The equipment used to drill a well.

Drill pipe—Steel pipe attached to collars or the bit to turn the bit and drill a well.

Drill stem test—A test to determine production potential of a particular formation interval using testing equipment installed in the drill string.

Drill string—The system of pipe, collars, subs and bit used to drill a well.

Dripstone—Cavern deposits formed by dripping mineral water and evaporation.

Drumlin—A smooth teardrop-shaped hill or mound of drift deposited by continental glaciation and tapered in the direction of ice flow.

Dry gas—Gas containing little or no liquid hydrocarbon.

Dry hole—An unsuccessful well.

Dune—A hill or mound of wind blown sand.

Dune field—An accumulation of dunes.

Dunite—A basic plutonic igneous rock containing mostly olivine and limited amounts of feldspar and ferromagnesian minerals.

Dynamic metamorphism—Metamorphism involving directed stress and deformation.

Earthquake—A sudden trembling or motion in the Earth accompanying strain release.

Earthquake wave—A seismic acoustic wave generated by an earthquake.

Earthtide—A semi-daily response of the earth to forces which produce ocean tides.

Effective porosity—The percent of a rock volume in which the pore spaces are connected to allow fluid flow.

Electric log—A well log obtained by passing electric current through rock formations penetrated by a well to illustrate diagnostic electrical characteristics.

Emergent coast—A coast formed by raising of the land or lowering of the sea.

Eogenetic burial—Burial of sediment between final sediment deposition and burial below the effects of near surface processes.

Eolian—Wind blown or wind related.

Eon—The longest geologic time unit: Phanerozoic Eon includes Paleozoic, Mesozoic and Cenozoic eras.

Epoch—A geologic time unit longer than an age and shorter than a period.

Equal area net—The Schmidt stereographic net used to prevent certain types of distortion in stereographic projection.

Era—A geologic time unit smaller than an eon.

Erathem—The largest formal time-rock unit above system.

Erosion—Removal of rock material to another place by one or several transportation agencies.

Esker—A sinuous, irregularly stratified ridge of subglacial stream detritus, deposited in a subglacial tunnel.

Estuary—A coastal marine drowned river valley.

Ethane—A natural gas of composition C_2H_6.

Eugeocline—A modification of "eugeosyncline" to demonstrate deposition and deposits like those deposited in a eugeosyncline.

Eugeosyncline—A linear depositional area in which rapid deposition, limited transportation, volcanism and siliceous environments are characteristic.

Evaporite—A rock or mineral deposited by precipitation during evaporation.

Evolution—Development and change of biota in an orderly progression over geologic time.

Exfoliation—The mechanical and chemical weathering process which produces rounded rock mass shapes. It results from plates or scales of rock striping or spalling from the exposed rock surface.

Exploration—The process of analyzing geological and geophysical data toward finding petroleum.

Exposure—A place on the Earth's surface where rock occurrences are visible.

Extrusive—Pertaining to igneous rock which flows or is ejected onto the Earth's surface to cool.

Facies—Rock type: lithology.

Facies change—Change of one rock type to another time-correlative rock type.

Fan—A gently sloping, commonly triangular mass of detrital material: alluvial fan, submarine fan.

Fault—A break in the Earth's crust along which there has been movement.

Fault block—A fault-bounded crustal unit.

Faulted anticline—An anticline disrupted or altered by faulting.

Fault trap—A fault-controlled petroleum accumulation.

Fauna—Animal population of an area or geologic time increment.

Faunal succession—The evolutionary sequence of fauna through geologic time.

Feldspar—A rock-forming aluminum silicate containing varying amounts of principally potassium, sodium or calcium.

Ferromagnesian mineral—A rock-forming silicate of iron and magnesium with varying amounts of calcium.

Fiord—A drowned, coastal, glacial valley with steep sides.

Fire flood—Combustion-drive secondary recovery process in which combustion of liquid petroleum in place is accomplished to reduce its surface tension by heat and increase its mobility toward a producing well.

Five spot—A secondary recovery well pattern with an injection well at each corner of a square and a producing well in the middle between them.

Flank—The limb of a structure where the strata dip in a consistent direction.

Flocculation—The process of clotting of colloidal particles in suspension allowing them to settle to form a sediment deposit.

Flood plain—The level land on either side of a river channel affected by flooding and constructed of river alluvium.

Flora—The plant population of an area or geologic time increment.

Flowing artesian well—An artesian well that flows freely without pumping.

Fluvial—Pertaining to rivers and streams.

Fluvio-glacial (sediments)—Detrital sediments deposited by glacial streams.

Fold—A bent or curved stratum, cleavage plane or foliation.

Fold axis—The line which when moved parallel to itself in space generates a fold.

Foliation—Planar development of structural or textural features especially in metamorphic rocks.

Foot wall—The underside wall rock of a fault, inclined vein or ore body.

Forearc basin—The shelf basin formed on the trench side of an active volcanic arc.

Forebar facies—Deposits on the seaward side of a barrier or baymouth bar.

Foreset beds—The inclined beds on an advancing or prograding deltaic slope or on the downstream slip-face side of wind or water current induced dunes and ripples.

Foreshore—The seaward part of the beach shore between high and low tide.

Formation—A discrete rock unit with characteristics suitable for distinctive study and mapping.

Formation water—Subsurface water found in individual rock units.

Fossil—Preserved plant or animal material.

Fracture (mineralogy)—How a mineral breaks; Can be a diagnostic physical property: conchoidal fracture.

Fracture (stress)—Stress-induced breaks in rock material occurring in conjugate sets.

Friable—Easily pulverized or crumbly rock or mineral material.

Fringing reef—A reef that borders a shore without being separated from it by a lagoon.

Frosted grain—A minutely impact-pitted or etched clastic grain, usually composed of quartz.

Frost wedging—Dislocation, prying and mechanical breakdown of fractured rock by expansion of ice in the fractures.

Gabbro—Mafic, plutonic igneous rock containing calcium feldspar, pyroxene and lesser amounts of olivine.

Gas—Natural petroleum hydrocarbon in vapor or gas form.

Gas cap—Free gas above liquid petroleum in a trapping reservoir.

Gas cap drive—Reservoir pressure augmented by or resulting from a gas cap.

Gas injection—The pumping of gas into a petroleum reservoir to maintain or re-establish its pressure to increase production.

Gas-oil contact—The interface between gas and underlying oil in a trap.

Gas-oil ratio—The number of cubic feet of gas produced per barrel of oil from an oil well.

Gas seep—Where gas escapes from ground surface exposures of a reservoir.

Gas show—A gas indication in well cuttings.

Gas-water contact—The interface between gas and underlying water in a trap in the absence of oil.

Generation (petroleum)—The process of converting organic material to migratable petroleum.

Geochemistry (geology)—The study of chemical elements in rocks and minerals of the earth's crust. Important in the study of petroleum source rocks and their depositional environments.

Geologic map—A map which illustrates geologic information including nature, age and distribution of rocks and structural features.

Geologic time—Earth history represented in the rock succession.

Geologic time scale—The chronologic sequence of geologic events which correspond to divisions of geologic time.

Geology—The study of the Earth.

Geomorphology—The study of the shape of the Earth's surface.

Geophysical survey—A study of the Earth and its crust by any of several geophysical methods.

Geophysics—Quantitative Earth study using physical techniques.

Geosyncline—A large, regional linear or basin like depressed area in which thick volcanic and sedimentary rocks accumulate. Related to plate margins and regional tectonism.

Geothermal gradient—The increase in temperature with depth below the Earth's surface: 1.5°F per 100 feet of depth is average.

Geyser—An intermittently erupting hot water spring or pool.

Glacial deposition—Deposition of sedimentary material (drift) by a glacier.

Glacial erosion—Glacial removal and transportation of rock material.

Glacial grooves/striae—Deep and shallow straight furrows cut into bedrock by glacially transported rocks and boulders.

Glacier—A slow moving gravity motivated ice accumulation.

Glacio-eolian sediments—Wind-blown glacially derived sediments.

Glacio-marine sediments—Marine sediments which contain glacial material.

Glassy texture—Igneous rock texture, caused by virtually instantaneous cooling, and so fine as to be without obvious crystal structure: amorphous.

Gneiss—High grade metamorphic rock comprising unicompositional bands of individual minerals.

Graded bed—A sedimentary stratum with coarse grains at the bottom which grade to fine grains at the top.

Graded stream—A stream at equilibrium in which erosion and deposition are balanced.

Gradient (stream)—The inclination or rate of rise or fall of a stream channel.

Grain—A fragment.

Grain size—How large or small a grain is.

Grainstone—A grain-supported carbonate rock with less than one percent intergranular mud.

Granite—An acidic plutonic igneous rock containing quartz, potassium feldspar, amphibole and mica.

Granularity—Being granular or made of grains.

Gravel—Unconsolidated detritus of larger than sand size particles.

Gravimeter—Used to measure differences in the Earth's gravitational field based upon variations in density.

Gravity (geophysical)—The Earth's gravitational field with density induced variations, measured by gravimetry for exploration purposes.

Gravity drive—Gravity oil drainage of a nonpressured reservoir.

Gravity map—A map contoured on gravity units (milligals) which illustrates areas of positive and negative gravity generated by density contrasts.

Gravity survey—An areal survey to illustrate gravity variations.

Ground moraine—A glacially deposited blanket of drift having no organization, structure or stratification.

Ground water—Water beneath the ground surface.

Group—A rock unit ranking above formation and comprising several formations.

Gypsum—Hydrous calcium sulfate.

Half life—The time required to radioactively break down a radioactive material to half its original radioactivity.

Halite—Rock salt which consists of sodium chloride.

Hanging valley—The higher level valley of a tributary to a lower level main valley formed by different rates of glacial erosion.

Hanging wall—The side or block above an inclined fault.

Hardness—A mineral physical property illustrated by Mohs' relative hardness scale from one (soft) to ten (hard).

Head land—A land projection into a valley, lake or ocean.

Head waters—The gathering upstream waters of a stream or river.

Heat flow—Flow of Earth heat measured as the geothermal gradient of a specific area.

Heave (fault)—The horizontal displacement of a fault.

Hexane—Liquid hydrocarbon of composition C_6H_{14}.

High energy depositional environment—High turbulence, high competence, high oxidation conditions including heavy waves and strong current.

High relief salt anticline—A salt-induced, salt-cored, large relief anticline with strongly draped, often faulted strata over its apex.

Hinge line—A line which connects hinge points of a folded layer.

Hinge point—The point of maximum curvature on a fold cross section.

Hinge surface—A surface which connects all of the hinge lines of a folded sequence.

Historical geology—The study of the geologic development of the Earth and its flora and fauna.

Hook (drilling)—A mechanical device, in a drilling system, which is raised and lowered by the traveling block and is used to attach to and raise and lower drilling tools.

Hook (deposition)—A sand spit recurved by the longshore and tidal currents that formed it.

Horn—An erosional spire or monolith formed by headward erosion of several glaciers.

Hornblende—A complex ferromagnesian mineral commonly associated with acidic igneous rocks.

Hornfels—A fine-grained contact metamorphic rock.

Horseshoe dune—A barchan dune with horns and slip face down wind.

Hot spring—A thermal spring with temperature over 98°F.

Hydraulic fracturing—Reservoir petroleum production enhancement by pumping high pressure fluid and sand into the reservoir to fracture it.

Hydrocarbon—Any organic compound comprising only carbon and hydrogen. Gas and oil are types of hydrocarbons.

Hydrochloric acid—Hydrogen chloride: A calcite and limestone solvent.

Hydrologic cycle—Circulation of water by precipitation, runoff, transpiration and evaporation in the atmosphere, on the Earth's surface, and in the subsurface responsible for erosion and land degradation.

Hydrolysis—Water-related decomposition. Water-related breakdown of some common silicate minerals: potassium feldspar to muscovite and quartz.

Hydrosphere—The water areas of the Earth: the oceans.

Hydrostatic pressure—The pressure caused at any point in a resting body of water.

Hypersaline—Extremely salty. Salinity significantly greater than that of sea water.

Ice cap—A radially flowing ice thickness with an area of less than 50,000 sq km that covers a large land mass but is smaller than a continental glacier.

Ice rafting—The process whereby sediment-laden icebergs derived from glaciers coming to the sea transport sediments to where the bergs melt and release the sediments to the sea bottom.

Igneous rock—Rock that has crystallized from magma (molten material).

Inclusion (igneous)—A fragment of older intruded rock in a younger igneous intrusion.

Index fossil—A fossil that is diagnostic of a specific geologic age or range.

Induration—Transformation of sediment to rock: lithification, diagenesis.

Inflection point—A point on a fold where limb curvature reverses itself.

In situ combustion—Combustion drive secondary recovery technique: fire flood.

Interdistributary swamp—The topographically low, marshy areas between delta distributary channels.

Intergranular—Between the grains of a rock.

Interior drainage—Surface runoff that collects in a topographic depression with no outlet.

Intermediate igneous rock—Igneous rock of composition between acidic and basic (mafic).

Intermediate principal compressive stress—Sigma 2 (σ_2). The principal compressive stress axis between maximum and least in the triaxial stress system. It is considered virtually neutral and is the line of intersection between conjugate shears.

Intermediate salt anticline—A salt-motivated anticline over which sediments are draped and faulted and which contains approximately 5000 feet of salt core.

Intracratonic basin—A basin that lies within the limits of a craton.

Intrusion (igneous)—An igneous body that penetrates older rocks before it cools.

Intrusive—Having the characteristics of an intrusion.

Isopach map—A contour map that illustrates the thickness variations of a rock unit or group of units.

Isoparaffin—A branched chain paraffinic hydrocarbon.

Isotope—One of two or more elemental species of the same element and having slightly different chemical and physical characteristics.

Juvenile water—Water from an igneous source. Less strictly defined it includes water from hot springs, hot pools, geysers, mud volcanoes, etc.

Kame—A sand and gravel terrace mound deposited during a sudden decrease in glacial stream velocity.

Karst topography—An irregular land form developed by solution of limestone terrain and named for the type locality at Karst, Yugoslavia.

Kelly—The usually four-sided hollow steel pipe that connects the drill pipe to the swivel and mud system through the kelly bushing of rotary drilling equipment.

Kelly bushing—The sheave system that fits into the rotary table and drives the kelly and the drill string of rotary drilling equipment.

Kerogen—An intermediate stage of thermally altered organic material between unaltered material and the migratable hydrocarbon, bitumen.

Kettle hole—A depression formed by the melting of a block of glacial ice surrounded by outwash.

Lacustrine—Pertaining to lakes and lake environments.

Lagoon—A sheltered, usually shallow body of water on the shore side of a barrier bar, barrier reef, barrier island, or baymouth bar.

Lagoon facies—Sediments deposited in a lagoon.

Lake—An enclosed nonmarine body of water.

Lake basin—The depression in which a lake resides.

Lake extinction—The elimination of a lake by erosion, infilling, vegetation, etc.

Lake sediments—Deposits formed in a lake.

Laminae—Thin layers of fine-grained sediment.

Land form—A specific topographic feature.

Land slide—Rapid down slope movement of unstable rock and soil material.

Lateral moraine—A ridge of glacial material along the floor of a glacial valley, formed by debris falling on the edges of the valley glacier from adjacent mountains.

Layer—A thickness of rock. A stratum.

Lava—Mobile, extrusive igneous rock.

Leaching—Chemical removal or alteration of rock and mineral constituents.

Least principal compressive stress—Sigma 3, (σ_3). The minimum principal compressive stress axis in the triaxial stress system. It represents the direction of elongation and extension.

Left-lateral strike-slip fault—A strike-slip fault along which the two blocks move to the left of or counter clockwise to each other.

Lense (sand)—An isolated sand body: river channel, barrier bar, barrier island.

Limb—A fold flank. One side of a fold.

Lime mud—Soft calcium carbonate precipitated from sea water.

Limestone—Lithified calcium carbonate.

Line-drive recovery pattern—A secondary recovery injection pattern which promotes linear petroleum migration to producing wells.

Lithification—Solidification of sediment to rock: induration, diagenesis.

Lithofacies map—A map which illustrates areal distribution and relationships of different rock types.

Lithology—Rock types: facies.

Lithosphere—The outermost layer of the Earth which includes continental and oceanic crust in its upper layers and the uppermost part of the mantle. It is up to 100 km thick.

Lithostatic pressure—Overburden pressure exerted by a thickness of rock.

Littoral zone—Marine zone between highest high tide and lowest low tide expressing maximum wave energy.

Loess—Windblown, wind-deposited, fine-grained glacially derived sediment.

Longitudinal dune—A long sand dune oriented parallel to a persistent, unidirectional wind.

Longitudinal profile—The shape of a stream valley from its headwaters to its mouth taken along its length.

Longshore current—A mostly meteorologically generated current parallel to the shore of an ocean or large body of water.

Longtooth bit—A rotary drill bit with long teeth for soft formation penetration.

Lost circulation—The loss of drilling mud into extremely porous formations during drilling.

Low energy depositional environment—A low oxygen, quiet water environment with no turbulence and low transportation capability.

Low relief salt pillow—A salt-motivated anticline over which sediments are gently draped and faulted with approximately 2000 feet of salt core.

Low relief salt ridge—A laterally extensive low relief salt pillow.

Luster—A physical property which describes the aspect of a mineral relative to metallic, submetallic or non-metallic appearance.

Lutite—Consolidated mud consisting of silt and/or clay.

Mafic igneous rock—Basic igneous rock lacking quartz and mica, which contains calcium-rich ferromagnesian silicates and feldspar.

Magma—Molten rock.

Magmatic differentiation—The process of temperature-controlled mineralogic separation of molten igneous material.

Magnetic survey—A survey process to measure variations in the magnetic field of the Earth caused by local or broad changes in rock magnetism.

Magnetic susceptibility—Ratio of induced magnetism to strength of the magnetic field inducing the magnetism.

Magnetometer—An instrument that measures regional and local changes in the magnetic field of the Earth.

Mantle—The second layer of the earth below the lithosphere, 2900 km thick, and with a specific gravity of 4.5–5.0.

Map—A diagram that represents geologic, geographic, topographic or other features on the Earth's surface.

Marble—Metamorphosed, coarse-textured limestone or dolomite.

Marine swamp—A seacoast swamp or marsh containing abundant vegetation in a low-lying salty or brackish environment in which coal often forms.

Mass wasting—Gravity induced downslope movement of weathered rock material without transportation by water, ice or other medium.

Matrix (igneous)—Finely crystalline or glassy ground mass between larger crystals in a porphyry.

Matrix (sedimentary)—Fine-grained silt or clay material infilling intergranular sedimentary pore space between coarse grains.

Maturation—The process of becoming mature involving petroleum source beds, coastlines, topography, streams, etc.

Mature coast—Characterized by a fairly straight, indented sea cliff and wave-cut terrace of medium width.

Mature stream—A moderately meandering stream with a flood plain, oxbow lakes, back swamps, approaching equilibrium between deposition and erosion.

Maximum principal compressive stress—The axis of maximum compression in the triaxial stress system which bisects a pair of conjugate shears and is represented by Sigma$_1$, (σ_1).

Meander—A bend in a river caused by the shift of a river course across a flood plain.

Mechanical weathering—Physical break down of rock material.

Medial moraine—Glacial material deposited by the coalescing of several lateral moraines in the middle portions of a glacial valley.

Medium tooth bit—A drilling bit with teeth between long and short used for penetrating medium hardness formations such as sandstone, hard shale, or soft limestone.

Member—A time-rock stratigraphic unit of which several can comprise a formation.

Mesogenetic burial—Overburden accumulation during the time between the formation of the depositional surface and subsequent erosional surface.

Mesozoic Era—An increment of geologic time of approximately 160 million years duration between the end of the Paleozoic Era 225 million years ago and the beginning of the Cenozoic Era 65 million years ago.

Metagenesis—Late stage diagenesis to early metamorphism corresponding to dry gas generation and thermally over-mature petroleum source sediments.

Metagenetic—Pertaining to late diagenesis and early metamorphism.

Metamorphic rock—Mineralogically, physically, chemically and structurally changed rock.

Metamorphism—The process of mineralogically, physically, chemically and structurally changing rock by temperature, pressure, shearing, and chemical means.

Meteor—A streak of light in the night sky produced by atmospheric friction on a fragment of celestial material entering the Earth's atmosphere.

Meteoric water—Water as liquid, snow, sleet, hail, mist, fog, etc., of atmospheric origin.

Meteorite—The solid celestial fragment that causes a meteor.

Methane—Odorless, colorless, inflammable natural gas.

Mica—A complex iron, magnesium silicate mineral with one perfect basal cleavage. Common forms include muscovite (white) and biotite (black).

Micrite—Translucent limestone matrix formed from carbonate mud precipitation with crystals less than 4 microns in diameter. A solidified lime mud.

Microcline—Acidic potassium feldspar. Orthoclase.

Mid-ocean ridge—A volcanic, seismically active, linear oceanic ridge from which seafloor spreading occurs and along which oceanic crust is generated.

Migration—Movement of petroleum through source and reservoir rocks.

Millidarcy—One thousandth of a darcy and a measure of permeability.

Mineral—A naturally occuring inorganic substance with a definite chemical composition, a definite crystal structure and physical properties that vary within predictable limits.

Mineralization—Introduction of mineral material to a rock by solution to increase its mineral content.

Mineralogy—The study of minerals.

Miogeocline—Shallow water sediment prograding seaward from a continental margin as in a miogeosyncline.

Miogeosyncline—The cratonward, non-volcanic, slowly subsiding portion of an orthogeosyncline.

Miscible drive—A secondary recovery production technique which injects a solvent into a reservoir to mobilize non-producible petroleum remaining after primary recovery.

Moderately rounded—Sediment grains between well and poorly rounded.

Moderately sorted—Sediment grains between well and poorly sorted.

Mohorovicic discontinuity—The seismic break between the crust and mantle below which acoustic wave velocities increase sharply.

Mohs hardness scale—Mineral relative hardness scale from softest to hardest: talc, gypsum, calcite, fluorite, apatite, orthoclase, quartz, topaz, corundum, diamond.

Monkey boards—The scaffolding attached approximately two thirds of the length of a drilling derrick above the derrick floor where the derrick man racks and unracks pipe during drilling operations.

Moraine—Till material deposited directly by glacial ice.

Mountain glacier—An alpine or valley glacier, usually at high elevation.

Mouth—Where a river flows into a lake, ocean or another stream.

Mud (drilling)—An aqueous mixture of a variety of fine-grained components used to lubricate drilling operations, recover cuttings and stabilize the drill hole.

Mud (sediment)—A mixture of silt and/or clay particles and water. Diagenetically transformed into siltstone and/or shale.

Mud cracks—Roughly polygonally distributed fractures caused by desiccation shrinkage of mud, clay or silt.

Mud flow—Rapid downslope movement of fine-grained earth materials in an aqueous, fluid system.

Mud pump—Equipment for circulating drilling mud.

Mudstone—Blocky or massive indurated mud without shale lamination or fissility.

Mud tank—Storage tank for drilling mud.

Natural gas—Petroleum in gaseous form at normal temperature and pressure.

Natural levee—A ridge of poorly sorted river sediment deposited by a river at flood on the flood plain at the edge of the channel.

Nearshore zone—An indefinite narrow zone extending seaward from the shoreline to a depth of about 10 m.

Neritic zone—Marine depth zone between lowest low tide and 200 m below sea level.

Neutral coast—A coast independent of submergence or emergence where no change relative to sea level occurs.

Nonassociated gas—Natural gas occuring in a reservoir without oil.

Nonconformity—An unconformity featuring sediment resting upon an eroded crystalline surface.

Nonconformity trap—A petroleum trap resulting from a nonconformity.

Noneffective porosity—Unconnected porosity resulting in no permeability.

Nonflowing well—A water well from which water does not flow.

Nonmarine deposition—Sedimentation which occurs under other than marine conditions.

Nonparallel fold—A fold in which the bedding ceases to be parallel because of overturning and flowage.

Normal fault—A fault representing the hanging wall down relative to the foot wall, a vertical σ_1, and horizontal σ_2 and σ_3.

Notch—An erosional re-entrant at the base of a sea cliff at its junction with the wave-cut terrace.

Oblique-slip—Fault slip other than dip-slip.

Oceanic crust—Crust under the ocean basins consisting of silicates of magnesium and iron: SIMA.

Offlap—Successive seaward deposition of conformable strata during marine regression.

Offshore zone—The shelf zone seaward from the nearshore at 10 meters of water depth to the shelf edge.

Oil field—Two or more closely related oil-producing areas on the same geologic feature.

Oil seep—An exposure of an oil reservoir or conduit where oil emerges at the ground surface.

Oil show—An indication of oil in well cuttings.

Oil-water contact—The interface between oil above and water below in a reservoir.

Old age coast—A hypothetical shoreline featuring a wide wave-cut terrace, regular unindented sea cliff and peneplaned coast.

Old age stream—A hypothetical slowly moving, strongly meandering graded stream with a very wide flood plain and very shallow gradient.

Olivine—A relatively unstable, mafic, rock-forming, iron and magnesium silicate mineral.

Onlap—Progressive onshore overlap or pinch out of transgressively deposited sediments.

Oolite—A sedimentary rock comprising concentrically precipitated calcium carbonate ooliths approximately one mm in diameter.

Oolitic limestone—A limestone, comprising calcareous ooliths, often having good reservoir potential.

Open marine environment—Marine environment seaward of the shore or barrier island environments.

Ore body—A continuous, economically important ore concentration.

Organic material—Animal or plant material, sometimes preserved as coal or petroleum.

Orthoclase—Acidic potassium feldspar: Microcline.

Oscillation wave—A water wave displaying circular particle motion but having no transportive or erosive capability.

Outcrop—A ground surface expression of a geologic structure or formation.

Outwash plain—A plain of stratified detrital material deposited in front of a glacier by subglacial and/or superglacial streams.

Overburden pressure—Accumulated rock pressure exerted on subsurface rocks by the overlying rock column.

Overthrust trap—A fault or fold trap lying above a thrust fault.

Overturned fold—A fold with an axial plane depressed below horizontal.

Oxbow lake—A river meander cut off as a lake by river channel straightening processes.

Oxidation—Chemical combination of oxygen with other substances resulting in compositional change and elimination of organic materials.

Oxide—A mineral compound in which primary elements are combined with oxygen.

Oxidizing environment—A depositional environment in which oxidation occurs. Chemical elimination of organic material occurs in an oxidizing environment.

Packstone—Granular carbonate rock, the grains of which occur in a self-supporting framework surrounded by some calcareous matrix.

Paleogeography—Study of the geography of the past.

Paleogeology—Study of the geology of the past.

Paleontology—Study of life of the past; study of fossils.

Paleozoic Era—Geologic time from the end of the Precambrian, 570 million years ago to the beginning of the Mesozoic, 225 million years ago.

Palynology—Study of pollen and spores. In geology, the study of fossil pollen and spores and their relation to stratigraphy.

Parabolic dune—A scoop-shaped sand dune, concave upwind, formed in strong wind and abundant sand conditions.

Paraffin oil—Crude oil with a high paraffin content.

Patch reef—A small, relatively isolated, organic, carbonate reef occurrence.

Peak oil generation—Maximum crude oil generation at most favorable conditions.

Peat—Altered, partly carbonized, water-saturated plant remains. A precursor to coal.

Pebble—A clastic fragment in the 4 millimeter to 64 mil-

limeter grain size range.

Pediment—A subaerially formed, rock-floored erosion surface formed in arid or semiarid areas, often at the base of an eroding mountain range front.

Pelagic deposits—Open ocean deposits.

Pelecypod—An aquatic bi-valved mollusk. A clam.

Pelletal limestone—Limestone containing abundant pellets comprising fecal and carbonate mud varieties.

Pellets—Rounded to spherical clasts, often carbonate and comprising fecal material from mollusks and worms, in well-sorted clastic carbonate rock.

Pentane—A paraffin hydrocarbon in petroleum.

Perched water table—A water table above a regional water table.

Percolation—Slow movement of water through small voids in a porous medium, as through the zone of aeration above a water table.

Peridotite—Coarse-grained plutonic rock of olivine and other mafic minerals and little or no feldspar.

Period—An increment of geologic time smaller than an era and larger than an epoch, i.e., the Cambrian period.

Permeability barrier—A barrier to transmission of fluid through a medium.

Permineralization—Deposition of mineral material in the pore spaces of bone or shell material.

Petrification—Molecular exchange of inorganic material for organic material usually by subsurface water, as in petrified or silicified wood.

Petroleum—Naturally occurring solid, liquid or gaseous hydrocarbons.

Petroleum accumulation—A concentration of naturally occurring hydrocarbons.

Petroleum migration—Movement of petroleum through permeable rock media from locality of generation.

Petroleum origin—The development of petroleum from naturally occurring organic raw materials.

Petroleum source rock—A sedimentary rock suitable for the generation of petroleum.

Phaneritic texture—Coarsely crystalline, usually plutonic igneous rock.

Phenocryst—Large crystal in, usually porphyritic, igneous rock.

Photogeology—Study of surface geologic features and structures using aerial photographs.

Phyllite—A foliated metamorphic rock of grade between slate and schist.

Physical properties—Nonchemical identifying physical characteristics of minerals, i.e., color, hardness, luster, etc.

Physical weathering—The mechanical, nonchemical

breakdown of rock material.

Physiography—Landform features of the surface of the Earth including topography, drainage, etc.

Piedmont glacier—An ice sheet of coalescing valley glaciers at the foot of a mountain range.

Pinnacle reef—A narrow or spire-shaped reef formed by vertical coral growth to accommodate rapid sea level rise.

Pitted outwash—Glacial outwash containing depressions left by melting of included ice blocks surrounded and covered by outwash sediments.

Plagioclase—A feldspar group containing varying amounts of sodium and calcium occurring in intermediate to basic composition igneous rock.

Planetary current—Oceanic currents generated by the rotation of the Earth.

Plate tectonics—A theory of crustal deformation involving the motions of lithospheric plates.

Playa lake—A desert or semiarid intermittent lake which dries up and leaves a playa or hard pan in the absence of rainfall.

Plucking—Glacial erosion resulting from freezing of glacial water seeping into cracks in subglacial rocks causing their displacement and removal by ice movement.

Plunging fold—A fold with a non-horizontal hinge line.

Pluton—An intrusive igneous body.

Plutonic—Pertaining to igneous intrusion.

Pneumatolytic metamorphism—Alteration and mineralization of rock terrain by fluid emanations from crystallizing igneous intrusions.

Point bar—Coarse-grained, often cross-bedded, river channel deposits on the inside, low velocity part of a meander.

Polymeric compound—Formed by combining two or more small molecules into larger molecules.

Polymorphic—Pertaining to the ability of a substance to crystallize in several crystal forms.

Poorly rounded—Sedimentary grains exhibiting angularity in lieu of rounding.

Poorly sorted—Clastic sediments containing a variety of grain sizes.

Pore space—Voids in rock materials.

Porosity—The amount of void space in rock materials.

Porphyry—An igneous rock containing large phenocrysts in a fine-grained matrix.

Post depositional history—Diagenetic, erosional deformational, etc., events affecting rocks after their deposition.

Pour point—The temperature at which crude oil will congeal or become liquid.

Precipitation—Rain, snow, sleet, hail, drizzle, etc., that falls upon the Earth's surface.

Precipitation—The process by which dissolved materials come out of solution.

Pressure gradient—Lithostatic or hydrostatic pressure that increases with depth below the Earth's surface.

Primary dolomite—Precipitated dolomite.

Primary earthquake wave—A compressive earthquake wave propagated in the vertical plane; P Wave; the first earthquake wave to arrive at a recording station.

Primary migration—Petroleum migration from source rock to reservoir.

Primary porosity—Sediment or particle porosity at time of deposition.

Primary recovery—Petroleum recovery prior to initial reservoir and/or pressure depletion.

Primitive (stereonet)—The outer circle or margin of a stereographic net.

Profile of equilibrium—The gently concave upward longitudinal profile of a stream along which deposition and erosion are balanced and the stream is at a grade: the gently concave upward seashore profile representing a balance between deposition and erosion.

Propane—A natural gas of the methane series.

"P" shear—A through-going, regional, strike-slip fault representing a primary displacement system and connecting shorter synthetic faults.

Pumping unit—A mechanical device used to extract petroleum from a reservoir.

Pyroclastic rock—Clastic rock of explosively volcanic origin.

Pyroxene—A dark, basic igneous, silicate mineral containing calcium, magnesium and iron: characteristic of gabbro.

Quartz—Silicon dioxide.

Quartzite—A general term applied to metamorphosed quartz sandstone.

Radial drainage pattern—An inwardly directed central drainage into a basin or an outwardly directed drainage from a dome.

Radiation—Sending out rays or particles.

Radioactive decay—The breakdown of substances by emission of radioactive particles.

Radioactivity—The emission of radioactive particles.

Radiometric age—A geologic age in years determined by measurement of elemental half-lives, radioactive emission rates and daughter product formation.

Radiometric time scale—The geologic time scale calibrated by the establishment of radiometric ages.

Recessional moraine—A terminal or lateral moraine built

behind the farthest advance of a glacier during its ultimate withdrawal.

Recovery program—A program of petroleum recovery from a reservoir: primary, secondary or tertiary recovery.

Recrystallization—The generation of new, usually larger, crystals in a rock.

Recumbent fold—A fold with a horizontal axial plane.

Red beds—Clastic sedimentary strata colored red or reddish brown by ferric iron oxide.

Reducing environment—A depositional or ecological condition in which oxygen is diminished or eliminated.

Reef—A bank, ridge or mound constructed by calcareous animals and plants.

Reef talus—Wave broken fragments from a reef and distributed at its foot.

Reef trap—A petroleum accumulation retained by reef permeability surrounded by adjacent rock impermeability.

Reflection seismic survey—A seismic survey in which the time of acoustic waves from ground surface to reflection horizon and back are measured or plotted to illustrate rock character and structure.

Refraction seismic survey—A seismic survey which measures velocities of acoustic waves through various rock types to illustrate rock character.

Regression—Withdrawal of the sea.

Regressive—Pertaining to sea withdrawal.

Relative geologic age—The geologic age of a geologic feature, event, fossil, structure, etc., compared to that of another.

Relative permeability—Comparison of the permeability of a medium at partial saturation to its permeability at 100% saturation.

Replacement—A subsurface water process involving solution exchange of one mineral for another in a rock medium.

Reservoir pressure—Pressure in a subsurface petroleum-bearing rock, due to overburden thickness, deformation, fluid column, etc.

Reservoir rock—A permeable subsurface rock unit which contains petroleum.

Resistivity—In electric logs, resistivity represents the measurable inabilty of a rock and its contained fluid to transmit an electrical charge.

Restricted marine environment—Where free circulation with the open sea is diminished.

Reverse fault—A fault the plane of which dips toward the upthrown block.

Ria—A drowned, long, narrow coastal inlet.

Right-lateral strike-slip fault—A strike-slip fault with right-handed or clockwise displacement.

Ripple marks—Current- or wave-generated ripples in sediments.

Rock—A naturally occurring aggregate of minerals of igneous, sedimentary, or metamorphic origin.

Rock competence—The ability of a rock to withstand stress.

Rock cycle—The interrelation of the three rock types and how geologic processes can change each of them into one or more of the others.

Rock fall—The freefall accumulation of weathered rock material at the base of a slope of cliff.

Rock-forming mineral—A mineral that in combination with other minerals forms rocks.

Rock glacier—A slowly moving mass of broken rock material sometimes containing interstitial ice to assist its downslope progress.

Rock sequence—A progression of layered rocks often illustrative of specific stratigraphic characteristics.

Rock slide—Sudden down slope movement of broken rock material accumulating at the slope base.

Rock type—Lithology, facies: kind of rock.

Rock unit—A specific rock increment illustrating certain recognizable characteristics.

Rounding—Reduction of rock fragment angularity.

Rubber sleeve core barrel—A core taking device which utilizes an interior rubber cylinder to increase core recovery.

Rudite—Clastic rock of particles larger than sand size.

Runoff—Atmospheric precipitation that moves downslope on land as streams and rivers.

Salinity—The concentrations of salts contained in a solution, i.e., sea water.

Salt (rock salt)—Rock deposits consisting of the mineral halite.

Salt anticline—A positive elongate structure containing a salt core.

Salt dome—A circular or elliptical, positive salt-cored structure which vertically penetrates or deforms the surrounding sediments.

Salt pillow—An early stage salt dome limited by available salt or deformation time.

Sand—A clastic particle size from 1/16 mm to 2 mm in diameter: an accumulation of such particles.

Sand bar—A ridge or linear sand feature built to or near the water surface by marine or fluvial waves and currents.

Sand dune—A wind-blown sand deposit.

Sandstone—A clastic sedimentary rock of sand-size particles.

Schist—A medium-grade metamorphic rock containing discrete mineral crystals along its foliation laminae.

Schmidt net—A graphic equal; area system of coordinates of latitude and longitude as small and great circles respectively for three-dimensional analysis of structural data.

Sea arch—A wave-formed bridge or arch formed by the meeting of two caves or a single throughgoing cave in a headland.

Sea cave—A wave-formed cave in a sea cliff.

Sea cliff—A wave-eroded coastal cliff.

Seal—An impermeable rock medium that isolates a reservoir and retains petroleum, i.e., shale, evaporite.

Seamount—A submarine mountain often eroded by the sea.

Seasonal laminations—Season-generated, climatic, lake sediment layers which change in color and organic content between winter and summer.

Secondary earthquake wave—Secondary shear or S waves that arrive after the faster primary or P waves.

Secondary migration—Petroleum migration within a reservoir.

Secondary porosity—Post-depositional rock porosity.

Secondary recovery—Petroleum recovery from a reservoir following primary recovery; achieved by various means of stimulation.

Sediment—Fragmental or precipitated material transported and deposited by gravity, water, wind, ice or precipitation.

Sedimentary rock—A rock comprising sediment.

Sedimentation—Accumulation of layered sediments.

Sediment overburden—The rock thickness that overlies a particular rock stratum.

Sediment source—The physical and/or chemical source of sediment deposited in a particular locality.

Seismic receiver—A geophone: a receiver of acoustic signals from the Earth.

Seismic record—A graphic display of acoustic waves recorded by a seismograph.

Seismic survey—A study of the acoustic properties of the Earth or a portion thereof.

Seismic wave—An acoustic or earthquake wave.

Seismograph—An instrument that detects, enhances and records acoustic signals.

Seismology—The study of the structure and seismicity of the Earth.

Self potential—Or spontaneous potential of an electric log determines shaly from non-shaly and porous from non-porous formations.

Series—A time-stratigraphic designation above stage and below system.

Serrated divide—A saw-toothed glacial ridge toward which galciers on both sides have eroded headward.

Shadow zone—An area blocked by refraction or non-transmission from receiving seismic waves.

Shale—A fine-grained, usually laminated, clastic rock of compacted clay or mud particles.

Shale shaker—A vibrating screen designed to screen drill cuttings from drilling mud.

Shale sloughing—The caving of shale from the sides of a well bore.

Shear—A stress-induced displacement along a zone or plane of failure.

Sheeting—A rock weathering phenomenon induced by relief of overburden pressure, resulting in thin rock slabs parallel to the rock surface.

Shelf sediments—Deposits, mostly neritic, on the continental shelf.

Shore—The narrow land zone parallel to a body of water.

Shoreface terrace—A wave-built sand and gravel terrace in the shore-face zone.

Shoreface zone—The seaward-sloping zone between the low-tide line and the offshore zone.

Short tooth bit—A tricone, hard formation, rotary bit.

Show—An indication of oil or gas in well cuttings or drilling mud.

Sidewall core—A formation sample from the wall of a well bore obtained by laterally firing a hollow projectile.

Silica—Silicon dioxide, quartz.

Silicate—Pertaining to the content of oxides of silicon.

Siliceous—A rock containing abundant free silica.

Sill—A sheet-like or tabular intrusive igneous body injected along the bedding planes or planar structure of the including rock.

Silt—A clastic sediment comprising fragments from 1/16 millimeter to 1/256 millimeter in diameter.

Siltstone—A rock made of silt.

SIMA—Oceanic crust comprising rocks rich in silica and magnesia.

Similar fold—A fold in which bed thickness is greater in the hinge than in the limbs.

Simple fault trap—A fault trap involving a single, simple fault.

Sink hole—A collapsed cavern formed by underground solution, commonly circular but often funnel shaped.

Slate—A low grade, fine-grained, foliated metamorphic rock commonly formed from shale.

Slaty cleavage—A low-grade metamorphic, platy mineral foliation perpendicular to maximum compressive stress commonly in slate.

Slide—A downslope mass movement of rock, earth, snow,

etc., resulting in a rockslide, landslide or snow and ice avalanche.

Slip—Displacement along a fault surface.

Slip direction—Direction of fault displacement.

Slump—A rotary downslope slide movement along a surface concave in the displacement direction.

Soft-sediment deformation—Deformation of incompetent, poorly consolidated and/or unstable sediments.

Solifluction—Slow downslope movement of water-saturated soil.

Solubility—The ability of a substance to dissolve in a solvent or solution.

Solution—Dissolving of mineral constituents by a solvent. A liquid (aqueous) medium in which minerals are dissolved.

Solution load—The dissolved load of a river.

Sorting—The degree of constancy of grain size in a clastic rock. Well-sorted rocks comprise grains of the same size. The dynamic process of achieving sorting of grains.

Source rock—A sedimentary rock in which petroleum forms.

Sparite—A limestone with crystalline cement more abundant than finer-textured precipitated micrite matrix.

Sparry cement—Crystalline cement.

Specific gravity—The ratio of the density of a substance to the density of water.

Spit—A long, narrow projection of land, or wave and current built sand, extending from shore into a body of water.

Spontaneous potential—See self potential.

Spreading center—A linear zone from which lithospheric plate extension occurs. A mid-ocean ridge/rift/rise system.

Spring—Where subsurface water flows from an aquifer onto the ground surface.

Spur—A ridge or land projection that extends laterally from a mountain or valley wall.

Stack—An isolated, eroded coastal remnant island detached from a land mass by wave action.

Stage—A time-stratigraphic unit ranking below series.

Stalactite—Dripstone attached to the ceiling of a cavern.

Stalagmite—Dripstone building upward from the floor of a cavern.

Static metamorphism—Metamorphism caused by the buildup of overburden or lithostatic pressure.

Steam flood—A secondary or tertiary petroleum recovery program in which steam is pumped into a reservoir to mobilize remaining petroleum by heating it.

Stereographic projection—Projection of solid or planar objects on a two dimensional surface consisting of parallels of latitude and meridians of longitude.

Stereonet—The graphic on which stereographic projection is made.

Strain—Deformation resulting from stress.

Stratigraphic cross-section—A section which illustrates stratigraphic relationships.

Stratigraphic range—The vertical or geologic age range of geologic phenomena.

Stratigraphic trap—A depositional, non-deformed petroleum trap.

Stratigraphic unit—A recognizable stratum or body of rock used for mapping, correlation and description.

Stratigraphy—The study of layered rocks.

Stratum—A layer of rock.

Streak—Color of powdered mineral streaked on unglazed porcelain.

Stream—A linear body of gravity motivated running water.

Stream channel—The bed of a stream.

Stream deposits—Sediments laid down by stream activity.

Stream erosion—The transportation of rock material by a stream.

Stream load—The sediment load transported by a stream.

Stream pattern—The geographic arrangement of a stream and its tributaries.

Stress—Applied force or pressure resulting in strain.

Striations—Geologically induced, usually parallel lines or scratches: glacial striations.

Strike—The horizontal line of intersection between a dipping surface and a horizontal plane.

Strike-slip fault—An essentially vertical fault demonstrating lateral displacement.

Striplog—A graphic, scaled log on which is plotted sample, drilling, percentage or other information.

Structural axis—A straight line around which a body or figure (fold) can be perceived to rotate.

Structural geology—The study of geologic structure.

Structural trap—A petroleum trap formed by deformation.

Structure—The result of crustal deformation. A deformed geologic feature.

Structure map—A map representation of geologic structure.

Subaerial—On the land surface; under the air.

Subangular grain—A slightly rounded angular grain.

Subaqueous—Under water.

Subduction—The descending of one crustal (lithospheric) plate beneath another. Oceanic crust is subducted beneath continental crust along subduction zones.

Subduction zone—A deep water trench zone along which subduction occurs: Japan Trench, Aleutian Trench, Java Trench.

Submarine canyon—A valley cut into the continental shelf or continental slope.

Submarine fan—A subaqueous fan-shaped, terrigenous clastic deposit located seaward of the mouths of rivers and canyons, or at the foot of a steep slope or the continental slope. Can be the result of turbidity current deposition in deep or shallow water.

Submarine slide—A downslope movement of unstable, unconsolidated sediment on the continental slope or other submarine slope.

Submergent coast—A drowned coast.

Subnormal pressure—A lower than normal subsurface pressure.

Subrounded grain—A partly rounded grain; one not completely rounded.

Subsidence—Downward movement of the Earth's surface.

Substitution—A ground water process involving exchange of one element for another in grain or cement material.

Subsurface geology—The study of drilling, logging, mining, etc., information to analyze stratigraphy and structure beneath the Earth's surface.

Subsurface water—Ground water.

Subthrust trap—A petroleum trap beneath a thrust fault.

Sulfur content—The amount of surfur in crude oil.

Superposition—The normal sequence of layered rocks; the oldest on the bottom and youngest on top.

Surface earthquake wave—An acoustic wave travelling along the Earth's surface.

Surface geology—Study of geology utilizing ground surface data.

Surface tension—The condition at the surface of a liquid caused by intermolecular attraction.

Surf zone—The marine edge zone between the farthest seaward breaking wave and the farthest landward wave swash.

Suspended load—The solid-carried load of a stream.

Swamp—A low, marshy, partly water-covered, vegetative area.

Swivel—The mechanical device attached to the upper end of the kelly to effect the transfer of drilling mud into the drill string.

Symmetric fold—A bilaterally symmetrical fold with equal flanks.

Syncline—A downwardly convex fold with limbs that dip toward the axis.

Synthetic fault—Minor faults of the same orientation and direction of displacement as associated major faults.

System—A worldwide time-stratigraphic rock unit ranking above series and below erathem.

Talus—Gravity accumulated, angular rock fragments at the base of a mountain slope.

Talus—Gravity accumulated, angular rock fragments at the foot of a reef.

Talus slope—A steep slope of accumulated rock fragments at the base of a cliff or scarp.

Tar—Thick, viscous petroleum.

Tarn—A lake or pool in a depression of a glacially eroded mountain valley.

Taste—A physical property useful in identification of some minerals. The taste of a mineral: as the taste of rock salt.

Taxonomy—Classification of plants and animals.

Tear fault—Strike-slip fault.

Tectonic—Pertaining to forces and resulting deformation in the Earth's crust.

Telogenetic—Describing the erosion and weathering of long buried carbonates and the associated porosity development.

Temperature current—A marine or lacustrine current that flows as a result of temperature variations.

Temperature gradient—The increase of temperature with depth in the Earth at $1.5°$ F per 100 feet.

Terminal moraine—The end moraine of a glacial system.

Terrace—A linear, level bank of sand and gravel, eroded and left higher than a down-cutting river.

Terrestrial—Pertaining to land.

Terrigenous—Derived from the land: terrigenous sediment.

Tertiary recovery—Petroleum recovery after primary and secondary production.

Thermal maturity—The degree of thermal alteration of sediments, source rocks, coal, fossils, etc.

Thermogenic gas—Gas derived from thermal alteration.

Three-point problem—A method of strike and dip determination using depth or elevation data from three points on a plane.

Throw—The vertical displacement of a fault.

Thrust fault—A reverse fault dipping at $30°$ ($45°$ by some accounts) or less.

Tidal current—A marine current induced by tidal activity.

Tide—The periodic, daily rise and fall of sea level due to gravitational effects of the moon and sun.

Till—Unsorted, unstratified drift deposited directly by a glacier.

Tilted fluid contact—A fluid contact in a reservoir tilted by deformation or hydrodynamics.

Time-rock unit—A chronostratigraphic unit of rock representing only a specific interval of geologic time.

Time unit—An increment of geologic time corresponding to the time of deposition of a specific corresponding time-rock or chronostratigraphic unit.

Tombolo—A sand or gravel feature connecting two islands or an island to the mainland.

Topographic cross-section—A cross-section showing topography.

Topographic map—A contour or other type of map illustrating topography.

Topography—Configuration of the land surface of the Earth's surface.

Topset beds—The flat-lying upper beds overlying the foreset beds of a cross-bedded or deltaic sequence.

Total porosity—The volume of rock porosity.

Transcurrent fault—A strike-slip fault.

Transform fault—A transverse fault system that changes one form of crustal motion to another: a regional strike-slip fault connecting two subduction zones, for example.

Transform plate margin—A lithospheric plate margin bounded by a transform fault: the boundary between the Caribbean and South American plates.

Transgression—Marine encroachment by rising sea level.

Transgressive—Pertaining to marine encroachment.

Translation fault—Strike-slip fault.

Translation wave—A breaking ocean or lake wave.

Transverse dune—A linear sand dune perpendicular to the wind direction.

Trap—A geologic feature in which petroleum can accumulate.

Travelling block—The system of contained sheaves through which pass steel cables which moves up and down in a drilling derrick and controls the drilling string.

Travel time—The time required for an acoustic wave to travel from one place to another in the Earth.

Trench—Subduction zone.

Triaxial stress—Stress comprising maximum, intermediate, and least principal compressive stress expressed along three mutually perpendicular axes.

Trilobite—A three-lobed marine arthropod ranging in age from Cambrian to Permian as an important index fossil.

Tubing—The pipe installed in a producing well through which petroleum is produced from a reservoir.

Tubing string—The length of tubing from reservoir to surface.

Tuff—Consolidated pyroclastic rocks.

Turbidite—A turbidity current-deposited, graded clastic sequence.

Turbidity current—A gravity-motivated current containing velocity suspended sediment which deposits graded sediments as its velocity decreases.

Type locality—The locality at which a geologic feature was first described or named; commonly the best-developed known locality of occurrence.

Unconformity—An interruption of the geologic record manifest as an erosion surface bounded by rocks which are not immediately chronologically successive.

Uplift—A redundant term to describe a structurally high portion of the crust.

Upthrown block—A structurally high crustal block of a set of two blocks separated by a fault: the opposite of downthrown block.

U-shaped valley—The typical shape of a glacial valley, differentiated from a V-shaped stream valley.

Valley fill—Valley detritus derived from erosion of adjacent higher topography. It is particularly common and often thick in arid and semi-arid areas.

Valve—One of two shells of a brachiopod or clam (pelecypod).

Varve—A seasonal sediment lamination common to glacial and temperate zone lakes.

Velocity—Stream velocity represents the downstream speed of the water in the channel.

Viscosity—The resistance to flow of a substance.

Vitrinite—Coalified humic (organic) material.

Vitrinite reflectance—The ability of vitrinite to reflect incident light and indicative of thermal maturity and petroleum source potential.

Volcanic—Extrusive, as a volcano.

Volcano—The commonly conical buildup of extrusive igneous material around a vent.

V-shaped valley—The typical shape of a youthful stream valley, differentiated from a U-shaped glacial valley.

Vug—A small void in a rock.

Vugular porosity—Or vuggy porosity, which describes rock porosity comprising vugs.

Wackestone—Mud-supported carbonate rock with more than 10% grains larger than 20 microns in diameter: calcarenite.

Washout—A larger diameter portion of a well bore caused by drilling mud removal of soft material.

Water drive—A reservoir pressure drive system involving natural water pressure to produce petroleum.

Water flood—A secondary or tertiary recovery program involving water injection to establish reservoir pressure to produce petroleum.

Water injection—Pumping of water into a reservoir to establish production pressure.

Water table—The upper surface of the ground water zone of saturation at the base of the zone of aeration.

Wave—An earthquake, seismic or acoustic wave of propagation.

Wave—An oscillation of water.

Wave base—The water depth at which wave action is no longer manifest.

Wave crest—The highest part of a wave.

Wave-cut bench (terrace)—A wave erosion platform at the base of a sea cliff.

Wave height—The vertical distance between a wave crest and the adjacent wave trough.

Wave length—The horizontal distance between two successive wave crests or troughs.

Wave period—The time between two successive moving crests or troughs.

Wave refraction—Wave bending caused by impingement of a wave against a shallowing bottom.

Wave trough—The lowest area between two wave crests.

Weathering—Mechanical and/or chemical breakdown of rock material.

Well—A hole drilled in the ground to produce water or petroleum.

Well bore—The three-dimensional, circular perforation that results from drilling a well.

Well cuttings—Rock fragments cut by drilling a well.

Well-rounded grains—Grains abraded to approach a rounded to spherical shape.

Well sample log—A graphic representation of symbols and descriptions of the rock sample sequence encountered in drilling a well.

Well-sorted grains—Grains of the same size.

Wet gas—Gas associated with liquid petroleum.

Wind deposition—Laying down of sediments by the wind.

Wind erosion—Removal of rock material by the wind.

Wrench fault—Strike-slip fault.

Wulff net—A stereographic projection net used in structural analysis.

Xenolith—An inclusion from surrounding rock in an igneous (intrusive) rock.

Youthful coast—A submergent, irregular, rugged coast.

Youthful stream—A fast-moving, relatively straight-channel stream in a V-shaped valley.

Zone of aeration—The ground water zone above the water table.

Zone of saturation—The ground water zone below the water table.

Index